南西日本菌類誌

軟質高等菌類

The fungal flora in southwestern Japan : Agarics and boletes

南西日本菌類誌

軟質高等菌類

寺嶋芳江 監修
Supervised by Yoshie Terashima

寺嶋芳江・高橋春樹・種山裕一 編著
Edited by Yoshie Terashima, Haruki Takahashi and Yuichi Taneyama

東海大学出版部

The fungal flora in southwestern Japan : Agarics and boletes

Supervised by Yoshie TERASHIMA
Tokai University Press, 2016
ISBN978-4-486-02085-1

General part 総論

Preface

Mushrooms are members of the higher fungi that produce fruiting bodies that can be seen clearly with the naked eye. Katsumoto (2010) generated the first comprehensive checklist of the fungi of Japan, and this list has subsequently been expanded by the Mycological Society of Japan. However, mushrooms from the Nansei Islands south of Kyushu Island, with Okinawa Island at their center, have not been examined in any great detail, and numerous taxa await description.

The mushrooms covered in this monograph were collected from the southwestern region of Japan, focusing on the Nansei Islands and the Kyushu region. Of the 35 mushroom taxa surveyed, 21 are new species.

Okinawa Prefecture consists of 49 inhabited islands and a number of uninhabited islands that extend over a vast area of about 1,000 km east-to-west and 400 km north-to-south. The prefecture consists of the Ryukyu Islands which connect Taiwan to the Satsunan Islands of Kagoshima Prefecture. Like Tokyo, Osaka and Fukuoka, Naha, the capital of Okinawa Prefecture has a warm humid climate; however, Naha is also considered to have a sub-tropical climate. The Sakishima Islands, which include Ishigaki and Iriomote Islands, are separated from the main island of Okinawa by about 410 km. The minimum average monthly temperature and precipitation on the Sakishima Islands is 18°C, which is warm enough for palm trees, and more than 60 mm, respectively.

According to the modified Köppen climate classification system, which is based on vegetation, temperature and precipitation, the Sakishima Islands are considered to have a tropical rainforest climate. The Ryukyu Archipelago extends in an arc that connected Japan to the Asian continent at the time of the last ice age 10,000 to 70,000 years ago. Consequently, the vegetation in southwestern Japan is dominated by broad-leaved tree species and floristic elements that are typical of the tropical regions on the Asian continent.

Given that Okinawa is a continental island, it has region-specific forests. Thus, the Sakishima Islands are characterized by primary, evergreen, broad-leaved forests of continental origin; areas that experience relatively large amounts of annual rainfall and a tropical marine climate.

Naturally, since mycorrhizal fungi obtain their carbon from trees, they are affected by vegetation. In the same way, the saprophytic mushrooms that decompose wood components are affected by the distribution of their host trees. Thus, the mushroom flora in southwestern Japan, including Okinawa, is highly affected in the combination of tropical and temperate trees growing together.

Surveys by Miyagi recoded a total of 62 mushroom species on the main island of Okinawa (Miyagi 1958, 1960, 1964b, 1967), and an additional 109 species were collected on surveys conducted by the Mycological Society of Japan. Similar surveys on the fungi of Ishigaki and Iriomote islands revealed a total of 53 species by Miyagi (Miyagi 1964a, 1971), 9 species by Fukiharu and Hongo (Fukiharu and Hongo 1995), 68 species by Hosoya et al. (Hosoya et al. 2011), and eight species by Murakami and Terashima (Murakami and Terashima 2011). Focusing on Ishigaki Island, Takahashi (http://www7a.biglobe.ne.jp/~har-takah/) has reported 39 species to date, and is busy describing more. Despite these studies, our current knowledge of the mushroom flora on the islands around Okinawa is incomplete.

In this monograph, 8 luminescent mushroom taxa are newly described. Of a global total of 77 luminescent species (Chew et al. 2014), 12 are distributed in Japan. Luminescent mushrooms typically inhabit tropical and subtropical regions. Although the mechanism underlying mushroom luminescence has not yet fully clarified, the mystique associated with this luminescence has attracted the attention of humans since the beginning of recorded history, and the mechanism involved could potentially be used to improve the lives of people in the future. It is my hope that this monograph will contribute meaningfully towards an understanding of the mushroom flora in southwestern Japan.

Terashima Yoshie

はじめに

　きのこ（高等菌類）は，肉眼で明確に確認できる大きさの子実体を形成する菌類を指す．日本全体を俯瞰すると，きのこリストは勝本（Katsumoto 2010）により日本産菌類集覧にまとめられ，その後の追記は日本菌学会により進められている．しかし，南西諸島（九州地方島嶼以南から沖縄県を中心とする地域）におけるきのこ相リストはもとより，記載についても遅々として進められてこなかった．本モノグラフに掲載したきのこの採取地域は，南西諸島を中心とした九州地域を含む日本の南西地域である．対象としたきのこは，35種であり，その中で新種としての記載は21種に上る．

　沖縄県は，49の有人島と多くの無人島から成り，東西約1,000 km，南北約400 kmと広大な県域を持つ．東京，大阪，福岡などは温暖湿潤気候区に属する．那覇は温帯の温暖湿潤気候区に分類されるが，亜熱帯気候区といわれることが多い．沖縄本島と先島諸島の石垣島・西表島とは約410 km離れている．先島諸島の最寒月平均気温はヤシの生育可能な18℃以上で，最少月降水量は60 mm以上ある．したがって，植生・気温・降水量に基づく改良ケッペンの気候区分では，熱帯の熱帯雨林気候区に属するといえる．

　沖縄県は，鹿児島県の薩南諸島から台湾をつなぐ琉球諸島に位置する．弧状に連なる島々の配置からも想像できるように，これらは7万年から1万年前の最終氷河期には一連の陸橋を形成し，大陸と地続きであった．植生をみると，沖縄本島の常緑広葉樹林では，西南日本の広葉樹を核として，大陸からの熱帯系の植物要素が混ざり，さらに，陸島となったために本地域固有の林相を呈している．先島諸島では，上記のように年間降水量は比較的多く，海洋性の熱帯性に近い気候環境のもとで，大陸由来の常緑広葉樹林を中心とした原生林が保たれている．

　樹木から炭素源の供給を受けている菌根性きのこは，当然植生の影響を受ける．また，木材成分を分解する腐生性きのこも寄主樹木の影響が大きい．沖縄県を含めた南西日本のきのこ相にはこのように，温帯性と熱帯性の樹木が混ざり合って生育している環境が大きく関与している．

　これまで沖縄本島のきのこ相については，宮城（Miyagi 1958, 1960, 1964b, 1967）による調査の結果62種が報告され，日本菌学会菌類採集会では109種が報告された．一方，石垣島・西表島に生息する菌類についての報告としては，宮城による53種（Miyagi 1964a, 1971），およびFukiharu and Hongo（1995）による9種，細矢ら（Hosoya et al. 2011）による68種および村上・寺嶋（Murakami and Terashima 2011）による8種のみである．また，高橋（http://www7a.biglobe.ne.jp/~har-takah/）により石垣島を中心とした39種のきのこ記載が順次進められてきた．しかし，沖縄県を中心とした地域のきのこ相に関する網羅的知見はいまだ十分ではない．

　本モノグラフでは，発光性きのこ8種が新種として記載されている．発光性きのこは世界では，77種（Chew et al. 2014）が報告され，日本では12種が確認されており，熱帯〜亜熱帯地域で採取される場合が多い．きのこ発光機構はまだ十分には解明されていないが，発光の神秘性は人を魅了し，発光機構は人々の生活に有効利用される可能性が高い．本モノグラフが南西日本のきのこ相の理解に貢献することを祈ってやまない．

寺嶋芳江

Materials and Methods

Morphological studies

Macroscopic features, including spore print colors, are all based on fresh materials. For microscopic observations of new species, the second editor mainly examined free-hand sections of dried materials mounted in Melzer's reagent, 30% NH4OH, 3% KOH, phloxine B, Congo red, or distilled water under a light microscope (Nikon Eclipse 50i, Nikon, Japan). Basidiospores were measured in side view, excluding the hilar appendix; ± standard deviation, Q range (length/width), and each average (mean length, mean width, mean Q) were statistically derived from a random selection of all basidiospores measured. Color names and notations in parentheses are taken from Kornerup and Wanscher (1983). Specimens presented here are deposited in the National Museum of Nature and Science in Tsukuba (TNS), the Kanagawa Prefectural Museum of Natural History, Japan (KPM), Osaka Museum of Natural History (OSA), and Herbarium of Natural History Museum and Institute, Chiba, Japan (CBM).

DNA extraction, PCR amplification and sequencing

For molecular analysis, DNA was extracted from fresh or dry specimens using a DNeasy Plant Mini Kit (Qiagen USA, Valencia, CA. http://www.qiagen.com/) or an FTA card (Whatman International Ltd., Maidstone, England). PCR amplification and cycle sequencing were performed to obtain sequences of nuclear ribosomal internally transcribed spacer regions (ITS) 1 and 2, and the 28S large sub unit (LSU). The primer pairs for PCR were ITS1F/ITS4 (White et al. 1990; Gardes and Bruns 1993) for the ITS region, and LR0R/LR5 (Vilgalys and Hester 1990) for the LSU. The PCR products of the ITS region were cloned into pGEM-T Easy Vectors (Promega, Madison, WI). Cloned colonies were PCR amplified using primer pair T7/SP6 and then screened using gel electrophoresis. The PCR products were sequenced on a 3130 Genetic Analyzer (Applied Biosystems, Foster City, CA) with a BigDye Terminator v3.1 Cycle Sequencing Kit using T7 and SP6 (ITS region) or LR0R and LR5 (LSU) promoter primers as sequencing primers. To confirm the species identity, the BLAST algorithm was used to compare the obtained sequences against those deposited in the NCBI sequence databases (Altschul et al. 1990).

Phylogenetic analysis

The sequences obtained in this study were aligned against sequences downloaded from GenBank using the Muscle function in MEGA5 (Tamura et al. 2011). The phylogenetic analyses were carried out using the Maximum Likelihood method (ML) implemented in MEGA5. The strength of the internal branches within the ML tree was statistically tested using 2000 bootstrap replications.

材料と方法

形態観察

　胞子紋を含む肉眼的データは全て生の標本に基づく．新種の顕微鏡観察に関しては主に第2編著者がデータの採取を行い，メルツァー溶液，30%アンモニア溶液，3%水酸化カリウム，フロキシンB，コンゴー赤，または蒸留水で封入した乾燥標本から作成された試料をNikon Eclipse 50i microscopeで観察した．担子胞子は嘴状突起を除く側面観を計測し，標準偏差，幅長比（Q），並びに各々の平均値（mean length, mean width, mean Q）は計測を行った全ての担子胞子から無作為に抽出したデータに基づき統計的に導出されている．括弧内に表記されている色票の名称はKornerup and Wanscher (1983)を用いた．記載文中に引用した標本は筑波国立科学博物館（TNS），神奈川県立生命の星・地球博物館（KPM），大阪市立自然史博物館（OSA）並びに千葉県立中央博物館（CBM）に保管されている．

DNA抽出，CR増幅および塩基配列決定

　分子生物学的解析には，生標本あるいは乾燥標本からDNeasy Plant Mini Kit（Qiagen USA, Valencia, CA. http://www.qiagen.com/），またはFTAカード（Whatman International Ltd, Maidstone, England）を用いてDNAを抽出し，PCR増幅およびサイクルシーケンス法により，核リボソームの内部転写領域（ITS）の1から2，および大サブユニット（LSU）の配列を読んだ．PCRプライマーは，ITS領域にはITS1FとITS4（White et al. 1990; Gardes and Bruns 1993），LSUにはLR0RとLR5（Vilgalys and Hester 1990）を用いた．ITS領域についてはPCR産物をpGEM-T Easy Vectors（Promega, Madison, WI, USA）内部でクローニングした．クローニングしたコロニーをプライマーT7とSP6でPCR増幅し，電気泳動ゲルで選別した．ITS領域にはT7とSP6，LSUにはLR0RとLR5をシーケンスプライマーとし，BigDye Terminator v3.1 Cycle Sequencing Kitを用いて，シーケンサー3130 Genetic Analyzer（Applied Biosystems, Foster City, CA, USA）にかけてPCR産物の塩基配列を決定した．得られた塩基配列は，NCBI塩基配列データベースで，BLASTアルゴリズム（Altschul et al. 1990）を用いて種相同性を確認した．

系統解析

　本研究で得られた塩基配列とGenBankからダウンロードした配列は，MEGA5（Tamura et al. 2011）に実装されたMuscleを用いてアラインメントし，最尤法（ML）により，系統解析を行った．系統樹における配列間の進化的距離は，2000回のブートストラップ繰り返しにより得られた．

Acknowledgements

Special thanks to Dr. Toshimitsu Fukiharu for providing several, highly valued, pre-publication reviews of this book. The authors are grateful to Dr. Tsuyoshi Hosoya (TNS), Dr. Kentaro Hosaka (TNS), Ms. Kanade Otsubo (KPM), and Dr. Takamichi Orihara (KPM) for allowing the specimens cited herein to be kept in the National Museum of Nature and Science in Tsukuba and the Kanagawa Prefectural Museum of Natural History. The first author thanks Ms. Kimiko Wada for her excellent photographs. Thanks also are owed to Mr. Yoshinori Nishino for provision of specimens of *Mycena lazulina*. This work was supported by JSPS KAKENHI, Grant-in-Aid for Publication of Scientific Research Results, Grant Number 15HP5222JSPS.

謝辞

本書を作成するにあたり，千葉県立中央博物館の吹春俊光博士には全体の構成と体裁につきまして御助言を賜りました．細矢剛博士と保坂健太郎博士には筑波国立科学博物館（TNS）において，また大坪奏氏と折原貴道博士には神奈川県立生命の星・地球博物館（KPM）において引用標本の登録と保管をしていただきました．兵庫きのこ研究会の和田貴美子氏には貴重な写真をご提供いただきました．西野嘉憲氏には，石垣島産コンルリキュウバンタケの標本をご提供いただきました．これらの方々に厚くお礼申し上げます．本研究は独立行政法人日本学術振興会科学研究費補助金研究成果促進費15HP5222の助成を受けたものです．

References 引用文献

Altschul SF, Gish W, Miller W, Myers EW, Lipman DJ. 1990. Basic local alignment search tool. J Mol Biol. 215 (3): 403-410.

Chew AL, Tan YS, Desjardin DE, Musa MY, Sabaratnam V. 2014. Four new bioluminescent taxa of *Mycena* sect. *Calodontes* from Peninsular Malaysia. Mycologia 106 (5): 976-988.

Fukiharu T, Hongo T. 1995. Ammonia fungi of Iriomote Island in the southern Ryukyus, Japan and a new ammonia fungus, Hebeloma luchuense. Mycoscience 36: 425-430.

Gardes M, Bruns TD. 1993. ITS primers with enhanced specifity for Basidiomycetes: application to identification of mycorrhizae and rusts. Mol Ecol 2: 113-118.

Hosoya T, Neda H, Hattori T, Hosaka K, Murakami Y, Fukiharu T, Kinjo K, Terashima Y. 2011. List of identified mushroom species during the Iriomote Island mushroom foray. The science bulletin of the Faculty of Agriculture, University of the Ryukyus 58: 21-28.

Katsumoto K. 2010. List of fungi record in Japan (in Japanese). The Kanto Branch of the Mycological Society of Japan, Tokyo.

Kornerup A, Wanscher JH. 1983. Methuen handbook of colour. Eyre Methuen, London.

Miyagi G. 1958. On the fungi of the Ryukyu Islands, vol 2 (in Japanese). Bulletin of Arts and Science Division, University of the Ryukyus, Mathematics and sciences 2: 35-40.

Miyagi G. 1960. Notes on luminous fungi, Filoboletus manipularis, on Okinawa, vol. 4 (in Japanese). Bulletin of Arts and Science Division, University of the Ryukyus, Mathematics and sciences 4: 77-87.

Miyagi G. 1964a. On a luminous fungus, Pleurotus lunaillustris from the Yaeyama Islands, vol. 7 (in Japanese). Bulletin of Arts and Science Division, University of the Ryukyus, Mathematics and sciences 7: 54-56.

Miyagi G. 1964b. Notes on the Agaricales of Okinawa Island (I) (in Japanese). Bulletin of Arts and Science Division, University of the Ryukyus, Mathematics and sciences. 7: 57-70.

Miyagi G. 1967. Notes on the Agaricales of Okinawa Island (II) (in Japanese). Bulletin of Arts and Science Division, University of the Ryukyus, Mathematics and sciences. 10: 38-45.

Miyagi G. 1971. Notes on the Agaricales of Iriomote Island and Ishigaki Island (1) (in Japanese). Biol Mag Okinawa 7: 33-37.

Murakami Y, Terashima Y. 2011. Mushroom species in Iriomote Island. The science bulletin of the Faculty of Agriculture, University of the Ryukyus 58: 29-34.

Tamura K, Peterson D, Peterson N, Stecher G, Nei M, Kumar S. 2011. MEGA5: Molecular Evolutionary Genetics Analysis using Maximum Likelihood, Evolutionary Distance, and Maximum Parsimony Methods. Molecular Biology and Evolution 28: 2731-2739.

Vilgalys R, Hester M. 1990. Rapid genetic identification and mapping of enzymatically amplified ribosomal DNA from several Cryptococcus species. Journal of Bacteriology 172: 4238-4246.

White TJ, Bruns TD, Lee S, Taylor J. 1990. Amplification and direct sequencing of fungal ribosomal RNA genes for phylogenetics. In: Innis MA, Gelfand DH, Sninsky JJ, White TJ eds. PCR protocols, a guide to methods and applications. San Diego, California: Academic Press. pp 315-322.

Contents
目次

General part 総論 ·· v
  Preface ·· v
  はじめに ·· vi
  Materials and Methods ··· vii
    Morphological studies ··· vii
    DNA extraction, PCR amplification and sequencing ·· vii
    Phylogenetic analysis ·· vii
  材料と方法 ·· viii
    形態観察 ·· viii
    DNA 抽出，CR 増幅および塩基配列決定 ·· viii
    系統解析 ·· viii
  Acknowledgements ··· ix
  謝辞 ··· ix
  References 引用文献 ··· x

Taxonomic part 分類記載 ··· 1
  1. ***Amanita chepangiana*** Tulloss & Bhandary ナンヨウシロタマゴタケ ···················· 3
  2. ***Amanita rubromarginata*** Har. Takah. フチドリタマゴタケ ································ 6
  3. ***Aureoboletus liquidus*** Har. Takah. & Taneyama, sp. nov. ヌメリアシナガイグチ ······ 17
  4. ***Boletus bannaensis*** Har. Takah. ナンヨウウラベニイグチ ································ 32
  5. ***Boletus virescens*** Har. Takah. & Taneyama, sp. nov. アオアザイグチ ··················· 45
  6. ***Chaetocalathus fragilis*** (Pat.) Singer ヒダフウリンタケ ··································· 54
  7. ***Crinipellis canescens*** Har. Takah. シラガニセホウライタケ ······························· 60
  8. ***Crinipellis rhizomorphica*** Har. Takah. ミドリニセホウライタケ ························· 68
  9. ***Cruentomycena orientalis*** Har. Takah. & Taneyama, sp. nov. ガーネットオチバタケ ··· 79
  10. ***Gymnopilus iriomotensis*** Har. Takah., Taneyama & Wada, sp. nov. ミナミホホタケ ····· 90
  11. ***Gymnopus albipes*** Har. Takah. & Taneyama, sp. nov. シロアシホウライタケ ·········· 98
  12. ***Gymnopus oncospermatis*** (Corner) Har. Takah. ヤシモリノカレバタケ ················ 105
  13. ***Gymnopus phyllogenus*** Har. Takah., Taneyama & Terashima, sp. nov.
      アシグロカレハタケ ··· 116
  14. ***Inocybe fuscomarginata*** Kühner フチドリトマヤタケ ······································ 124
  15. ***Inocybe humilis*** (J. Favre & E. Horak) Esteve-Rav. & Vila コカブラアセタケ ········· 126
  16. ***Leccinellum rhodoporosum*** (Har. Takah.) Har. Takah., comb. nov. ウラベニヤマイグチ ···· 129
  17. ***Marasmiellus arenaceus*** Har. Takah., Taneyama & Wada, sp. nov. シロスナホウライタケ ···· 141

18. *Marasmiellus lucidus* Har. Takah., Taneyama & S. Kurogi, sp. nov. ヒメホタルタケ ………… 155
19. *Marasmiellus venosus* Har. Takah., Taneyama & A. Hadano, sp. nov. ヒメヒカリタケ ………… 165
20. *Micropsalliota cornuta* Har. Takah. & Taneyama, sp. nov. ダイダイツノハラタケ ………… 174
21. *Mycena comata* Har. Takah. & Taneyama, sp. nov. キジムナハナガサ ………… 184
22. *Mycena flammifera* Har. Takah. & Taneyama, sp. nov. モリノアヤシビ ………… 197
23. *Mycena lazulina* Har. Takah., Taneyama, Terashima & Oba, sp. nov.
    コンルリキュウバンタケ ………… 209
24. *Mycena luxfoliata* Har. Takah., Taneyama & Terashima, sp. nov. カレハヤコウタケ ………… 219
25. *Mycena stellaris* Har. Takah., Taneyama & A. Hadano, sp. nov. ホシノヒカリタケ ………… 227
26. *Pleurotus nitidus* Har. Takah. & Taneyama, sp. nov. シロヒカリタケ ………… 238
27. *Psilocybe capitulata* Har. Takah. ナンヨウシビレタケ ………… 250
28. *Psilocybe definita* Har. Takah. & Taneyama, sp. nov. ハマシビレタケ ………… 263
29. *Pulveroboletus brunneoscabrosus* Har. Takah. ウロコキイロイグチ ………… 274
30. *Resinomycena fulgens* Har. Takah., Taneyama & Oba, sp. nov. ギンガタケ ………… 284
31. *Rubinoboletus monstrosus* Har. Takah. ダルマイグチ ………… 295
32. *Strobilomyces brunneolepidotus* Har. Takah. & Taneyama, sp. nov. チャオニイグチ ………… 303
33. *Strobilurus luchuensis* Har. Takah., Taneyama & Pham, sp. nov.
    リュウキュウマツカサキノコ ………… 313
34. *Termitomyces intermedius* Har. Takah. & Taneyama, sp. nov. シロアリシメジ ………… 324
35. *Tylopilus obscureviolaceus* Har. Takah. スミレニガイグチ ………… 336

Scientific name Index 学名索引 ………… 347
Japanese name Index 和名索引 ………… 348

Taxonomic part
分類記載

1. *Amanita chepangiana* Tulloss & Bhandary ナンヨウシロタマゴタケ

Mycotaxon 43: 25 (1992) [MB#355483].

Macromorphological characteristics (Fig. 2): Pileus up to 125 mm in diameter, convex to nearly plane, with a long sulcate-striate margin; surface glabrous, dry, pale greyish to light yellowish brown at the center, almost pure white toward the margin. Context soft, relatively thick, white; odor and taste indistinct. Lamellae free, crowded, white; edges fimbriate, concolorous. Stipe 140–170 × 17–19 mm, almost equal, slightly tapering upward, central, terete, at first stuffed then becoming hollow above; surface dry, fibrous-scaly to lacerate-scaly, pure white; annulus thin, membranous, attached toward the stipe apex, pendant, white with slightly yellowish tinge; volva up to 77 × 46 mm, large, saccate, thick, membranous, lobed, white, sometimes stained dingy brownish.

Micromorphological characteristics (Fig. 1): Basidiospores 9.3–11.8 × 8.8–11.0 μm (N = 40, averaging 10.7–11.0 × 9.6–10.0 μm, Q = 1.0–1.3, mean Q = 1.10–1.12), subglobose to broadly ovoid, inamyloid, colorless, smooth, thin-walled. Basidia 20–39 × up to 14.5 μm, 4-spored, narrowly clavate, thin-walled. Marginal cells of lamellae (remnants of the annulus) 23–50 × 12.5–16.3 μm, narrowly cylindrical, clavate to ellipsoid with a short pedicellate base, colorless, thin-walled. Clamp connections present.

Habitat and phenology: Solitary or scattered on ground in evergreen broad-leaved forests (dominated by *Castanopsis sieboldii* (Makino) Hatus. ex T. Yamaz. et Mashiba), May to November.

Known distribution: Okinawa (Iriomote Island), Nepal, northern Thailand, southwestern China, and Korean Peninsula.

Specimens examined: TNS-F-61368, on ground in an evergreen broad-leaved forest, Uehara, Iriomote Island, Yaeyama-gun, Okinawa Pref., 26 Nov. 2013, leg. Takah. Kobayashi; KPM-NC0023866, Oomiya road Park, on ground in an evergreen broad-leaved forest, Iriomote Island, Yaeyama-gun, Okinawa Pref., 24 May 2014, leg. Hoang ND Pham.

Japanese name: Nayou-shiro-tamagotake (named by T. Kobayashi).

Comments: This species belongs to the genus *Amanita*, subgenus *Amanita*, section *Caesareae* Singer ex Singer (Singer 1986) because of the inamyloid basidiospores, the well-developed annulus, the distinctly long sulcate-striate margin of the pileus, and the stipe having a saccate volva instead of a basal bulb.

The macro- and micromorphological characteristics of the present collection correspond well with *A. chepangiana* originally described from Nepal (Tulloss and Bhandary 1992; Yang 1997, 2000, 2005; Sanmee, et al. 2008).

Amanita caesarea var. *alba* Gillet (≡ *A. caesarea* f. *alba* (Gillet) E.-J. Gilbert) from Europe (Gillet 1878; Neville and Poumarat 2004) is somewhat similar to *A. chepangiana* in appearance, but it forms shorter marginal striations on the pileus, a much shorter stipe, and short ellipsoid, significantly narrower basidiospores: 10–12 × 7–8.5 μm (Neville and Poumarat 2004).

References 引用文献

Gillet CC. 1878. Les Hyménomycètes ou Description de tous les Champignons qui Croissent en France 1: 1–176.

Neville P, Poumarat S. 2004. Amaniteae. *Amanita, Limacella* & *Torrendia*. Fungi Europaei 9: 1–1120. Candusso, Alassio.

Sanmee R, Tulloss RE, Lumyong P, Dell B, Lumyong S. 2008. Studies on *Amanita* (Basidiomycetes: Amanitaceae) in Northern Thailand. Fung Diver 32: 97–123.

Singer R. 1986. The Agaricales in modern taxonomy, 4th edn. Koeltz, Koenigstein.

Tulloss RE, Bhandary HR. 1992. *Amanita chepangiana* - a new species from Nepal. Mycotaxon 43: 25–31.

Yang ZL. 1997. Die *Amanita*-Arten von Sudwestchina. Bibl Mycol 170: 1-240.
Yang ZL. 2000. Revision of the Chinese *Amanita* collections deposited in BPI and CUP. Mycotaxon 75: 117-130.
Yang ZL. 2005. Amanitaceae (in Chinese). Flora Fungorum Sinicorum 27: 1-258, Beijing.

1. ナンヨウシロタマゴタケ（小林孝人新称）*Amanita chepangiana* Tulloss & Bhandary
Mycotaxon 43: 25 (1992) [MB#355483].

肉眼的特徴（Fig. 2）：傘は径125 mmまで，まんじゅう型～ほぼ平ら，周縁部に放射状の長い溝線を表す；表面は平滑，粘性を欠き，中央部は淡灰色～淡黄褐色，周縁部に向かって純白色を呈する．肉は軟質，相対的に厚く，純白色；特別な味や臭いはない．ヒダは離生，密，白色；縁部は長縁毛状，同色．柄は140-170×17-19 mm，ほぼ上下同大，上部に向かいわずかに細くなり，中心性，最初中実まもなく上部が中空になる；表面は粘性を欠き，繊維状小鱗片で密に被われ，ささくれ状～多少段だら模様を形成し，純白色；ツバは薄い膜質，柄の上部に垂れ下がるように付着し，白色でわずかに黄色の色彩を帯びる；ツボは77×46 mmまで，大形，袋状，厚い膜質，裂け目を生じ，白色，時にくすんだ帯褐色の斑点を表す．

顕微鏡的特徴（Fig. 1）：担子胞子は9.3-11.8×8.8-11.0 μm（N＝40，平均10.7-11.0×9.6-10.0 μm，Q＝1.0-1.3，mean Q＝1.10-1.12），亜球形～広卵形，非アミロイド，無色，平滑，薄壁．担子器は20-39×-14.5 μm，4胞子性，狭棍棒状，薄壁．ヒダの縁部の細胞（ツバの残存組織）は23-50×12.5-16.3 μm，狭円柱状，棍棒状～短い小柄のある楕円形，ほぼ無色，薄壁．かすがい状突起を有する．

生態および発生時期：スダジイを主体とする広葉樹林内地上に孤生または散生，5月～11月．

分布：沖縄（西表島），ネパール，タイ北部，中国南西部，朝鮮半島．

供試標本：TNS-F-61368，常緑広葉樹林内地上，沖縄県八重山郡西表島上原，2013年11月26日，小林孝人採集；KPM-NC0023866，常緑広葉樹林内地上，沖縄県八重山郡西表島大見謝ロードパーク，2014年5月24日，Hoang ND Pham 採集．

コメント：本種は柄の基部が球根状膨大部の代わりに袋状のツボを形成し，発達したツバを有し，傘周縁部に明瞭な長い溝状条線を表し，担子胞子が非アミロイドの特徴に基づき，テングタケ属 *Amanita*，テングタケ亜属 subgenus *Amanita*，タマゴタケ節 section *Caesareae* Singer ex Singer（Singer 1986）に分類される．

　本試料の肉眼的および顕微鏡的特徴は最初ネパールから記載された *A. chepangiana*（Tulloss and Bhandary 1992; Yang 1997, 2000, 2005; Sanmee, et al. 2008）に一致する．

　欧州産 *Amanita caesarea* var. *alba* Gillet（Gillet 1878; Neville and Poumarat 2004）は本種に外観がやや類似するが，傘周縁部の条線はより短く，柄が短形で，担子胞子の幅がより狭い点で有意差が認められる．

1. *Amanita chepangiana* ナンヨウシロタマゴタケ ── 5

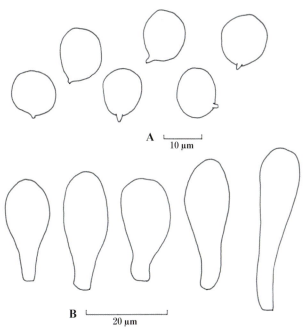

Fig. 1 – Micromorphological features of *Amanita chepangiana* (TNS-F-61368): **A.** Basidiospores. **B.** Marginal cells of the lamellae. Illustrations by Kobayashi, T.
ナンヨウシロタマゴタケの顕鏡図（TNS-F-61368）：**A.** 担子胞子．**B.** ヒダ縁部の細胞．**C.** ツボを構成する菌糸と末端細胞．図：小林孝人．

Fig. 2 – Basidiomata of Amanita chepangiana (TNS-F-61368) on ground in an evergreen broad-leaved forest, 26 Nov. 2013, Uehara, Iriomote Island. Photo by Kobayashi, T.
亜熱帯性常緑広葉樹林内地上に発生したナンヨウシロタマゴタケの子実体（TNS-F-61368）．2011年5月25日，西表島上原．写真：小林孝人．

2. *Amanita rubromarginata* Har. Takah. フチドリタマゴタケ

Mycoscience 45 (6): 372 (2004) [MB#373370].

Etymology: The specific epithet means "red marginate", referring to the reddish orange marginate lamellae.

English name: "Red-Skirted Slender Caesar" (named by Dr. Rodham E. Tulloss).

Macromorphological characteristics (Figs. 4–11): Pileus 40–80 mm in diameter, at first cylindrical-campanulate, then expanding to nearly plane to slightly concave and subumbonate, long sulcate-striate from the margin toward the center; surface glabrous, subviscid when wet, at first orange (6B7–8) to brownish orange (6C7–8 to 7C7–8) overall, then reddish yellow (4B7–8) toward the margin, in age greyish yellow (4C7) at the center. Flesh soft, 3–7 mm thick in the center of the pileus, yellowish white, deeper yellow below the surface, odor and taste mild. Stipe 60–120 × 5–16 mm, subcylindrical or slightly tapering upward, central, terete, hollow; surface dry, silky fibrillose, pale yellow, usually covered with reddish orange (7B7–8), appressed, indefinite squamules forming irregular transverse zones; annulus 10–15 mm wide, thin, membranous, attached toward the stipe apex, pendant, reddish orange (7B7–8), striate; volva 20–40 × 10–25 mm, large, saccate, thick, membranous, bilobate, white, sometimes stained dingy brownish. Lamellae free, very close (55–70 reaching the stipe), with 1–3 series of lamellulae, up to 10 mm broad, pale yellow; edges fimbriate, reddish orange (7B7–8).

Micromorphological characteristics (Fig. 3): Basidiospores 8–9 × 5.5–7 μm (n = 20 spores per two specimens, Q = 1.28–1.45), broadly ovoid to short ellipsoid, smooth, colorless, inamyloid, thin-walled, with a large refractive guttule. Basidia 25–30 × 8–10 μm, clavate, 4-spored; basidioles clavate. Marginal cells of lamellae (remnants of the annulus) 30–60 × 8–15 μm, gregarious to scattered, narrowly clavate to pyriform, colorless or with dark orange (5A8) to orange (5A6) intracellular pigment, thin-walled. Subhymenium ramose-inflated, 20–40 μm thick, with 1-3 layers of subglobose to ellipsoid cells, 5–20 × 5–14 μm. Hymenophoral trama narrow, bilateral; element hyphae cylindrical, clavate to subfusiform. Pileipellis a cutis of narrow cylindrical cells 4–8 μm wide, with orange (5A6) to deep orange (5A8) intracellular pigment, occasionally with clamped septa, thin-walled. Hyphae of pileitrama 5–14 μm wide, parallel to the pileipellis elements, cylindrical, colorless, thin-walled. Stipitipellis a cutis of parallel, repent hyphae 4–14 μm wide, cylindrical, with reddish yellow (4A7) to deep yellow (4A8) vacuolar pigment, thin-walled. Stipe trama composed of longitudinally running, cylindrical hyphae 8–38 μm wide, colorless, thin-walled. Volval tissue composed of 4–8 μm wide, cylindrical hyphae and scattered, oblong, broadly clavate, ellipsoidal to subglobose cells 35–75 × 12–40 μm, occasionally with clamped septa, thin-walled.

Habitat and phenology: Solitary or scattered on ground in evergreen broad-leaved forests (dominated by *Quercus miyagii* Koidz. and *Castanopsis sieboldii* (Makino) Hatus. ex T. Yamaz. et Mashiba), May to September.

Known distribution: Okinawa (Ishigaki Island).

Specimens examined: KPM-NC0011979 (holotype), on ground in a *Castanopsis-Quercus* forest, Banna Park, Ishigaki-shi, Okinawa Pref., 15 Sep., 2003, coll. Takahashi, H.; KPM-NC0010087, same location, 8 Jun., 2002, coll. Takahashi, H.; KPM-NC0023873, same location, 25 May 2011, coll. Takahashi, H.

Japanese name: Fuchidori-tamagotake.

Comments: Distinctive features of this species are found in the orange to brownish orange then reddish yellow pileus; the pale yellow, squamulose stipe with a thin, membranous, reddish-orange annulus and a thick, saccate volva; the reddish orange marginate lamellae; and the habitat of subtropical evergreen broad-leaved forests.

The long sulcate-striate pileus, the presence of a membranous annulus and a thick, saccate volva, and the inamyloid basidiospores suggest that this species is a member of the section *Caesareae* Singer ex Singer in the genus *Amanita*, subgenus *Amanita* as defined by Singer (Singer 1986). Within the section, this species seems to be closely allied with *Amanita hemibapha* (Berk. & Broome) Sacc. var. *hemibapha*, originally described from Sri Lanka (Berkeley and Broome 1871; Saccardo 1887; Hongo 1987), *Amanita caesareoides* Lj. N. Vassiljeva from the Russian maritime region (Vassiljeva 1950; Endo et al. 2013), and *Amanita javanica* (Corner & Bas) T. Oda, C. Tanaka & Tsuda, originally described from Java (Boedijn 1951; Oda et al. 1999). These species are distinct in producing a different-colored pileus, a yellow annulus, and not marginate lamellae. Moreover, *A. javanica* forms wholly orange-yellow to ochre-yellow basidiomata. *Amanita hemibapha* var. *ochracea* Z. L. Yang from southwestern China (Yang 1997) also resembles *A. rubromarginata* to some degree, though it has an ochraceous pileus and yellow marginate, white lamellae.

In the circumscription of Corner & Bas (Corner and Bas 1962), *A. javanica* has been relegated to *A. hemibapha* as a morphological subspecies on the basis of the color of basidiomata. As for *A. rubromarginata*, we recognize this taxon as distinct based on the morphological characteristics and comparison of ITS sequence data derived from the present taxon with the nucleotide sequences of taxa available in GenBank (Dr. Naoki Endo's personal communication).

References 引用文献

Berkeley MJ, Broome CE. 1871. On some species of *Agaricus* from Ceylon. Trans Linn Soc London 27: 149-152.
Boedijn KB. 1951. Notes on Indonesian fungi. The genus *Amanita*. Sydowia 5: 317-327.
Corner EJH, Bas C. 1962. The genus *Amanita* in Singapore and Malaya. Persoonia 2: 241-304.
Endo N, Gisusi S, Fukuda M, Yamada A. 2013. In vitro mycorrhization and acclimatization of *Amanita caesareoides* and its relatives on *Pinus densiflora*. Mycorrhiza 23 (4): 303-15.
Hongo T. 1987. Amanitaceae. In: Imazeki R, Hongo T (eds) Colored Illustrations of Mushrooms of Japan I (in Japanese). Hoikusha, Osaka, pp 115-136.
Oda T, Tanaka C, Tsuda M. 1999. Molecular phylogeny of Japanese *Amanita* species based on nucleotide sequences of the internal transcribed spacer region of nuclear ribosomal DNA. Mycoscience 40 (1): 57-64.
Saccardo PA. 1887. Sylloge Hymenomycetum, Vol. I. Agaricineae. Sylloge Fungorum 5: 1-1146.
Singer R. 1986. Agaricales in modern taxonomy, 4th edn. Koeltz, Koenigstein.
Takahashi H. 2004. Two new species of Agaricales from southwestern islands of Japan. Mycoscience 45 (6): 372-376.
Vassiljeva LN. 1950. Species novae fungorum. Notulae Systematicae e Sectione Cryptogamica Instituti Botanici Nomeine V.L. Komarovii Academiae Scientificae USSR 6: 188-200.
Yang ZL. 1997. Die *Amanita*-arten von Südwestchina. Bibl Mycol 170: 1-240.

2. フチドリタマゴタケ *Amanita rubromarginata* Har. Takah.

Mycoscience 45 (6): 372 (2004) [MB#373370].

英語名："Red-Skirted Slender Caesar"（Rodham E. Tulloss 博士により命名）

肉眼的特徴（Figs. 4-11）：傘は径40-80 mm，最初釣り鐘形，のちほぼ平に開き，中央部がやや突出し，周縁部から中心に向かって放射状の長い溝線を表す；表面は平滑，湿時やや粘性があり，初め全体に橙褐色を呈するが，のち周縁部から黄色を帯び，老成時は中央部が灰黄色を呈する．肉は軟質で，傘中央部の厚さは3-7 mm，淡黄色〜類白色，表皮直下は黄色，味および臭いは温和．柄は60-120×5-16 mm，類円柱形または頂部に向かってやや細くなり，中心生，中空；表面は粘性を欠き，絹状〜繊維状，淡黄色，通常橙赤色〜帯褐橙色の圧着した薄い膜質の鱗片がだんだら模様を形成する；ツバは幅10-15 mm，薄い膜質，柄の上部に垂れ下がるように付着し，

橙赤色～帯褐橙色，条線を表す；ツボは20-40×10-25 mm，大形，厚い膜質，袋状，裂け目を生じ，白色，時にくすんだ帯褐色の斑点を表す．ヒダは離生，密（柄に到達するヒダは55-70），1-3の小ヒダを伴い，幅10 mm以下，淡黄色；縁部は長縁毛状，橙赤色～帯褐橙色に縁取られる．

顕微鏡的特徴（Fig. 3）：担子胞子は8-9×5.5-7 μm（n = 20 spores per two specimens, Q = 1.28-1.45），広卵形～短楕円形，平坦，無色，非アミロイド，薄壁．担子器は25-30×8-10 μm，こん棒形，4胞子性；偽担子器はこん棒形．ヒダの縁部を構成する細胞（ツバの残存組織）は30-60×8-15 μm，群生～散生，亜こん棒形～洋梨形，無色または暗橙色～橙色の色素が細胞内に存在し，薄壁．子実下層は厚さ20-40 μm，1-3の類球形～楕円形の細胞（5-20×5-14 μm）からなる．子実層托実質は狭く，両側型；菌糸は円柱形，こん棒形～類紡錘形．傘表皮組織は円柱形の菌糸（幅4-8 μm）からなる平行菌糸被を成し，橙色の色素が細胞内に存在し，時にクランプを持ち，薄壁．傘実質の菌糸は幅5-14 μm，傘表皮の細胞と平行に配列し，円柱形，無色，薄壁．柄表皮組織は平行菌糸被を成し，菌糸は幅4-14 μm，円柱形，黄色の色素が液胞内に存在し，薄壁．柄の実質は縦に沿って配列した円柱形の菌糸（幅8-38 μm）からなり，無色，薄壁．ツボの組織は幅4-8 μmの円柱形の菌糸からなり，広こん棒形または楕円形～類円柱形の細胞（35-75×12-40 μm）が混在し，時にクランプを持ち，薄壁．

生態および発生時期：スダジイ，オキナワウラジロガシを主体とする照葉樹林内地上に孤生または散生，5月～9月．

分布：沖縄（石垣島）．

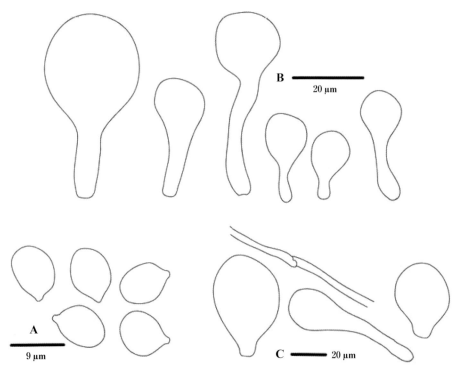

Fig. 3 – Micromorphological features of *Amanita rubromarginata* (KPM-NC0011979): **A**. Basidiospores. **B**. Marginal cells of the lamellae. **C**. Hypha and terminal cells of the volval tissue. Illustrations by Takahashi, H.
フチドリタマゴタケの顕鏡図（KPM-NC0011979）：**A**. 担子胞子．**B**. ヒダ縁部の細胞．**C**. ツボを構成する菌糸と末端細胞．図：高橋春樹．

供試標本：KPM-NC0011979（正基準標本），常緑広葉樹林内地上，沖縄県石垣市バンナ公園，2003年9月15日，高橋春樹採集；KPM-NC0010087，同上，2002年6月8日，高橋春樹採集；KPM-NC0023873，同上，2011年5月25日，高橋春樹採集．

主な特徴：傘表面は最初帯褐橙色のち帯赤黄色；柄表面は黄色地に橙色の段だら模様を表し，上部に薄い膜質で帯赤橙色〜帯褐橙色のツバを形成し，根元に厚い袋状のツボがある；ヒダは帯赤

Fig. 4 − Basidiomata of *Amanita rubromarginata* (KPM-NC0023873) on ground in a *Castanopsis-Quercus* forest, 25 May 2011, Banna Park, Ishigaki Island. Photo by Takahashi, H.
　スダジイ−オキナワウラジロガシ林内地上に発生したフチドリタマゴタケの子実体（KPM-NC0023873），2011年5月25日，石垣島バンナ公園．写真：高橋春樹．

Fig. 5 – Mature basidiomata of *Amanita rubromarginata* on ground in a *Castanopsis-Quercus* forest, 11 Aug. 2010, Banna Park, Ishigaki Island. Photo by Takahashi, H.
スダジイ-オキナワウラジロガシ林内地上に発生したフチドリタマゴタケの成熟した子実体. 2010年8月11日, 石垣島バンナ公園. 写真：高橋春樹.

Fig. 6 – Mature basidiomata of *Amanita rubromarginata* on ground in a *Castanopsis-Quercus* forest, 11 Aug. 2010, Banna Park, Ishigaki Island. Photo by Takahashi, H.
スダジイ-オキナワウラジロガシ林内地上に発生したフチドリタマゴタケの子実体，2010年8月11日，石垣島バンナ公園．写真：高橋春樹．

橙色に縁取られる：スダジイ，オキナワウラジロガシを主体とする亜熱帯性常緑広葉樹林内地上に発生．

コメント：傘周縁部に放射状の長い溝線を表し，膜質のツバと厚い袋状のツボを形成し，そして非アミロイドの担子胞子を持つ性質は本種がテングタケ属 Amanita，テングタケ亜属 subgenus Amanita，タマゴタケ節 section Caesareae Singer ex Singer (Singer 1986) に位置することを示唆している．

　節内においてスリランカから最初報告されたタマゴタケ Amanita hemibapha (Berk. & Broome) Sacc. var. hemibapha (Berkeley and Broome 1871; Saccardo 1887; Hongo 1987)，旧ソビエトの沿海州から新種記載された Amanita caesareoides Lj. N. Vassiljeva (Vassiljeva 1950; Endo et al. 2013)，ジャワ島から新種記載されたキタマゴタケ Amanita javanica (Corner & Bas) T. Oda, C. Tanaka & Tsuda (Boedijn 1951; Oda et al. 1999) はフチドリタマゴタケに最も近縁と思われるが，傘の色が異なり，ツバは黄色で，ヒダにフチドリを欠く．また最近中国南西部から報告された Amanita hemibapha var. ochracea Z. L.Yang (Yang 1997) もフチドリタマゴタケに似るが，傘は黄土色で，黄色に縁取られた白色のヒダを持つ点で異なる．

　Corner & Bas (Corner and Bas 1962) は主に色の相違に基づきキタマゴタケを A. hemibapha の形態的亜種として扱っている．フチドリタマゴタケに関してはITS領域の解析結果においてデータベース上に相同性が極めて近い種類はなく，独立種として扱うのは妥当とされている (信州大遠藤直樹氏からの私信)．

Fig. 7 – Underside view of the basidioma of *Amanita rubromarginata* (KPM-NC0023873), 25 May 2011, Banna Park, Ishigaki Island. Photo by Takahashi, H.
フチドリタマゴタケのヒダ (KPM-NC0023873)．2011年5月25日．石垣島バンナ公園．写真：髙橋春樹．

Fig. 8 – Mature basidioma of *Amanita rubromarginata* (KPM-NC0023873), 25 May 2011, Banna Park, Ishigaki Island. Photo by Takahashi, H.
フチドリタマゴタケの成熟した子実体（KPM-NC0023873），2011年5月25日，石垣島バンナ公園．写真：高橋春樹．

Fig. 9 – Mature basidioma of *Amanita rubromarginata* (KPM-NC0023873), 25 May 2011, Banna Park, Ishigaki Island. Photo by Takahashi, H.
フチドリタマゴタケの成熟した子実体（KPM-NC0023873），2011年5月25日，石垣島バンナ公園．写真：高橋春樹．

Fig. 10 – Immature basidioma of *Amanita rubromarginata* (KPM-NC0010087), Jun. 8, 2002, Banna Park, Ishigaki Island. Photo by Takahashi, H.
 フチドリタマゴタケの幼菌（KPM-NC0010087）．2002年6月8日．石垣島バンナ公園．写真：高橋春樹．

Fig. 11 – Aged basidiomata of *Amanita rubromarginata* (KPM-NC0023873) on ground in a *Castanopsis-Quercus* forest, 25 May 2011, Banna Park, Ishigaki Island. Photo by Takahashi, H.
スダジイ-オキナワウラジロガシ林内地上に発生したフチドリタマゴタケの老成した子実体（KPM-NC0023873），2011年5月25日，石垣島バンナ公園．写真：高橋春樹．

3. *Aureoboletus liquidus* Har. Takah. & Taneyama, **sp. nov.** ヌメリアシナガイグチ

MycoBank no.: MB 809926.

Etymology: The specific epithet "*liquids*" comes from the Latin word for "liquiform, transparent, pure", referring to the glutinosity of the basidiomata.

Distinctive features of this species are found in basidiomata shrouded by an entirely glutinous sheath; a reddish brown, irregularly rugose pileus; an elongate stipe with evanescent, white annulus; greyish yellow hymenophore, ellipsoidal basidiospores ornamented with prominent longitudinal ridges; subcylindrical to narrowly subfusoid, sinuous pleurocystidia; clavate-pedicellate, often subcapitate terminal elements in the outermost layer of pileipellis and stipitipellis; and a habitat of *Castanopsis-Quercus* forests.

Macromorphological characteristics (Figs. 18–23): Pileus 45–70 mm in diameter, hemispherical when young, expanding to convex, at first smooth then irregularly rugose; surface remarkably glutinous, innately fibrillose, reddish brown (8D7–8) overall; margin attached with white, appendiculate veil when young. Flesh up to 9 mm, gelatinous, whitish, unchanging where bruised; odor and taste not distinctive. Stipe 100–170 × 8–17 mm, cylindrical but somewhat thickened toward the base, slender, tall, central, terete, solid, longitudinally rugulose, non-reticulate; surface strongly glutinous, greyish brown (8D3) to reddish brown (8D4–5); annulus 10–14 mm wide, evanescent, thick, membranous, glutinous, white, attached to stipe apex; base shrouded in viscid, whitish mycelial tomentum. Tubes 9–13 mm deep, adnexed, abruptly depressed around the stipe, greyish yellow (3B5–7), unchanging when bruised; pores up to 1.5 mm, angular, concolorous with the tubes.

Micromorphological characteristics (Figs. 13–17): Basidiospores (10.3–) 11.9–13.3 (–15.0) × (6.8–) 8.3–9.5 (–11.7) μm (n = 107, mean length = 12.56 ± 0.70, mean width = 8.92 ± 0.62, Q = (1.23–) 1.33–1.49 (–1.67), mean Q = 1.41 ± 0.08) in internal line, (12.8–) 14.2–15.8 (–17.2) × (8.8–) 10.5–11.8 (–13.9) μm (n = 99, mean length = 15.04 ± 0.80, mean width = 11.15 ± 0.68, Q = (1.20–) 1.29–1.42 (–1.61), mean Q = 1.35 ± 0.06) in external line, inequilateral with an obscure suprahilar depression in profile, broadly ellipsoid to ovoid-ellipsoid in face view, ornamented with longitudinal ridges over the entire surface, orange-yellow (4B7–8) in water, yellowish brown in KOH, distinctly dextrinoid, thick-walled (0.5–1 μm); ridges 15–19 per spores, occasionally anastomosing, (1.1–) 1.2–1.5 (–1.8) μm high (n = 47, mean width = 1.31 ± 0.15). Basidia (36.0–) 38.5–45.9 (–53.1) × (13.8–) 15.2–17.1 (–17.8) μm (n = 33, mean length = 42.20 ± 3.68, mean width = 16.17 ± 0.94) in main body, (4.8–) 5.6–6.8 (–7.7) × (2.4–) 2.8–3.6 (–3.9) μm (n = 46, mean length = 6.21 ± 0.62, mean width = 3.16 ± 0.41) in sterigmata, clavate to broadly clavate, four-spored. Basidioles broadly clavate. Cheilocystidia gregarious when young, becoming inconspicuous in age, (14.7–) 16.4–24.7 (–32.7) × (8.0–) 9.2–11.5 (–13.4) μm (n = 52, mean length = 20.57 ± 4.17, mean width = 10.35 ± 1.14), subclavate to subfusiform, occasionally septate, smooth, hyaline, colorless in KOH, inamyloid, thin-walled. Pleurocystidia scattered, (46.4–) 59.2–75.0 (–80.3) × (10.2–) 11.3–15.5 (–19.8) μm (n = 41, mean length = 67.10 ± 7.87, mean width = 13.43 ± 2.10), subcylindrical to narrowly subfusoid, somewhat flexuous, smooth, hyaline, colorless in KOH, inamyloid, thin-walled. Hymenophoral trama bilateral-divergent of the *Boletus*-subtype; elements (5.6–) 6.8–9.6 (–11.5) μm wide (n = 35, mean width = 8.19 ± 1.40) in lateral stratum, (3.5–) 3.9–5.4 (–6.2) μm wide (n = 28, mean width = 4.63 ± 0.77) in mediostratum, cylindrical, smooth, hyaline, colorless in KOH, inamyloid, thin-walled. Outermost layer of pileipellis an ixotrichoderm consisting of loosely interwoven hyphae (2.8–) 3.9–5.0 (–5.8) μm wide (n = 78, mean width = 4.40 ± 0.55), at first vertically and compactly arranged then disrupted with age, cylindrical, smooth or occasionally thinly encrusted with hyaline crystals, colorless in KOH, inamyloid, thin-walled; terminal elements (18.9–) 26.8–48.0 (–70.4) × (4.3–) 5.4–7.3 (–8.7) μm (n = 37, mean length = 37.44 ± 10.61, mean width = 6.36 ± 0.97), clavate-pedicellate, often with a subcapitate apex, smooth or thinly encrusted with hyaline crystals, thin-walled. Innermost layer of pileipellis composed of more or less prostrate, loosely arranged hyphae (3.3–) 4.6–7.9 (–10.2) μm wide (n = 61, mean width = 6.27 ± 1.66), subcylindrical, with intracellular (vacuolar) brownish pigment (in water), yellowish in KOH, inamyloid or dextrinoid, smooth, not incrusting, thin-walled. Pileitrama of subcylindrical, subparallel hyphae (3.9–) 5.9–12.6 (–17.7) μm wide (n = 65, mean width = 9.23 ± 3.36), smooth, gelatinized, hyaline, colorless in KOH, inamyloid, thin-walled. Stipitipellis similar to the outermost layer of pileipellis; terminal elements (31.9–) 37.4–60.2 (–75.6) × (5.5–) 6.1–7.5 (–8.3) μm (n = 38, mean length = 48.80 ± 11.40, mean width = 6.77 ± 0.70), colorless, smooth, not encrusted, thin-walled; underlying stratum consisting of prostrate, parallel, subcylindrical hyphae (3.8–) 4.4–6.3 (–7.2) μm wide (n = 56, mean width = 5.33 ± 0.94), with intracellular (vacuolar) brownish pigment (in water), yellowish in KOH, inamyloid or dextrinoid, smooth, thin-walled. Stipe trama composed of longitudinally running, cylindrical hyphae (4.8–) 6.9–11.4 (–16.5) μm wide (n = 89, mean width = 9.13 ± 2.25), unbranched, smooth, hyaline, brownish in KOH, inamyloid, with slightly thickened walls (0.8–) 0.9–1.3 (–1.4) μm (n = 33, mean thickness = 1.12 ± 0.19). Elements of basal mycelium (2.9–) 3.3–4.5 (–5.8) μm wide (n = 44, mean width = 3.88 ± 0.60), hyaline, colorless pale yellow in KOH, inamyloid, thin-walled. Elements of annulus (3.0–) 3.7–4.7 (–5.3) μm wide (n = 77, mean width = 4.21 ± 0.48), cylindrical, occasionally branched, loosely interwoven, septate, gelatinized, hyaline, colorless in KOH, inamyloid, thin-walled. Clamp connections absent.

Habitat and phenology: Solitary to scattered on ground in evergreen broad-leaved forests, from late May to September.

Known distribution: Okinawa (Ishigaki Island), Honshu (Hiroshima, Yamaguchi, Fukui, Gifu).

Holotype: TNS-F-39710, on ground in an evergreen broad-leaved forest dominated by *Quercus miyagii* Koidz. and *Castanopsis sieboldii* (Makino) Hatus. ex T. Yamaz. et Mashiba, Mt. Banna, Ishigaki-shi, Okinawa Pref., 11 Aug. 2011, coll. Takahashi, H.

Extralimital specimens examined: TNS-F-52264, on the ground under *C. sieboldii*, Myojin-cho, Tsuruga-shi, Fukui Pref., 17 Sep. 2008, coll. Kuribayashi, Y.; TNS-F-52265, on ground under *C. sieboldii*, Hiroshima Ryokka Center, Hiroshima Pref., 7 Sep. 2011, coll. Kawakami, Y.; TNS-F-52266, on ground under *C. sieboldii*, Ohara-Lake, Tokudi, Yamaguchi-shi, Yamaguchi Pref., 11 Sep. 2011, coll. Kaijo, K.; TNS-F-52267, Agi, Nakatsugawa-shi, Gifu Pref., 13 Sep. 2013, coll. Yamada, H.

Gene sequenced specimens and GenBank accession numbers: TNS-F-39710, AB968238 (ITS); TNS-F-52265, AB972884 (LSU); TNS-F-39710, AB972885 (LSU); TNS-F-52267, AB972886 (LSU).

Japanese name: Numeri-asinagaiguchi (named by H. Takahashi & Y. Taneyama).

Comments: The entirely glutinous basidiomata with a distinct, white annulus, the greyish yellow hymenophore, and the basidiospores with prominent longitudinal ridges suggest that the new species is closely related to the genus *Boletellus*, section *Ixocephali* Singer (Singer 1945, 1986). However, Singer's definition of the genus *Boletellus* based on basidiospore morphology is considered to be rather artificial and polyphyletic. Based on a comprehensive evaluation of macromorphological and ecological characteristics such as the entirely glutinous basidiomata, yellow hymenophore, and laurel forest habitat, we propose that this new species and the section *Ixocephali* are accommodated in the genus *Aureoboletus* (Pouzar 1957; Klofac 2010). Molecular phylogenetic trees (Fig. 12) based on ribosomal large subunit (LSU) sequences also support the present fungus is within the *Aureoboletus* clade.

Table 1. Comparison of basidiospore measurements.
担子胞子の比較測定結果.

Specimen	Reference	Country	Spore size
Boletellus longicollis RSNB5836 (neotype)	Horak 2011	Malaysia	11–12 × 6.8–8 μm
Boletellus longicollis	Zhishu et al. 1993	China	10.5–14 × 9–10.5 μm
Boletellus longicollis	Nagasawa E 1987	Japan	12.5–16 × 10–12 μm
Boletellus longicollis 6527 (ex Herb. Hongo, ZT)	Horak 2011	Japan	12.5–15 × 8.5–9.5 μm
Aureoboletus liquidus TNS-F-39710	This study	Ishigaki, Japan	(12.8–) 14.2–15.8 (–17.2) × (8.8–) 10.5–11.8 (–13.9) μm
Aureoboletus liquidus TNS-F-52264	This study	Fukui, Japan	(12.1–) 13.7–15.1 (–16.3) × (9.2–) 10.4–11.7 (–12.4) μm
Aureoboletus liquidus TNS-F-52265	This study	Hiroshima, Japan	(12.6–) 13.7–15.8 (–16.9) × (9.1–) 9.9–11.8 (–13.0) μm
Aureoboletus liquidus TNS-F-52266	This study	Yamaguchi, Japan	(12.4–) 13.2–15.0 (–17.0) × (9.6–) 10.3–11.5 (–12.4) μm
Aureoboletus liquidus TNS-F-52267	This study	Gifu, Japan	(12.5–) 13.4–15.1 (–16.5) × (6.5–) 9.7–11.3 (–12.5) μm

"*Boletellus*" *longicollis* (Ces.) Pegler & T.W.K. Young from Malaysia (Cesati 1879; Corner 1972; Pegler and Young 1981; Horak 2011) appears to be most closely related to *A. liquidus*, but it has significantly smaller basidiospores (Table 1): 11–12 × 6.5–8 μm (Horak 2011) and subventricose, much shorter pleurocystidia: –45 × 6–9 μm (Corner 1972). *Aureoboletus liquidus* also comes in the alliance of "*Boletellus*" *singeri* Gonz.-Velázq. & R. Valenz. from Mexico and Belize (Gonzáles-Velázquez and Valenzuela 1995; Ortiz-Santana et al. 2007), which differs in possessing a pale yellow pileus, somewhat narrower basidiospores: 11.2–18.4 × 7.2–9.6 μm, mean Q = 1.74 (Ortiz-Santana et al. 2007), and mostly fusoid to ventricose, significantly shorter pleurocystidia: 49–64 × 12–20 μm (Gonzáles-Velázquez and Valenzuela 1995).

New combinations 新組み合わせ

Aureoboletus* section *Ixocephali (Singer) Har. Takah. & Taneyama, **comb. nov.** ヌメリコウジタケ属ヌメリアシナガイグチ節（高橋春樹 & 種山裕一新称）

 MycoBank no.: MB 813806.

Basionym: *Boletellus* section *Ixocephali* Singer, Farlowia 2: 135 (1945).

Type species: *Boletopsis singaporensis* Pat. & C.F. Baker, J Straits Brch R Asiat Soc 78: 69. 1918 (= *Boletus longicollis* Ces.).

Aureoboletus longicollis (Ces.) Har. Takah. & Taneyama, **comb. nov.**

 MycoBank no.: MB 813821.

Basionym: *Boletus longicollis* Ces., Atti Accad Sci fis mat Napoli 8 (3): 4 (1879).

Aureoboletus singeri (Gonz.-Velázq. & R. Valenz.) Har. Takah. & Taneyama, **comb. nov.**

 MycoBank no.: MB 813863.

Basionym: *Boletellus singeri* Gonz.-Velázq. & R. Valenz. [as '*singerii*'], Mycotaxon 55: 400 (1995).

3. *Aureoboletus liquidus* ヌメリアシナガイグチ

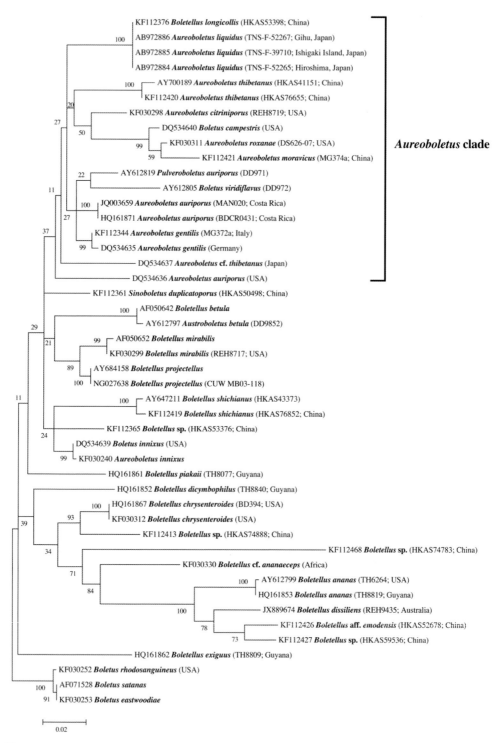

Fig. 12 – Molecular Phylogenetic anaylsis by Maximum Likelihood method based on nuc-rLSU sequences. The trees showed that all of the *A. liquidus* sequences cluster together with a Chinese specimen identified as "*Boletellus longicollis*", belonging to "*Aureoboletus* clade". These results corroborate those of Wu et al. (Wu et al. 2014).

　核リボソーム LSU を用いた最尤法による系統樹．解析結果によればヌメリアシナガイグチは，中国産の *Boletellus longicollis* と同定された標本と同一のクレードを形成し，ヌメリコウジタケ属分岐群（*Aureoboletus* clade）に含まれる事が示唆された．この結果は Wu ら（Wu et al. 2014）に準じている．

References 引用文献

Cesati V. 1879. Mycetum in itinere Borneensi lectorum. Atti Accad Sci fis mat Napoli 8 (3): 1-28.
Corner EJH. 1972. *Boletus* in Malaysia. Government Printing Office, Singapore.
Gonzáles-Velázquez A, Valenzuela R. 1995. A new species of *Boletellus* (Basidiomycotina, Agaricales: Boletaceae) from Mexico. Mycotaxon 55: 399-404.
Horak E. 2011. Revision of Malaysian Species of Boletales s.l. (Basidiomycota) Described by E.J.H. Corner (1972, 1974). Malayan forest records 51: 1-283, Forest Research Institute Malaysia.
Klofac W. 2010. Die Gattung *Aureoboletus*, ein weltweiter Überblick. Ein Beitrag zu einer monographischen Bearbeitung [The genus *Aureoboletus*, a world-wide survey. A contribution to a monographic treatment]. Öst Z Pilzk 19: 133-74.
Nagasawa E. 1987. Strobilomycetaceae. In: Imazeki R, Hongo T (eds), Colored illustrations of mushrooms of Japan I. Hoikusha, Osaka, pp 273-285 (in Japanese).
Ortiz-Santana B, Lodge DJ, Baroni TJ, Both EE. 2007. Boletes from Belize and the Dominican Republic. Fungal Diversity 27 (2): 247-416.
Pegler DN, Young TWK. 1981. A natural arrangement of the Boletales, with reference to spore morphology. Trans Brit Mycol Soc 76 (1): 103-146.
Pouzar Z. 1957. Nova genera macromycetum I. Ceská Mykologie 11 (1): 48-50.
Singer R. 1945. The Boletineae of Florida with notes on extralimital species. I. The Strobilomycetaceae. Farlowia 2: 97-141.
Singer R. 1986. The Agaricales in modern taxonomy, 4th edn. Koeltz, Koenigstein.
Wu G, Feng B, Xu J, Zhu XT, Li YC, Zeng NK, Hosen MI, Yang ZL. 2014. Molecular phylogenetic analyses redefine seven major clades and reveal 22 new generic clades in the fungal family Boletaceae, Fung Diver 69 (1): 93-115.
Zhishu B, Guoyang Z, Taihui L. 1993. The Macrofungus Flora of China's Guangdong Province. The Chinese University Press, Hong Kong, p 461.

3. ヌメリアシナガイグチ（新種；高橋春樹 & 種山裕一新称）*Aureoboletus liquidus* Har. Takah. & Taneyama, sp. nov.

肉眼的特徴（Figs. 18-23）：傘は径 45-70 mm，最初半球形のち饅頭形になり，最初平坦まもなく不規則な皺状隆起を表す；表面は著しい粘液に被われ，埋生繊維状，全体に赤褐色；縁部は未熟な時白色の縁片膜を付着する．肉は厚さ9 mm 以下，類白色，ゼラチン化し，空気に触れても変色せず，特別な味や匂いはない．柄は100-170×8-17 mm，円柱形，根元に向かってやや太くなり，痩せ型で背が高く，中心生，中実，縦に沿って浅い縦皺を表し，網目模様を欠く；表面は著しい粘液に被われ，灰褐色〜赤褐色を帯びるが傘より淡色；ツバは幅10-14 mm，消失性，厚い膜質，粘液に被われ，白色，柄の頂部に付着する；根元の菌糸体は白色．管孔は長さ9-13 mm，上生，柄の周囲において急激に嵌入し，黄色〜暗オリーブ黄色；孔口は角形，1.5 mm 以下，管孔と同色，管孔及び孔口は傷を受けても変色しない．

顕微鏡的特徴（Figs. 13-17）：担子胞子は内輪郭（畝状隆起の内側）において (10.3-) 11.9-13.3 (-15.0) × (6.8-) 8.3-9.5 (-11.7) μm (n = 107, mean length = 12.56 ± 0.70, mean width = 8.92 ± 0.62, Q = (1.23-) 1.33-1.49 (-1.67), mean Q = 1.41 ± 0.08)，外輪郭（畝状隆起の外側）において (12.8-) 14.2-15.8 (-17.2) × (8.8-) 10.5-11.8 (-13.9) μm (n = 99, mean length = 15.04 ± 0.80, mean width = 11.15 ± 0.68, Q = 1.29-1.42, mean Q = 1.35 ± 0.06)，下部側面に不明瞭でなだらかな凹みがあり（イグチ型），広楕円形〜卵状楕円形，表面に縦に走る畝状隆起を表し，水封で橙黄色，アルカリ溶液において黄褐色，強偽アミロイド，厚壁 (0.5-1 μm)；畝状隆起は1胞子中15-19 本，ところどころ分岐し，高さ1.2-1.5 μm．担子器は (36.0-) 38.5-45.9 (-53.1) × (13.8-) 15.2-17.1 (-17.8) μm（本体），(4.8-) 5.6-6.8 (-7.7) × (2.4-) 2.8-3.6 (-3.9) μm（ステリグマ），こん棒形〜広こん棒形，4胞子性．縁シスチジアは幼時群生し成熟とともに不明瞭となり，(14.7-) 16.4-24.7 (-32.7) ×

(8.0-) 9.2-11.5 (-13.4) μm，類こん棒形〜類紡錘形，しばしば隔壁を有し，平滑，無色，非アミロイド，薄壁．側シスチジアは（46.4-) 59.2-75.0 (-80.3)×(10.2-) 11.3-15.5 (-19.8) μm，散生，類円柱形〜狭紡錘形，ところどころやや曲がりくねり，平滑，無色，非アミロイド，薄壁．子実層托実質はイグチ亜型；菌糸は側層において幅 (5.6-) 6.8-9.6 (-11.5) μm，中層において幅 (3.5-) 3.9-5.4 (-6.2) μm，平滑，無色，非アミロイド，薄壁．傘表皮組織の最外層は最初密集した粘性毛状被を成すがのち次第に構造がくずれて緩く錯綜する；菌糸は幅 (2.8-) 3.9-5.0 (-5.8) μm，円柱形，しばしば無色の結晶物が薄く凝着し，アルカリ溶液において無色，非アミロイド，薄壁；末端細胞は (18.9-) 26.8-48.0 (-70.4)×(4.3-) 5.4-7.3 (-8.7) μm，長い柄を持つこん棒形，しばしばやや頂部頭状形になり，薄壁．傘表皮組織の内層を構成する菌糸は幅 (3.3-) 4.6-7.9 (-10.2) μm，円柱形，匍匐性，緩く錯綜し，淡褐色の色素が細胞の液胞内に存在し，非アミロイドまたは偽アミロイド，平滑で凝着物は見られない．傘実質の菌糸はやや並列または緩く錯綜し，幅 (3.9-) 5.9-12.6 (-17.7) μm，円柱形，平滑，ゼラチン化し，無色，非アミロイド，薄壁．柄表皮組織は傘に類似する；末端細胞は (31.9-) 37.4-60.2 (-75.6)×(5.5-) 6.1-7.5 (-8.3) μm，無色，平滑，凝着物を欠き，薄壁；表皮下層の菌糸は幅 (3.8-) 4.4-6.3 (-7.2) μm，匍匐性で並列し，類円柱形，アルカリ溶液において淡黄色，非アミロイドまたは偽アミロイド，平滑，薄壁．柄の実質の菌糸は幅 (4.8-) 6.9-11.4 (-16.5) μm，縦に沿って配列し，円柱形，無分岐，平滑，無色，アルカリ溶液において帯褐色，非アミロイド，やや厚壁 (0.9-1.3 μm)．根元の菌糸体を構成する菌糸は幅 (2.9-) 3.3-4.5 (-5.8) μm，無色，アルカリ溶液において無色または淡黄色，非アミロイド，薄壁．被膜を構成する菌糸は幅 (3.0-) 3.7-4.7 (-5.3) μm，円柱形，しばしば分岐し，緩く錯綜し，隔壁を持ち，ゼラチン化し，無色，非アミロイド，薄壁．全ての菌糸はクランプを欠く．

生態および発生時期：常緑広葉樹林内地上に孤生または散生，5月下旬〜9月．

分布：沖縄（石垣島），本州（広島，山口，福井，岐阜）．

供試標本：TNS-F-39710（正基準標本），スダジイ，オキナワウラジロガシを主体とする常緑広葉樹林内地上，沖縄県石垣市バンナ岳，2011年8月11日，高橋春樹採集．

地域外供試標本：TNS-F-52264，スダジイの樹下，福井県敦賀市明神町，2008年9月17日，栗林義宏採集；TNS-F-52265，スダジイの樹下，広島県広島緑化センター，2011年9月7日，川上嘉章採集；TNS-F-52266，スダジイの樹下，山口県山口市徳地大原湖，2011年9月11日，海上和江採集；TNS-F-52267，岐阜県中津川市阿木，2013年9月13日，山田弘採集．

分子解析に用いた標本並びに GenBank 登録番号：TNS-F-39710，AB968238（ITS）；TNS-F-52265，AB972884（LSU）；TNS-F-39710，AB972885（LSU）；TNS-F-52267，AB972886（LSU）．

主な特徴：子実体は全体に粘液に被わる；傘は赤褐色，不規則な皺状隆起を表す；柄は長く伸び，消失性の白色のツバを持つ；子実層托（管孔）は黄色；担子胞子は楕円形で，縦に走る畝状隆起を表す；側シスチジアは類円柱形〜狭紡錘形で，やや曲がりくねる；傘と柄の上表皮層の末端細胞はこん棒形またはやや頂部頭状形；スダジイ，オキナワウラジロガシを主な構成樹種とする常緑広葉樹林内地上に発生．

コメント：全体に粘液に被われた子実体，膜質の被膜の存在，黄色の子実層托，そして縦に走る皺状構造物を表す担子胞子は本種がキクバナイグチ属 *Boletellus*，イクソケファリ節 section *Ixocephali* Singer（Singer 1945, 1986）に近縁であることを示唆しているが，担子胞子の表面構造に基づく *Boletellus* 属の分類概念は多系統で人為的な感は否めない．ここでは粘液に被われた子実体，黄色の子実層托，そしてシイ・カシ林に発生する肉眼的，生態的性質を総合的に判断し，本種並びにイクソケファリ節についてヌメリコウジタケ属 *Aureoboletus*（Pouzar 1957; Klofac 2010）との組み合わせを提唱する．核リボソームの大サブユニット（LSU）の配列に基づく分子

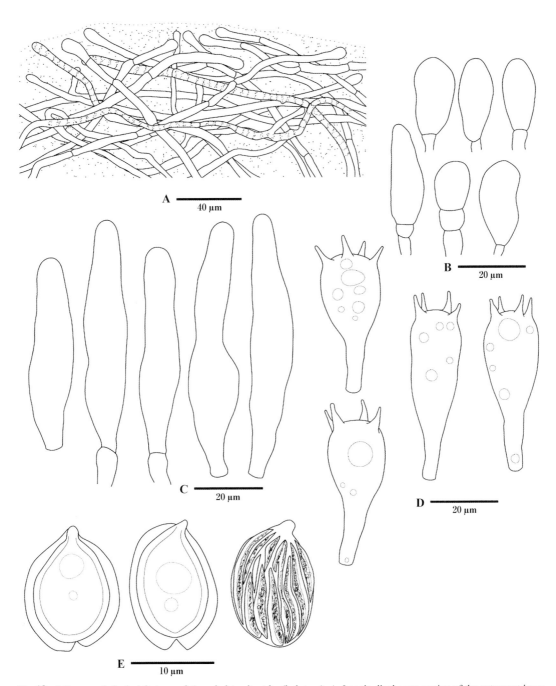

Fig. 13 – Micromorphological features of *Aureoboletus liquidus* (holotype): **A**. Longitudinal cross section of the outermost layer of pileipellis. **B**. Cheilocystidia. **C**. Pleurocystidia. **D**. Basidia. **E**. Basidiospores. Illustrations by Taneyama, Y.
ヌメリアシナガイグチの顕鏡図（正基準標本）：**A**. 傘の上表皮層の縦断面. **B**. 縁シスチジア. **C**. 側シスチジア. **D**. 担子器. **E**. 担子胞子. 図：種山裕一.

24 —— 3. *Aureoboletus liquidus* ヌメリアシナガイグチ

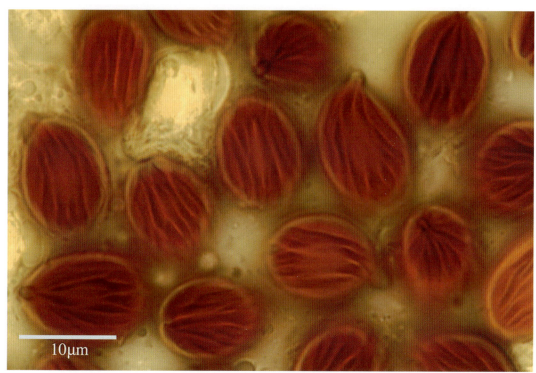

Fig. 14 – Basidiospores of *Aureoboletus liquidus* (in Melzer's reagent, holotype). Photo by Taneyama, Y.
ヌメリアシナガイグチの担子胞子（メルツァー溶液で封入，正基準標本）．写真：種山裕一．

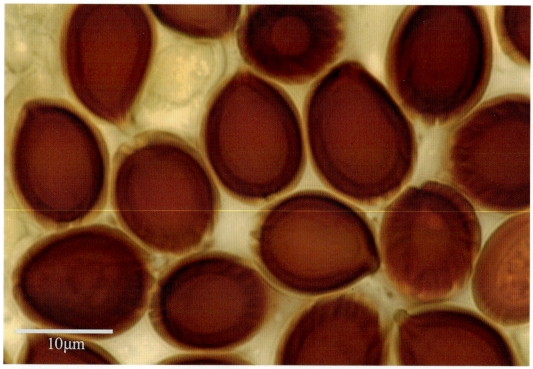

Fig. 15 – Basidiospores of *Aureoboletus liquidus* (in Melzer's reagent, holotype). Photo by Taneyama, Y.
ヌメリアシナガイグチの担子胞子（メルツァー溶液で封入，正基準標本）．写真：種山裕一．

3. *Aureoboletus liquidus* ヌメリアシナガイグチ ―― 25

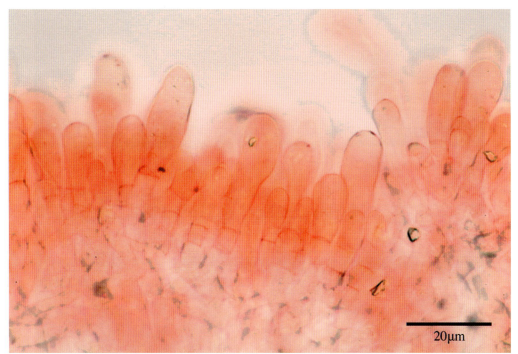

Fig. 16 – Longitudinal cross section of the pore in *Aureoboletus liquidus* (in 3%KOH and Congo red stain, TNS-F-52265). Photo by Taneyama, Y.
ヌメリアシナガイグチの孔口の縦断面（3％水酸化カリウム溶液で封入した後コンゴー赤染色，TNS-F-52265）．写真：種山裕一.

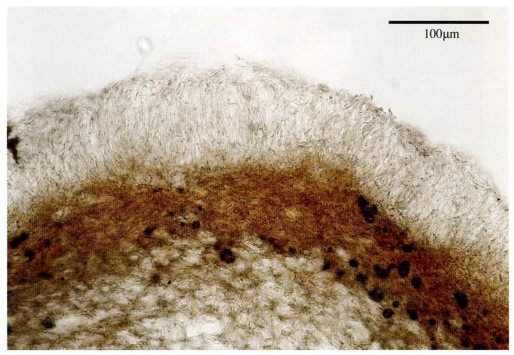

Fig. 17 – Longitudinal cross section of the pileipellis of *Aureoboletus liquidus* (from an immature basidioma in 3% KOH, TNS-F-52265). Photo by Taneyama, Y.
ヌメリアシナガイグチの傘上表皮層の縦断面（3％水酸化カリウムで封入した未熟な子実体，TNS-F-52265）．写真：種山裕一.

Fig. 18 – Basidiomata of *Aureoboletus liquidus* occurring on ground in an evergreen broad-leaved forest dominated by *Q. miyagii* and *C. sieboldii* (holotype), 11 Aug. 2011, Mt. Banna, Ishigaki Island. Photo by Takahashi, H.
　スダジイ，オキナワウラジロガシを主体とする常緑広葉樹林内に発生したヌメリアシナガイグチの子実体（正基準標本），2011年8月11日，石垣島バンナ岳．写真：高橋春樹．

Fig. 19 – Basidioma of *Aureoboletus liquidus* on ground in an evergreen broad-leaved forest dominated by *Q. miyagii* and *C. sieboldii* (holotype), 11 Aug. 2011, Mt. Banna, Ishigaki Island. Photo by Takahashi, H.
スダジイ，オキナワウラジロガシを主体とする常緑広葉樹林内に発生したヌメリアシナガイグチの子実体（正基準標本），2011年8月11日，石垣島バンナ岳．写真：高橋春樹．

Fig. 20 – Basidioma of *Aureoboletus liquidus* on ground in an evergreen broad-leaved forest dominated by *Q. miyagii* and *C. sieboldii* (holotype), 11 Aug. 2011, Mt. Banna, Ishigaki Island. Photo by Takahashi, H.
スダジイ，オキナワウラジロガシを主体とする常緑広葉樹林内に発生したヌメリアシナガイグチの子実体（正基準標本），2011年8月11日．石垣島バンナ岳．写真：高橋春樹．

Fig. 21 – Basidioma of *Aureoboletus liquidus*, 18 May 2002, Mt. Banna, Ishigaki Island. Photo by Takahashi, H.
ヌメリアシナガイグチの子実体，2002年5月18日，石垣島バンナ岳．写真：高橋春樹．

系統樹（Fig. 12）は本種がヌメリコウジタケ属分岐群（*Aureoboletus* clade）に位置することを支持している．

　マレーシア産 "*Boletellus*" *longicollis*（Ces.）Pegler & T. W. K. Young from Malaysia（Cesati, V. 1879；Corner 1972；Horak 2011；Pegler & T. W. K. Young 1981）は本種に最も近縁と思われるが，より小形の担子胞子（表1）：11–12×6.5–8 µm（Horak 2011）および片膨れ状でより短形の側シスチジア：–45×6–9 µm（Corner 1972）を有する．本種はメキシコおよびベリーズから報告された "*Boletellus*" *singeri* Gonz.-Velázq. & R. Valenz.（Gonzáles-Velázquez and Valenzuela 1995；Ortiz-Santana et al. 2007）にも類似する．しかし後者は淡黄色の傘を持ち，やや幅の狭い担子胞子：11.2–18.4×7.2–9.6 µm，mean Q = 1.74（Ortiz-Santana et al. 2007）および紡錘形〜片膨れ状でより短形の側シスチジア：49–64×12–20 µm（Gonzáles-Velázquez and Valenzuela 1995）を形成する．

Fig. 21 – Basidioma of *Aureoboletus liquidus*, 18 May 2002, Mt. Banna, Ishigaki Island. Photo by Takahashi, H.
　ヌメリアシナガイグチの子実体，2002年5月18日，石垣島バンナ岳．写真：高橋春樹．

Fig. 23 – Underside view of the mature basidioma of *Aureoboletus liquidus*, 18 May 2002, Mt. Banna, Ishigaki Island. Photo by Takahashi, H.
ヌメリアシナガイグチの成菌の孔口，2002年5月18日，石垣島バンナ岳．写真：高橋春樹．

4. *Boletus bannaensis* Har. Takah. ナンヨウウラベニイグチ

Mycoscience 48 (2): 90 (2007) [MB#529510].

Etymology: The specific epithet refers to a toponym of the type locality, Mt. Banna (Ishigaki Island).

Macromorphological characteristics (Figs. 25-34): Pileus 30-110 mm in diameter, at first hemispherical, expanding to broadly convex, with inrolled then straight margin; surface dry, minutely velutinous to subglabrous, often minutely rimose-areolate when mature, brownish grey (5C2 to 5D2) or brownish orange (5C3) to greyish brown (5D3), often partially or at times entirely dull red (10B3-4 to 11B3-4) when young, in places with blackish spots in age. Flesh up to 25 mm thick, firm, whitish in the pileus, pale yellowish in the stipe, faintly staining light blue (23A4-5) when cut; odor and taste indistinct. Stipe 40-100 × 10-25 mm, subequal or somewhat enlarged toward the base, central, terete, solid, usually finely reticulated above or only at the extreme apex by a thin-veined, concolorous or greyish red (10C5) to brownish red (10C6) reticulum with meshes subequal (above) to longitudinally elongated (below); surface dry, subglabrous to subtomentose, entirely pastel yellow (2-3A4) to light yellow (2-3A5) when young, then darker toward the base, slowly staining blue where handled; base covered with whitish strigose mycelium. Tubes 5-9 mm deep, relatively short in comparison with the size of the pileus, shallowly depressed around the stipe or subdecurrent, pastel yellow (2-3A4) to light yellow (2-3A5), slowly staining blue when cut; pores small (2-3 per mm), subcircular, concolorous or greyish red (7B6) to reddish orange (7B7), slowly staining blue where handled.

Micromorphological characteristics (Fig. 24): Basidiospores 6.5-9 × 3.5-4 µm (n = 63 spores of 5 basidiocarps, Q = 1.8-2.3), inequilateral with a shallow suprahilar depression in profile, oblong ellipsoid in face view, smooth, melleous in H_2O, thick-walled (up to 1 µm). Basidia 18-30 × 6-9 µm, clavate, four-spored. Cheilocystidia gregarious, 28-31 × 5-6.5 µm, fusiform, smooth, reddish in water, thin-walled. Pleurocystidia scattered, similar in shape to the cheilocystidia but colorless. Hymenophoral trama divergent, composed of hyphae 4-10 µm wide, cylindrical, smooth, hyaline, thin-walled. Pileipellis a trichoderm of loosely interwoven hyphae, 2-5 µm wide, cylindrical, smooth, hyaline or with intracellular brown pigment (in water), thin-walled. Pileitrama of cylindrical, loosely interwoven hyphae 5-8 µm wide, smooth, colorless, thin-walled. Stipe trama composed of longitudinally arranged, cylindrical cells 4-10 µm wide, unbranched, smooth, colorless, thin-walled. Clamps absent in all tissues.

Habitat and phenology: Solitary to scattered on ground in evergreen broad-leaved forests dominated by *Quercus miyagii* Koidz. and *Castanopsis sieboldii* (Makino) Hatus. ex T. Yamaz. et Mashiba, May to October.

Known distribution: Okinawa (Ishigaki Island).

Specimens examined: KPM-NC0010094 (holotype), on ground in an evergreen broad-leaved forest dominated by *Q. miyagii* and *C. sieboldii,* Banna Park, Ishigaki-shi, Okinawa Pref., 30 May 2002, coll. Takahashi, H.; KPM-NC0023874, same location, 20 Oct. 2011, coll. Takahashi, H.

Japanese name: Nanyou-urabeni-iguchi.

Comments: Distinctive features of this species include the medium-sized, boletoid basidiomata; the greyish brown pileus occasionally with pinkish tinge; the cyanescent flesh; the finely reticulate, light yellow stipe; the minute, light yellow pores often discolored red; and the habitat in subtropical evergreen broad-leaved forests.

The *Boletus*-habit of the basidiomata with the minute pores discolored red suggests that the present species belongs to the genus *Boletus*, section *Luridi* Fr. as defined in Singer (Singer 1986). Within this section, *B. bannaensis* seems to be closely related to *Boletus satanas* Lenz, originally described from

Europe (Singer 1967; Alessio 1985; Bessette et al. 2000), *Boletus firmus* Frost from North American (Snell and Dick 1970; Bessette et al. 2000), and *Boletus quercinus* Hongo from Japan (Hongo 1967; Hongo and Nagasawa 1975). These taxa are distinct from *B. bannaensis* mainly by having a persistently greyish pileus not tinged with dull red, and a reddish or greyish brown (not entirely yellowish) stipe. Moreover, *B. satanas* has a massive bulbous stipe, and *B. quercinus* forms a pileipellis with intercellular, incrusting pigments.

References 引用文献
Alessio CL. 1985. *Boletus* Dill.ex L. (sensu lato). Fungi Europaei 2. Biella Giovanna, Saronno.
Bessette AE, Roody WC, Bessette AR. 2000. North American Boletes. A color guide to the fleshy pored mushrooms. Syracuse University Press, New York.
Hongo T. 1967. Notulae Mycologicae (6). Mem Shiga Univ 17: 89-95.
Hongo T, Nagasawa E. 1975. Notes on some boleti from Tottori. Rept Tottori Mycol Inst (Japan) 12: 31-40.
Singer R. 1967. Die Röhrlinge II. Verlag Julius Klinkhardt, Bad Heilbrum Obb.
Singer R. 1986. Agaricales in modern taxonomy, 4th edn. Koeltz, Koenigstein.
Snell WH, Dick EA. 1970. The boleti of northeastern North America. Cramer, Vaduz.
Takahashi H. 2007. Five new species of the Boletaceae from Japan. Mycoscience 48 (2): 90-99.

4. ナンヨウウラベニイグチ *Boletus bannaensis* Har. Takah.
Mycoscience 48 (2): 90 (2007) [MB#529510].

肉眼的特徴（Figs. 25-34）：傘は径30-110 mm，最初半球形，のち饅頭形～ほぼ平開し，表面は乾性，密綿毛状，成熟するとしばしば部分的に細かい亀甲状のひび割れを生じ，帯褐灰色～灰褐色，幼時しばしば部分的にまたは時に全面にバラ色を呈し，老成すると所々黒褐色の染みを表す．肉は厚さ25 mm以下（傘中央部），傘は類白色，柄は淡黄色，空気に触れると徐々に弱く青変し，特別な味や臭いはない．柄は40-100×10-25 mm，ほぼ上下同大または下方に向かってやや太くなり，中心生，中実，通常上部または頂部のみに帯褐赤色または表面と同色の繊細な網目を表す；表面は乾性，ほぼ平滑～やや密綿毛状，最初全体に淡黄色を呈し，のち下部に向かって暗色を帯び，触れると徐々に青変する；根元は類白色剛毛状の菌糸体に被われる．管孔は肉の厚さと比較して相対的に短く長さ5-9 mm，柄の周囲において僅かに嵌入するかまたはやや垂生し，淡黄色，空気に触れると徐々に青変する；孔口は小型（2-3 per mm），類円形，管孔と同色または橙赤色を帯び，傷を受けると徐々に青変する．

顕微鏡的特徴（Fig. 24）：担子胞子は$6.5-9×3.5-4$ μm（n = 63 spores of 5 basidiocarps, Q = 1.8-2.3），下部側面に不明瞭でなだらかな凹みがあり（イグチ型），長楕円形，平坦，蜜色，厚壁．担子器は18-30×6-9 μm，こん棒形，4胞子性．縁シスチジアは群生し，28-31×5-6.5 μm，紡錘形，平坦，帯赤色（水封），薄壁．側シスチジアは散在し，縁シスチジアに似るが無色．子実層托実質は散開型，菌糸は幅4-10 μm，円柱形，平滑，無色，薄壁．傘の表皮は毛状被を形成し，菌糸は幅2-5 μm，緩く錯綜し，円柱形，平滑，無色または細胞内に帯褐色の色素を有し，薄壁．傘実質の菌糸は幅5-8 μm，緩く錯綜し，円柱形，平滑，無色，薄壁．柄実質の菌糸は幅4-10 μm，縦に沿って配列し，円柱形，無分岐，平滑，無色，薄壁．全ての組織において菌糸はクランプを欠く．

生態および発生時期：スダジイ，オキナワウラジロガシを主体とする常緑広葉樹林内地上に孤生または散生，5月～10月．

分布：沖縄（石垣島）．

供試標本：KPM-NC0010094（正基準標本），常緑広葉樹林内地上，沖縄県石垣市バンナ公園，2002年5月30日，高橋春樹採集；KPM-NC0023874，同上，2011年10月20日，高橋春樹採集．

34 —— 4. *Boletus bannaensis* ナンヨウウラベニイグチ

主な特徴：子実体は中型のヤマドリタケ型；傘は灰褐色または時にバラ色を帯びる；肉は青変性を持つ；柄は淡黄色の地に繊細な網目模様を表す；孔口は微細で，淡黄色またはしばしば橙赤色を帯びる；スダジイ，オキナワウラジロガシを主体とする常緑広葉樹林内地上に発生．

コメント：ヤマドリタケ型の子実体の類型並びに微細で，橙赤色を帯びる孔口を持つ性質は本種がヤマドリタケ属 *Boletus*，ウラベニイロガワリ節 section *Luridi* Fr.（Singer 1986）に属することを示唆している． 節内では欧州産 *Boletus satanas* Lenz（Singer 1967；Alessio 1985；Bessette et al. 2000），北米産 *Boletus firmus* Frost（Snell and Dick 1970；Bessette et al. 2000），日本産ナガエノウラベニイグチ *Boletus quercinus* Hongo（Hongo 1967；Hongo and Nagasawa 1975）に最も近縁と思われる．しかしながらこれらの分類群は傘が常に灰褐色を呈し，赤色を帯びないこと，また柄が赤色または灰褐色を呈し，全体に淡黄色を帯びることがない点でナンヨウウラベニイグチと異なる．さらに *B. satanas* は柄の基部が球根状に膨らみ，ナガエノウラベニイグチは傘表皮組織において細胞外凝着色素を有する．

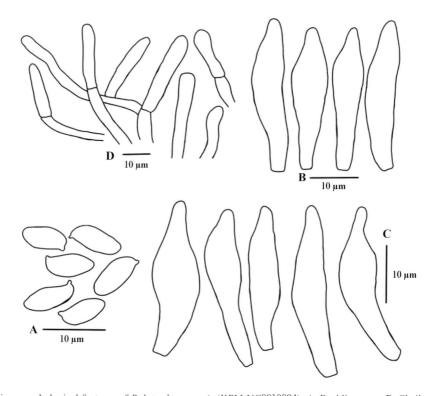

Fig. 24 – Micromorphological features of *Boletus bannaensis* (KPM-NC0010094): **A**. Basidiospores. **B**. Cheilocystidia. **C**. Pleurocystidia. **D**. Elements of the pileipellis. Illustrations by Takahashi, H.
ナンヨウウラベニイグチの顕微鏡図（KPM-NC0010094）：**A**. 担子胞子．**B**. 縁シスチジア．**C**. 側シスチジア．**D**. 傘表皮組織の菌糸．図：高橋春樹．

4. *Boletus bannaensis* ナンヨウウラベニイグチ —— 35

Fig. 25 – Basidiomata of *Boletus bannaensis* (KPM-NC0023874) on ground in a *Castanopsis-Quercus* forest, 20 Oct. 2011, Banna Park, Ishigaki Island. Photo by Takahashi, H.
スダジイ-オキナワウラジロガシ林内地上に発生したナンヨウウラベニイグチの子実体（KPM-NC0023874），2011年10月20日，石垣島バンナ公園．写真：高橋春樹．

4. *Boletus bannaensis* ナンヨウウラベニイグチ

Fig. 26 – Mature basidioma of *Boletus bannaensis* (KPM-NC0023874) on ground in a *Castanopsis-Quercus* forest, 20 Oct. 2011, Banna Park, Ishigaki Island. Photo by Takahashi, H.
スダジイ-オキナワウラジロガシ林内地上に発生したナンヨウウラベニイグチの成熟した子実体（KPM-NC0023874），2011年10月20日，石垣島バンナ公園．写真：高橋春樹．

Fig. 27 − Immature basidiomata of *Boletus bannaensis* (KPM-NC0023874) on ground in a *Castanopsis-Quercus* forest, 20 Oct. 2011, Banna Park, Ishigaki Island. Photo by Takahashi, H.
スダジイ-オキナワウラジロガシ林内地上に発生したナンヨウウラベニイグチの幼菌（KPM-NC0023874），2011年10月20日．石垣島バンナ公園．写真：高橋春樹．

4. *Boletus bannaensis* ナンヨウウラベニイグチ

Fig. 28 – Immature basidiomata of *Boletus bannaensis* (KPM-NC0023874), 20 Oct. 2011, Banna Park, Ishigaki Island. Photo by Takahashi, H.
スダジイ-オキナワウラジロガシ林内地上に発生したナンヨウウラベニイグチの幼菌（KPM-NC0023874），2011年10月20日，石垣島バンナ公園. 写真：高橋春樹.

Fig. 29 – Immature basidioma of *Boletus bannaensis* (KPM-NC0023874), 20 Oct. 2011, Banna Park, Ishigaki Island. Photo by Takahashi, H.
ナンヨウウラベニイグチの幼菌（KPM-NC0023874），2011年10月20日，石垣島バンナ公園．写真：高橋春樹．

Fig. 30 – Immature basidioma of *Boletus bannaensis* (KPM-NC0023874), 20 Oct. 2011, Banna Park, Ishigaki Island. Photo by Takahashi, H.
ナンヨウウラベニイグチの幼菌（KPM-NC0023874），2011年10月20日．石垣島バンナ公園．写真：高橋春樹．

Fig. 31 – Mature basidioma of *Boletus bannaensis* (KPM-NC0023874), 20 Oct. 2011, Banna Park, Ishigaki Island. Photo by Takahashi, H.
ナンヨウウラベニイグチの成熟した子実体（KPM-NC0023874），2011年10月20日，石垣島バンナ公園．写真：高橋春樹．

42 —— 4. *Boletus bannaensis* ナンヨウウラベニイグチ

Fig. 32 – Pileus surface of *Boletus bannaensis* (KPM-NC0023874), 20 Oct. 2011, Banna Park, Ishigaki Island. Photo by Takahashi, H.
ナンヨウウラベニイグチの傘表面 (KPM-NC0023874), 2011年10月20日, 石垣島バンナ公園. 写真：高橋春樹.

Fig. 33 – Underside view of the basidioma of *Boletus bannaensis* (KPM-NC0023874), 20 Oct. 2011, Banna Park, Ishigaki Island. Photo by Takahashi, H.
ナンヨウウラベニイグチの子実体（孔口側，KPM-NC0023874），2011年10月20日，石垣島バンナ公園．写真：高橋春樹．

44 —— 4. *Boletus bannaensis* ナンヨウウラベニイグチ

Fig. 34 – Longitudinal cross section of the basidioma of *Boletus bannaensis* (KPM-NC0023874), 20 Oct. 2011, Banna Park, Ishigaki Island. Photo by Takahashi, H.
ナンヨウウラベニイグチの子実体断面 (KPM-NC0023874), 2011年10月20日, 石垣島バンナ公園. 写真：高橋春樹.

5. *Boletus virescens* Har. Takah. & Taneyama, **sp. nov.** アオアザイグチ

MycoBank no.: MB 809941.

Etymology: The specific epithet means "virescent" or "greenish in color".

Distinctive features of this species are found in relatively small, xerocomoid basidiomata stained with greenish-blue spots in places; minutely rimose, rugulose pileus; ellipsoidal basidiospores; absence of cheilocystidia; ventricose-rostrate pleurocystidia; a pileipellis composed of a trichodermial palisade with more or less inflated, catenulate elements; narrowly fusoid-ventricose or subclavate to subcylindrical caulocystidia; and a habitat in evergreen broad-leaved forests.

Macromorphological characteristics (Figs. 36-40): Pileus 32-40 mm in diameter, at first hemispherical, then expanding to convex, irregularly rugulose, often marked with numerous, superficial, minute clefts or cracks; surface subtomentose, dry, pale yellow (4A3-4) in ground color, often stained with light turquoise (24A4-5) spots in places that become reddish brown (8E5-6) in age; margin not appendiculate. Flesh up to 5 mm, whitish, at times stained greyish red (7B5-6) at the base of stipe and right beneath the surfaces of the pileus and stipe, occasionally with light turquoise (24A4-5) spots here and there, unchanging where bruised; odor not distinctive, taste undetermined. Stipe 50-55 × 5-6 mm, cylindrical but somewhat enlarged toward the apex, slender, central, terete, solid, longitudinally rugulose, sometimes slightly reticulated by a thin-veined, concolorous reticulum at the apex; surface dry, glabrous, concolorous with the pileus above, reddish brown (8E6-7) downward, occasionally with light turquoise (24A3-4) to greyish turquoise (24D4-5) stains especially near the apex; base shrouded by whitish mycelial tomentum. Tubes 3-5 mm deep, adnexed to adnate, often with subdecurrent teeth, greyish orange (5B3-4) around the pore, dull yellow (3B3-4) near the pileus, sometimes stained with light turquoise (24A4-5) spots in places, unchanging when bruised; pores up to 1.5 mm, angular, concolorous with the tubes. Veil absent.

Micromorphological characteristics (Fig. 35): Basidiospores (7.2-) 8.4-9.6 (-11.0) × (3.7-) 4.4-5.0 (-5.6) μm (n = 134, mean length = 8.99 ± 0.59, mean width = 4.71 ± 0.28, Q = (1.63-) 1.80-2.03 (-2.49), mean Q = 1.91 ± 0.11), inequilateral with a shallow suprahilar depression in profile, ellipsoid to oblong-ellipsoid in face view, with a broadly rounded or subacute apex, smooth, pale greenish grey in water, melleous in KOH, inamyloid, with slightly thickened walls (up to 0.5 μm). Basidia (27.1-) 31.2-36.8 (-38.7) × (9.7-) 10.2-12.0 (-13.2) μm (n = 38, mean length = 34.00 ± 2.78, mean width = 11.13 ± 0.92) in main body, (3.9-) 4.2-5.3 (-6.0) × (1.5-) 1.9-2.6 (-2.9) μm (n = 26, mean length = 4.75 ± 0.51, mean width = 2.23 ± 0.33) in sterigmata, clavate, four-spored. Cheilocystidia none (edge of pore fertile). Pleurocystidia infrequent, (34.1-) 35.7-50.6 (-52.4) × (8.4-) 8.3-11.6 (-11.7) μm (n = 4, mean length = 43.14 ± 7.48, mean width = 9.97 ± 1.62), ventricose-rostrate, smooth, hyaline, thin-walled. Hymenophoral trama with parallel (mature stage), cylindrical hyphae (3.9-) 5.4-7.2 (-8.1) μm wide (n = 45, mean width = 6.30 ± 0.87) in lateral strata, (3.0-) 3.6-4.7 (-6.1) μm wide (n = 43, mean width = 4.13 ± 0.57) in a mediostratum, smooth, hyaline, thin-walled. Pileipellis of a compacted trichodermial palisade with more or less inflated, catenulate elements; constituent hyphal cells (11.6-) 19.5-41.1 (-64.3) × (7.6-) 10.3-16.4 (-24.6) μm (n = 51, mean length = 30.28 ± 10.78, mean width = 13.36 ± 3.07), subcylindrical to ovate, sometimes subfusoid, distinctly constricted at the septa, often encrusted with reddish brown granules and patches (in water), yellowish brown in KOH, thin-walled; terminal elements (22.6-) 30.5-53.4 (-71.7) × (8.0-) 9.7-14.6 (-18.5) μm (n = 36, mean length = 41.93 ± 11.46, mean width = 12.15 ± 2.41), subcylindrical, subfusiform or beaked, not encrusted. Pileitrama of cylindrical, subparallel hyphae (4.2-) 5.7-9.6 (-13.0) μm wide (n = 73, mean width = 7.64 ± 1.93), smooth, hyaline, thin-walled. Stipitipellis a cutis of parallel, repent hyphae (2.5-) 3.2-4.5 (-5.1) μm wide (n = 35, mean width = 3.83 ± 0.67), cylindrical, smooth, with reddish brown intracellular pigment in water, yellowish brown in KOH, thin-walled; caulocystidia scattered, (32.1-) 30.7-40.7 (-45.3) × (7.4-) 7.6-9.1 (-9.6) μm (n = 6, mean length = 35.73 ± 5.02, mean width = 8.34 ± 0.74) at the apex of stipe, (20.2-) 21.4-29.8 (-33.1) × (5.0-) 6.0-8.4 (-9.3) μm (n = 21, mean length = 25.63 ± 4.19, mean width = 7.24 ± 1.21) at the middle portion of stipe, (28.6-) 36.6-63.6 (-74.9) × (3.4-) 4.8-7.1 (-8.7) μm (n = 26, mean length = 50.11 ± 13.50, mean width = 5.98 ± 1.13) at the base of stipe, narrowly fusoid-ventricose or subclavate to subcylindrical, smooth, pale yellow in water and KOH, thin-walled; caulobasidia scattered, (24.0-) 27.0-37.8 (-48.6) × (6.6-) 7.5-9.2 (-10.1) μm (n = 16, mean length = 32.40 ± 5.37, mean width = 8.37 ± 0.84), clavate, yellowish brown in water, pale yellow in KOH, 1-2-3-4-spored. Stipe trama composed of longitudinally running, cylindrical hyphae (5.4-) 7.3-11.7 (-15.3) μm wide (n = 38, mean width = 9.47 ± 2.19), unbranched, smooth, hyaline, thin-walled. Elements of basal mycelium (2.4-) 2.8-4.3 (-5.0) μm wide (n = 50, mean width = 3.52 ± 0.73), cylindrical, smooth, hyaline, thin-walled. All tissues inamyloid, pale yellow or colorless in KOH, without clamp connections.

Habitat and phenology: Solitary or scattered on the ground in evergreen broad-leaved forests, June.

Known distribution: Okinawa (Ishigaki Island).

Holotype: TNS-F-48227, on ground in an evergreen broad-leaved forest, Fukai, Ishigaki-shi, Okinawa Pref., 10 Jun.2012, coll. Taneyama, Y.

Other specimens examined: TNS-F-48226, on the ground in an evergreen broad-leaved forest, Fukai, Ishigaki-shi, Okinawa Pref., 10 Jun. 2012, coll. Taneyama, Y; TNS-F-48228, same location, 10 Jun. 2012, coll. Taneyama, Y; TNS-F-48250, same location, 10 Jun. 2012, coll. Taneyama, Y.

Japanese name: Aoaza-iguchi (named by H. Takahashi & Y. Taneyama).

Comments: Judging from the small, xerocomoid basidiomata and the greenish brown basidiospores, the new species seems to be best accommodated in the genus *Boletus*, section *Subtomentosi* Fries sensu Smith and Thiers (Smith and Thiers 1971).

The xerocomoid, virescent basidiomata and the catenulate pileipellis elements of the new species are comparable with *Xerocomus sulcatipes* Heinem. & Gooss.-Font. from Congo (Heinemann 1951, 1954). The African species can be differentiated from *B. virescens* by much larger basidiomata: 40-70 mm in diameter in the pileus (Heinemann 1951) and significantly longer elongate-subfusiform basidiospores: 12.0-14.7 × 4.3-5.3 μm (Heinemann 1951).

References 引用文献

Heinemann P. 1951. Champignons récoltés au Congo Belge par Mme M. Goossens-Fontana. I. Boletineae. Bull Jard Bot Brux 21: 223-346.

Heinemann P. 1954. Boletineae. Flore Iconographique des Champignons du Congo 3: 49-80.

Smith AH, Thiers HD. 1971. The Boletes of Michigan. University of Michigan Press, Ann Arbor.

5. アオアザイグチ（新種：高橋春樹 & 種山裕一新称）*Boletus virescens* Har. Takah. & Taneyama, **sp. nov.**

肉眼的特徴（Figs. 36-40）：傘は径 32-40 mm，最初半球形のち饅頭形になり，しばしば多数の細かい亀裂を生じ，不規則な皺状隆起を表す；表面はやや密綿毛状，粘性を欠き，淡黄色の地にしばしば所々に緑青色の染みを生じ，老成すると赤褐色を帯びる；縁部は膜状に突出しない．肉は厚さ5 mm 以下，類白色，柄の根元および傘と柄の表皮直下において灰赤色を帯び，空気に

触れても変色せず，特別な匂いはなく，味は不明．柄は50-55×5-6 mm，円柱形，頂部に向かってやや拡大し，痩せ型，中心生，中実，縦に沿って浅い縦皺を表し，明瞭な網目模様を欠くが時に頂部において僅かな網目を表す；表面は粘性を欠き，平滑，上部は傘と同色，下方に向かって赤褐色を帯び，しばしば頂部付近に緑青色の染みを生じる；根元は白色の綿毛状菌糸体に被われる．管孔は長さ2-3 mm，上生〜直生またはしばしば僅かに垂生し，柄の周囲において浅く嵌入し，孔口の周囲では灰橙色，傘の近くでは鈍黄色を呈する；孔口は角形，1.5 mm 以下，管孔と同色，管孔並びに孔口は傷を受けても変色しない．被膜を欠く．

顕微鏡的特徴（Fig. 35）：担子胞子は (7.2-) 8.4-9.6 (-11.0)×(3.7-) 4.4-5.0 (-5.6) μm（n = 134, mean length = 8.99±0.59, mean width = 4.71±0.28, Q = (1.63-) 1.80-2.03 (-2.49), mean Q = 1.91±0.11），下部側面に浅いなだらかな凹みがあり（イグチ型），楕円形〜長楕円形，頂部は鈍頭またはやや尖り，平坦，水封で灰緑色，アルカリ溶液において蜜色，非アミロイド，壁は厚さ0.5 μm以下．担子器は (27.1-) 31.2-36.8 (-38.7)×(9.7-) 10.2-12.0 (-13.2) μm（本体），(3.9-) 4.2-5.3 (-6.0)×(1.5-) 1.9-2.6 (-2.9) μm（ステリグマ），こん棒形，4胞子性．縁シスチジアはない（孔口の縁部は稔性）．側シスチジアは稀で，(34.1-) 35.7-50.6 (-52.4)×(8.4-) 8.3-11.6 (-11.7) μm，片膨れ状で頂部は嘴状に伸び，平滑，無色，薄壁．子実層托実質の菌糸は成熟時平列し，円柱形，側層において幅 (3.9-) 5.4-7.2 (-8.1) μm，中層において幅 (3.0-) 3.6-4.7 (-6.1) μm，平滑，無色，薄壁．傘表皮組織は隙間なく並んだ柵状毛状被で多少膨大した菌糸細胞が鎖状につながる；菌糸は (11.6-) 19.5-41.1 (-64.3)×(7.6-) 10.3-16.4 (-24.6) μm，円柱形〜卵形，時に類紡錘形，隔壁の周囲で著しく収縮し，しばしば赤褐色（水封）の粒状色素が菌糸細胞の外側に凝着し，アルカリ溶液において黄褐色，薄壁；末端細胞は (22.6-) 30.5-53.4 (-71.7)×(8.0-) 9.7-14.6 (-18.5) μm，亜円柱形，亜紡錘形または嘴形，凝着色素を欠く．傘実質の菌糸は緩く錯綜し，幅 (4.2-) 5.7-9.6 (-13.0) μm，平滑，無色，薄壁．柄表皮組織は平行菌糸被；菌糸は幅 (2.5-) 3.2-4.5 (-5.1) μm，円柱形，平滑，水封において赤褐色の細胞内色素を有し，薄壁；柄シスチジアは散生し，(32.1-) 30.7-40.7 (-45.3)×(7.4-) 7.6-9.1 (-9.6) μm（頂部），(20.2-) 21.4-29.8 (-33.1)×(5.0-) 6.0-8.4 (-9.3) μm（中部），(28.6-) 36.6-63.6 (-74.9)×(3.4-) 4.8-7.1 (-8.7) μm（根元），幅の狭い片膨れ状紡錘形または亜こん棒形〜亜円柱形，平滑，淡黄色（水封およびアルカリ溶液），薄壁；柄胞子器は散在し，(24.0-) 27.0-37.8 (-48.6)×(6.6-) 7.5-9.2 (-10.1) μm，こん棒形，黄褐色（水封），淡黄色（アルカリ溶液），1-2-3-4-胞子性．柄実質の菌糸は幅 (5.4-) 7.3-11.7 (-15.3) μm，縦に沿って配列し，円柱形，無分岐，平滑，無色，薄壁．根元の綿毛状菌糸体を構成する菌糸は幅 (2.4-) 2.8-4.3 (-5.0) μm，円柱形，平滑，無色，薄壁．全ての組織は非アミロイド，アルカリ溶液において淡黄色または無色，クランプを欠く．

生態および発生時期：常緑広葉樹林内地上に孤生または散生，6月．

分布：沖縄（石垣島）．

供試標本：TNS-F-48227（正基準標本），常緑広葉樹林内地上，石垣市樫海，2012年6月10日，種山裕一採集；TNS-F-48226，同上，2012年6月10日，種山裕一採集；TNS-F-48228，同上，2012年6月10日，種山裕一採集；TNS-F-48250，同上，2012年6月10日，種山裕一採集．

主な特徴：子実体は小形のアワタケ型で，所々に青色の染みを生じる；傘は細かい亀裂と小皺を表す；担子胞子は楕円形；縁シスチジアを欠く；側シスチジアは片膨れ状紡錘形で頂部が嘴状に伸びる；傘表皮組織は多少とも膨大した菌糸細胞が鎖状につながり，柵状毛状被をなす；柄シスチジアは狭紡錘形または類こん棒形〜類円柱形；常緑広葉樹林内地上に発生．

コメント：小形のアワタケ型子実体およびオリーブ褐色の担子胞子を有する性質から，本種は Smith and Thiers（Smith and Thiers 1971）の分類概念によるヤマドリタケ属 *Boletus*，アワタケ節 section *Subtomentosi* Fries に位置すると考えられる．

48 —— 5. *Boletus virescens* アオアザイグチ

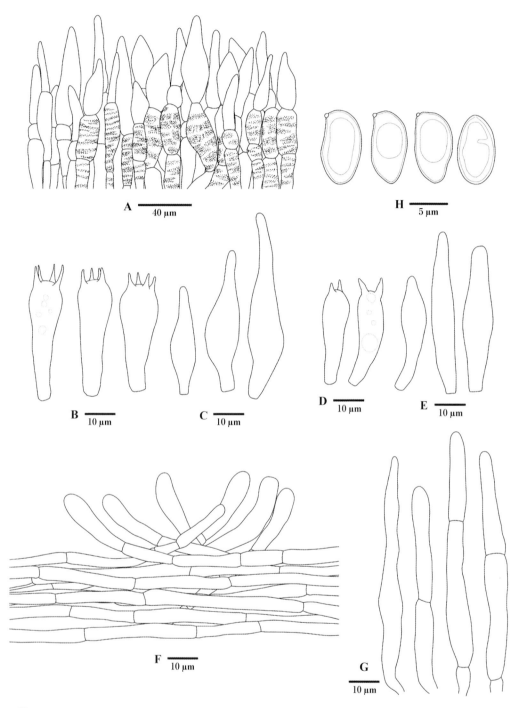

Fig. 35 – Micromorphological features of *Boletus virescens* (holotype): **A**. Longitudinal cross section of the pileipellis. **B**. Basidia. **C**. Pleurocystidia. **D**. Caulobasidia. **E**. Caulocystidia from the very apex of stipe. **F**. Longitudinal cross section of the stipitipellis (from the middle portion of stipe). **G**. Caulocystidia from the lower portion of stipe. **H**. Basidiospores. Illustrations by Taneyama, Y.

アオアザイグチの顕微鏡図（正基準標本）：**A**. 傘の表皮組織の縦断面．**B**. 担子器．**C**. 側シスチジア．**D**. 柄シスチジア．**E**. 柄頂部の柄シスチジア．**F**. 柄表皮組織の断面（柄中部）．**G**. 柄下部の柄シスチジア．**H**. 担子胞子．図：種山裕一．

5. *Boletus virescens* アオアザイグチ

Fig. 36 − Basidiomata of *Boletus virescens* (holotype) occurring on ground in an evergreen broad-leaved forest, 10 Jun. 2012, Fukai, Ishigaki Island. Photo by Taneyama, Y.
常緑広葉樹林内地上から発生したアオアザイグチの子実体（正基準標本），2012年6月10日，石垣島桴海．写真：種山裕一．

　青色の染みを生じるアワタケ型の子実体並びに柵状毛状被からなる傘表皮組織はアフリカ（コンゴ）産 *Xerocomus sulcatipes* Heinem. & Gooss.-Font.（Heinemann 1951, 1954）と共通するが，アフリカ産種はより大型の子実体：傘の径40-70 mm（Heinemann 1951）と長紡錘形でより長い担子胞子：12.0-14.7×4.3-5.3 µm（Heinemann 1951）を形成する点で本種と異なる．

5. *Boletus virescens* アオアザイグチ

Fig. 37 – Basidioma of *Boletus virescens* (TNS-F-48226) occurring on ground in an evergreen broad-leaved forest, 10 Jun. 2012, Fukai, Ishigaki Island. Photo by Taneyama, Y.
常緑広葉樹林内地上から発生したアオアザイグチの子実体 (TNS-F-48226), 2012年6月10日, 石垣島桴海. 写真：種山裕一.

5. *Boletus virescens* アオアザイグチ —— 51

Fig. 38 – Basidioma of *Boletus virescens* occurring on ground in an evergreen broad-leaved forest, 10 Jun. 2012, Fukai, Ishigaki Island. Photo by Taneyama, Y.
常緑広葉樹林内地上から発生したアオアザイグチの子実体，2012年6月10日，石垣島桴海．写真：種山裕一．

Fig. 39 – Underside view of the basidioma of *Boletus virescens* (holotype), 10 Jun. 2012, Fukai, Ishigaki Island. Photo by Taneyama, Y.
アオアザイグチの子実体（管孔側，正基準標本），2012年6月10日．石垣島桴海．写真：種山裕一．

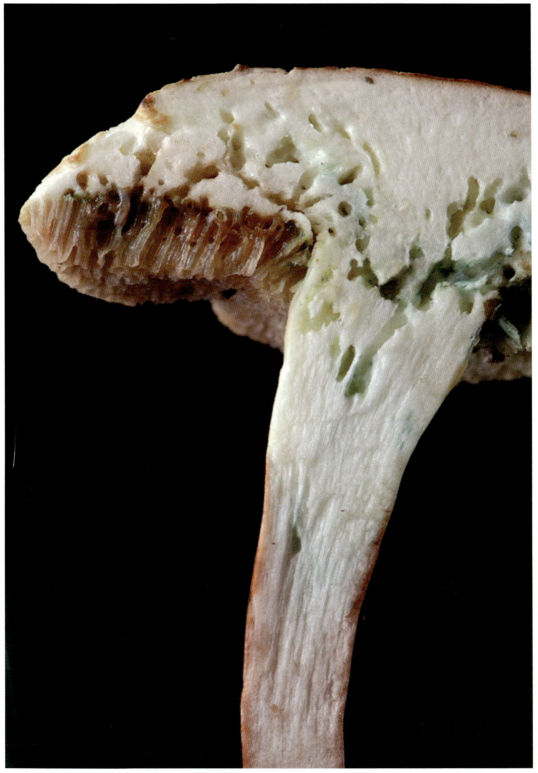

Fig. 40 – Longitudinal cross section of the basidioma of *Boletus virescens* (holotype), 10 Jun. 2012, Fukai, Ishigaki Island. Photo by Taneyama, Y.
アオアザイグチの子実体断面（正基準標本），2012年6月10日，石垣島桴海．写真：種山裕一．

6. *Chaetocalathus fragilis* (Pat.) Singer ヒダフウリンタケ
Lilloa 8: 520 (1943) [MB#285125].
≡ *Crinipellis fragilis* Pat., Philipp J Sci, C, Bot 10: 97 (1915) [MB#234871].
≡ *Marasmius fragilis* (Pat.) Sacc. & Trotter, Syll Fung 23: 153 (1925) [MB#146692].
≡ *Lachnella fragilis* (Pat.) Locq., Bull trimest Soc mycol Fr 68: 166 (1952) [MB#299245].

Macromorphological characteristics (Figs. 43–45): Pileus (0.5–) 1–2 (–2.5) mm in diameter, 0.5–1 mm tall, very small, membranous, astipitate and dorsally attached to substratum, discoid-cyphelloid, with pilose margin; surface not hygrophanous, appressed fibrillose to woolly-tomentose, pure white overall. Flesh very thin (up to 0.2 mm), white. Odor and taste not distinctive. Rudiment of stipe (columella) 0.1–0.2 mm in diameter, white, reduced to a central small papilla, situated in the underside of the pileus and not directly attached to a substratum. Lamellae adnexed to almost free, subdistant (13–18) with 1–2 series of lamellulae, medium broad (0.2–0.6 mm), thin, white, not intervenose in the interstices; edges even, concolorous.

Micromorphological characteristics (Figs. 41, 42): Basidiospores $8-8.5 \times 6-7$ µm (n = 33, Q = 1.2–1.3), subglobose to shortly ovoid-ellipsoid, smooth, colorless, inamyloid, thin-walled. Basidia $16-22 \times 4-6$ µm, clavate, 4-spored; basidioles subclavate to subfusoid. Cheilocystidia $15-23 \times 6-12$ µm, gregarious, forming a compact sterile edge, projecting from the hymenium, metuloid-type, clavate to subclavate, usually with one or two cylindrical apical appendages 5–11 µm long, with an obtuse apex, colorless in water, hyaline in KOH, with strongly dextrinoid, moderately thickened walls (up to 1 µm thick), often in the upper portion encrusted with coarse hyaline crystals. Pleurocystidia scattered, similar to cheilocystidia. Hyphae of hymenophoral trama 2–9 µm wide, entangled, cylindrical, smooth, colorless, inamyloid, thin-walled, occasionally with clamped septa. Pileipellis a hypotrichial layer of loosely interwoven, subcylindrical hyphae 3–6 µm wide, with smooth, weakly dextrinoid, colorless walls up to 1 µm thick, occasionally with clamped septa; hairs of pileus $200-300 \times 2-5$ µm, arising directly from the hypotrichium, densely entangled, cylindrical, with a subacute apex, sometimes flexuous, hyaline in KOH, with smooth, hyaline walls 1–2 µm thick, strongly dextrinoid, without a secondary septum. Hyphae of pileitrama similar to those of the hymenophoral trama.

Habitat and phenology: Usually abundant on dead twigs (unidentified substrata) in evergreen broad-leaved forests, June to August.

Known distribution: Okinawa (Ishigaki Island), Philippines, Hawaii and Thailand.

Specimens examined: KPM-NC0017520, on dead twigs in an evergreen broad-leaved forest, Banna Park, Ishigaki-shi, Okinawa Pref., 25 Jul. 2010, coll. Takahashi, H. & Terashima, Y.; KPM-NC0017521, same location, 28 Jul. 2010, coll. Takahashi, H.; KPM-NC0017522, same location, 11 Aug. 2010, coll. Takahashi, H; KPM-NC0017523, same location, 6 Jun. 2009, coll. Takahashi, H.; KPM-NC0017524, same location, 8 Jun. 2009, coll. Takahashi, H.; KPM-NC0017525, same location, 15 Jun. 2009, coll. Takahashi, H.; TNS-F-48304, same location, 11 Jun. 2012, coll. Taneyama, Y.

Japanese name: Hida-fuurintake.

Comments: The diagnostic features of this species are the minute (1–2 mm in diameter on average), white, membranous, discoid-cyphelloid basidiomata having a central, small, papillate columella; the subglobose to shortly ovoid-ellipsoid basidiospores; the dextrinoid hymenial metuloids with an obtuse apex; the strongly dextrinoid hairs covering the pileus surface; and the gregarious habit on dead twigs in evergreen broad-leaved forests. The presence of dextrinoid hymenial metuloids suggests placement of this species in the genus *Chaetocalathus*, section *Holocystis* Singer (Singer 1986).

The Japanese material fully conforms to *C. fragilis* originally described from Philippines (Patouillard 1915; Saccardo and Trotter 1925; Singer 1943; Locquin 1952). At the collection locality in Japan (Ishigaki Island, Okinawa Pref.), the species commonly occurs on dead twigs in evergreen broad-leaved forests in summer. The present species seems to be widely distributed in eastern Asia and the Pacific region.

Among the white *Chaetocalathus* species, Southeast Asian *Chaetocalathus semisupinus* (Berk. & Broome) Pegler, redescribed by Pegler from Sri Lanka (Pegler 1986), seems to be the closest species to *C. fragilis* as it has relatively small pileus: 1-3 mm in diameter (Pegler 1986). The former species, however, does not conform to *C. fragilis* as it has white to tan colored lamellae acquiring a reddish tinge, somewhat longer, broadly ovoid basidiospores: 8-9.5 μm long (Pegler 1986), and setiform hymenial cystidia with an acute apex. The appearance of the basidiomata in *C. fragilis* is also analogous to *Chaetocalathus ehretiae* Neda from Japan (Neda and Doi 1998), which differs in forming much larger pileus: 2-4 mm in diameter (Neda and Doi 1998), pale brown to yellow lamellae, globose to subglobose, somewhat larger basidiospores: 8.5-9×7-8 μm (Neda and Doi 1998), and seta-like cheilocystidia. Unfortunately, the features of pleurocystidia and amyloid reaction of cheilocystidia were not mentioned in the protologue of *C. ehretiae*.

References 引用文献

Locquin M. 1952. Sur la non-validité de quelques genres d'Agaricales. Bull trimest Soc mycol Fr 68: 165-169.
Neda H, Doi Y. 1998. Notes on Agarics in Kyushu District. Mem Natn Sci Mus, Tokyo 31: 89-95.
Patouillard NT. 1915. Champignons des Philippines communiqués par C.F.Baker, II. Philipp J Sci, C, Bot 10: 85-98.
Pegler DN. 1986. Agaric flora of Sri Lanka. Kew Bull, Addit Ser 12: 1-519.
Saccardo PA, Trotter A. 1925. Supplementum Universale, Pars X. Basidiomycetae. Syll fung 23: 1-1026.
Singer R. 1943. A monographic studies of the genera of *Crinipellis* and *Chaetocalathus*. Lilloa 8: 441-534.
Takahashi H. 2011. Two new species of Agaricales and a new Japanese record for *Chaetocalathus fragilis* from Ishigaki Island, a southwestern island of Japan. Mycoscience 52 (6): 392-400.

6. ヒダフウリンタケ *Chaetocalathus fragilis* (Pat.) Singer

Lilloa 8: 520 (1943) [MB#285125].
≡ *Crinipellis fragilis* Pat., Philipp J Sci, C, Bot 10: 97 (1915) [MB#234871].
≡ *Marasmius fragilis* (Pat.) Sacc. & Trotter, Syll Fung 23: 153 (1925) [MB#146692].
≡ *Lachnella fragilis* (Pat.) Locq., Bull trimest Soc mycol Fr 68: 166 (1952) [MB#299245].

肉眼的特徴（Figs. 43-45）：傘は径 (0.5-) 1-2 (-2.5) mm，高さ0.5-1 mm，微小で膜質，中央部が基質に背着し，柄を欠き，背着生，フウリンタケ型〜チャワンタケ型，平坦，縁部は長毛に被われる；表面は非吸水性，圧着した繊維状〜密綿毛状，全体に純白色を呈する．肉は非常に薄く，厚さ 0.2 mm 以下，純白色，特別な味や臭いはない．痕跡柄（柱軸）は径 0.1-0.2 mm，子実層托側の中央に微小な乳頭状突起を形成し，基質に直接接続しない．ヒダは上生〜ほぼ離生し，やや疎（柄に到達するヒダは13-18），1-2本の小ヒダを交え，幅 0.2-0.6 mm，純白色，連絡脈はない；縁部は全縁で，同色．

顕微鏡的特徴（Figs. 41, 42）：担子胞子は8-8.5×6-7 μm（n = 33, Q = 1.2-1.3），類球形，平坦，無色，非アミロイド，薄壁．担子器は16-22×4-6 μm，こん棒形，4胞子性；偽担子器は亜こん棒形〜亜紡錘形．縁シスチジアは15-23×6-12 μm，群生し，縁部に不稔帯を形成し，子実層から突出し，メチュロイド（冠石灰のう体）型，類こん棒形〜こん棒形，しばしば頂部に1-2本の円柱形付属糸（長さ5-11 μm）を具え，無色（水封），苛性カリ水溶液中で無色，強偽アミロイドに染

まり，やや厚壁（厚さ1 μm 以下），しばしば付属糸は無色透明な結晶物に被覆される．側シスチジアは散在し，縁シスチジアに似る．子実層托実質の菌糸細胞は幅2-9 μm，もつれて錯綜し，円柱形，平滑，無色，非アミロイド，薄壁，時に隔壁にクランプを有する．傘表皮組織の菌糸は幅3-6 μm，円柱形，緩く錯綜し，無色，壁は厚さ1 μm 以下で弱偽アミロイド，時に隔壁にクランプを具える；毛状細胞は200-300×2-5 μm，密集して絡み合い，円柱形，頂部はやや尖り，時に屈曲し，無色，苛性カリ水溶液において無色，厚壁（厚さ1-2 μm），強偽アミロイド，二次隔壁は存在しない．傘実質の菌糸細胞は子実層托実質と共通する．

分布：沖縄（石垣島），フィリピン，ハワイ，タイ．

生態および発生時期：常緑広葉樹林内の枯れ枝上に群生〜散生，6月〜8月．

供試標本：KPM-NC0017520，常緑広葉樹林内枯れ枝上，沖縄県石垣市バンナ公園，2010年7月25日，採集者：高橋春樹 & 寺嶋芳江；KPM-NC0017521，同上，2010年7月28日，採集者：高橋春樹；KPM-NC0017522，同上，2010年8月11日，採集者：高橋春樹；KPM-NC0017523，同上，2009年6月6日，採集者：高橋春樹；KPM-NC0017524，同上，2009年6月8日，採集者：高橋春樹；KPM-NC0017525，同上，2009年6月15日，採集者：高橋春樹；TNS-F-48304，同上，2012年6月11日，種山裕一採集．

主な特徴：1) 子実体は微小（平均径1-2 mm）なチャワンタケ形〜フウリンタケ形で柄を欠き，傘の中央部が直接基質に背着し，全体に純白色，子実層托側の中央部に微小な乳頭状突起（痕跡柄）を具える．2) 担子胞子は類球形．3) メチュロイド型シスチジアが子実層に存在する．4) 傘は偽アミロイドに染まる毛状細胞に被われる．5) 常緑広葉樹林内の枯れ枝上に群生する．

コメント：偽アミロイドに染まるメチュロイドを有する性質において，本種は Singer (Singer 1986) の分類概念によるケカゴタケ属 *Chaetocalathus*，ヒダフウリンタケ節（高橋春樹新称）section *Holocystis* Singer に属すると考えられる．

　沖縄産の標本はフィリピンから新種記載された *C. fragilis*（Patouillard 1915；Saccardo and Trotter 1925；Singer 1943；Locquin 1952）とよく一致する．本種は東南アジア〜太平洋地域に渡り広く分布しているものと思われる．

　Pegler (Pegler 1986) の記載によるスリランカ産 *Chaetocalathus semisupinus*（Berk. & Broome）Pegler は微小で白色の傘を形成する点で本種に最も近縁と思われるが，ヒダが淡黄褐色〜赤色を帯び，担子胞子は本種に比べてやや長形：8-9.5 μm long (Pegler 1986) で広卵形になり，先端部が尖った剛毛体様シスチジアを有する．九州から報告されたケカゴタケ *Chaetocalathus ehretiae* Neda (Neda and Doi 1998) も純白色の子実体および類球形の担子胞子を持つ点で本種に似るが，子実体はより大型：傘の径2-4 mm (Neda and Doi 1998) で，淡褐色〜黄色のヒダを有し，やや大型でより球形に近い担子胞子：8.5-9×7-8 μm (Neda and Doi 1998) を形成する．残念ながらケカゴタケの原記載において側シスチジアの特徴と縁シスチジアのアミロイド反応に関するデータは言及されなかった．

6. *Chaetocalathus fragilis* ヒダフウリンタケ ——57

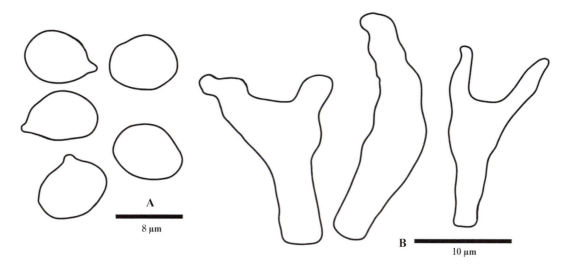

Fig. 41 – Micromorphological features of *Chaetocalathus fragilis* (KPM-NC0017520): **A**. Basidiospores. **B**. Cheilocystidia. Illustrations by Takahashi, H.
ヒダフウリンタケの顕鏡図（KPM-NC0017520）：**A**. 担子胞子．**B**. 縁シスチジア．図：高橋春樹．

Fig. 42 – Cheilocystidia (metuloid) of *Chaetocalathus fragilis* (in Melzer's reagent, KPM-NC0017520). Photo by Takahashi, H.
ヒダフウリンタケの縁シスチジア（メチュロイド）（メルツァー溶液でマウント，KPM-NC0017520）．写真：高橋春樹．

6. *Chaetocalathus fragilis* ヒダフウリンタケ

Fig. 43 – Basidiomata of *Chaetocalathus fragilis* (TNS-F-48304) on a dead twig in an evergreen broad-leaved forest, 11 Jun. 2012, Banna Park, Ishigaki Island. Photo by Taneyama, Y.
常緑広葉樹林内の枯れ枝上に発生したヒダフウリンタケの子実体（TNS-F-48304），2012年6月11日，石垣島バンナ公園．写真：種山裕一．

Fig. 44 – Basidiomata of *Chaetocalathus fragilis* (TNS-F-48304) on a dead twig in an evergreen broad-leaved forest, 11 Jun. 2012, Banna Park, Ishigaki Island. Photo by Taneyama, Y.
常緑広葉樹林内の枯れ枝上に発生したヒダフウリンタケの子実体（TNS-F-48304），2012年6月11日，石垣島バンナ公園．写真：種山裕一．

6. *Chaetocalathus fragilis* ヒダフウリンタケ —— 59

Fig. 45 – Basidiomata of *Chaetocalathus fragilis* (KPM-NC0017520) on a dead twig in an evergreen broad-leaved forest, 25 Jul. 2010, Banna Park, Ishigaki Island. Photo by Takahashi, H.
常緑広葉樹林内の枯れ枝上に発生したヒダフウリンタケの子実体（KPM-NC0017520），2010年7月25日，石垣島バンナ公園．写真：高橋春樹．

7. *Crinipellis canescens* Har. Takah. シラガニセホウライタケ

Mycoscience 41 (2): 171 (2000) [MB#464656].

≡ *Moniliophthora canescens* (Har. Takah.) Kerekes & Desjardin, Fung Diver 37: 137 (2009) [MB#513037].

Etymology: The specific epithet refers to the hoary fibrils enveloping the pileus and stipe.

Macromorphological characteristics (Figs. 47–51): Pileus 6–15 mm in diameter, at first hemispherical with incurved margin, then convex to plane, sometimes with an obtuse umbo or depressed center, radially sulcate-striate up to the disk; surface dry, covered overall with white, fine, pubescent hairs, at first reddish brown (8F7–8, 9F7–8) under hoary fibrils, then brown (7E7–8, 8E7–8) at center, brownish yellow (5C7–5C8) to yellowish brown (5D7–5D8) toward the ciliate-villose margin, orange (6B8) to brownish orange (6C8) when wet. Flesh very thin (up to 0.3 mm), concolorous with the surface, pliant, tough, odor and taste none. Stipe 10–15 × 0.5–1 mm, almost equal or slightly enlarged at the base, central, slender, terete, hollow, at first white overall, then brown (7E7–8, 8E7–8) toward the base, white pruinose above, villose to hispid below, attached to a white, appressed mycelial pad over the substratum. Lamellae adnexed to subcollariate, distant (14–18 reach the stipe), with 0–1 lamellulae, up to 1 mm broad, slightly intervenose, concolorous with the pileus; edges even, concolorous. Spore print pure white.

Micromorphological characteristics (Fig. 46): Basidiospores 9.2–10.4 × 3.7–4.3 µm (n = 34, Q = 2.41–2.48), ellipsoid, smooth, colorless, inamyloid, thin-walled. Basidia 26.4–37.3 × 6.4–8.8 µm, clavate, four-spored; basidioles clavate. Cheilocystidia 16–21.6 × 5–8.5 µm, forming a compact sterile edge, subclavate to irregularly shaped, with one or several finger-like appendages 2.4–4.2 µm wide, or shortly lobed or knobbed, colorless or pale melleous, inamyloid, thin-walled. Pleurocystidia none. Hymenophoral trama regular; element hyphae similar to those of the pileitrama. Pileipellis a hypotrichial layer (cutis) of subcylindrical cells 30–60 × 6–12 µm, often inflated, with orange (6B8) to brownish orange (6C8) walls up to 1 µm thick, encrusted with brownish orange (6C8) pigment, brownish red (11C8) in Melzer's reagent (dextrinoid); hairs of pileus 230–1200 × 4–7 µm, arising directly from the hypotrichium, repent or erect, cylindrical, tapering to an obtuse apex or with a broadly rounded apex, sometimes flexuous, smooth, brownish violet (11D8) to violet-brown (11E8) in Melzer's reagent (strongly dextrinoid), with colorless or pale melleous walls 1.6–3.2 µm thick, without a secondary septum. Hyphae of pileitrama 6–16 µm wide, parallel, cylindrical to subfusiform, often inflated, smooth, with pale melleous cytoplasmic pigments, brownish red (11C8) in Melzer's reagent (dextrinoid), thin-walled. Stipitipellis of fasciculate hairs arising from cutis hyphae; hairs of stipe similar to those of the pileus. Stipe trama composed of longitudinally running, cylindrical hyphae 5.3–8 µm wide, smooth, with pale melleous to brownish orange (6C8) cytoplasmic pigments, brownish red (11C8) in Melzer's reagent (dextrinoid), thin-walled. Clamps present in all tissues.

Habitat and phenology: Solitary or scattered on dead shrubs in evergreen broad-leaved forests, April to November, common.

Known distribution: Okinawa (Yaeyama Islands: Ishigaki Island and Iriomote Island), Malaysia (Kerekes and Desjardin 2009).

Specimens examined: KPM-NC0005014 (holotype), on dead twigs of various kinds of low trees in an evergreen broad-leaved forest, along the Urauchi River, Iriomote Island, Okinawa Pref., 2 Jun. 1999, coll. Takahashi, H.; CBM-FB-24120, same location, 1 Oct. 1996, coll. Takahashi, H.; CBM-FB-24121, same location, 15 May 1997, coll. Takahashi, H.; CBM-FB-24122, same location, 6 Jun. 1998, coll. Takahashi, H.; TNS-F-48181, on dead twigs of various kinds of low trees in an evergreen broad-leaved forest, Omoto-yama, Ishigaki-shi, Okinawa Pref., 9 Jun. 2012; KPM-NC0023876, on dead twigs of

Ardisia quinquegona Blume, Banna-dake, Ishigaki-shi, Okinawa Pref., 2 Jun. 2013, coll. Takahashi, H. Japanese name: Shiraga-nise-houraitake.

Comments: This species is characterized by the radially sulcate-striate, canescent pileus colored brownish yellow at maturity; the white or brown, villose to hispid stipe attached by white, appressed mycelial pad to the substratum; the distant lamellae concolorous with the pileus; the subclavate or irregularly shaped cheilocystidia with one or several finger-like appendages; the dextrinoid hairs on the pileus and stipe; and the basidiome formation on dead twigs of various kinds of low trees in evergreen broad-leaved forests.

The lack of bright pigments, the relatively long (more than 7 mm) and central stipe, the ellipsoid basidiospores less than 9 μm wide, and the absence of pleurocystidia suggest that this species belongs in the genus *Crinipellis*, section *Crinipellis*, subsection *Crinipellis* (Singer 1942, 1986).

Within the subsection, *C. canescens* appears to be closely related to several neotropical taxa that have a small pileus (less than 10 mm) and cheilocystidia with several appendages, such as *Crinipellis pseudostipitaria* Singer (Pegler 1983; Singer 1942), *Crinipellis septotricha* Singer (Pegler, 1983; Singer 1942), and *Crinipellis stupparia* (Berk. & M.A.Curtis) Pat. (Pegler 1983). The latter three taxa are distinct in having a pileus with a dark papillate umbo and in lacking a basal mycelium. Moreover, *C. pseudostipitaria* grows on grass debris, and *C. septotricha* and *C. stupparia* form hairs with secondary ladder-septations. *Crinipellis canescens* is also similar to *Crinipellis omotricha* (Berk.) D.A. Reid, redescribed by Pegler (Pegler 1986) from material collected in Sri Lanka, except that the latter has simple cheilocystida and a graminicolous habitat. *Crinipellis hepatica* Corner (Corner 1996), recently described from Malaysia, appears to be the closest ally. Unlike *C. canescens*, however, the former has a reddish brown pileus, an insititious stipe, subfusiform basidiole, and subventricose to subfusiform cheilocystidia without appendages.

Based on the ITS sequence data for the Malaysian *C. canescens* materials, Kerekes and Desjardin (Kerekes and Desjardin 2009) proposed relegating the species to the anamorphic genus *Moniliophthora* (Evans et al. 1978; Aime and Phillips-Mora 2005). However, they did not base this recommendation on topotypic materials, and they did not specify the morphological differences between *Crinipellis* and *Moniliophthora*. We therefore prefer to retain the species in the genus *Crinipellis* as defined by Singer (Singer 1986).

References 引用文献

Aime MC, Phillips-Mora W. 2005. The causal agents of witches' broom and frosty pod rot of cacao (chocolate, *Theobroma cacao*) form a new lineage of Marasmiaceae. Mycologia 97 (5): 1012-1022.

Corner EJH. 1996. The agaric genera *Marasmius, Chaetocalathus, Crinipellis, Heimiomyces, Resupinatus, Xerula* and *Xerulina* in Malesia. Beih Nova Hedwigia 111: 1-175.

Evans HC, Stalpers JA, Samson RA, Benny GL. 1978. On the taxonomy of *Monilia roreri*, an important pathogen of *Theobroma cacao* in South America. Can J Bot 56 (20): 2528-2532.

Kerekes JF, Desjardin DE. 2009. A monograph of the genera *Crinipellis* and *Moniliophthora* from Southeast Asia including a molecular phylogeny of the nrITS region. Fung Diver 37: 101-152.

Pegler DN. 1983. Agaric flora of the Lesser Antilles. Kew Bull, Addit Ser 9: 1-668.

Pegler DN. 1986. Agaric flora of Sri Lanka. Kew Bull, Addit Ser 12: 1-519.

Singer R. 1942. A monographic study of the genera *Crinipellis* and *Chaetocalathus*. Lilloa 8: 441-534 (published in 1943).

Singer R. 1986. The Agaricales in modern taxonomy, 4th edn. Koeltz, Koenigstein.

Takahashi H. 2000. Three new species of *Crinipellis* found in Iriomote Island, southwestern Japan, and central Honshu, Japan. Mycoscience 41 (2): 171-182.

7. シラガニセホウライタケ *Crinipellis canescens* Har. Takah.
Mycoscience 41 (2): 171 (2000).
≡ *Moniliophthora canescens* (Har. Takah.) Kerekes & Desjardin, Fung Diver 37: 137 (2009) [MB#513037].

肉眼的特徴 (Figs. 47-51)：傘は径 6-15 mm，最初縁部は内側に巻き，半球形，のち饅頭形〜平らに開き，時に中丘を具えるかまたは中央部が凹み，放射状に溝線を表す；表面は乾性，全体に白い微毛に被われ，微毛の下は最初赤褐色，のち中央部は褐色で周縁部に向かって黄褐色を帯び，湿時橙褐色を呈する．肉は非常に薄く（0.3 mm 以下），傘の表面と同色，弾力性があり，質は強靱，特別な味や匂いはない．柄は 10-15×0.5-1 mm，ほぼ上下同大または基部において僅かに拡大し，中心生，やせ型，中空，最初は全体に白色，のち根本にかけて褐色を帯び，上部は白色粉状，下部は微毛を帯び，根元は基質を被う発達した白色菌糸体に接続する．ヒダは上生〜不完全な襟帯を形成し，疎（柄に到達するヒダは14-18），小ヒダは0-1，幅1 mm 以下，僅かに連絡脈を形成し，傘と同色．胞子紋は白色．

顕微鏡的特徴 (Fig. 46)：担子胞子は 9.2-10.4×3.7-4.3 µm (n = 34, Q = 2.41-2.48)，楕円形，平坦，無色，非アミロイド，薄壁．担子器は 26.4-37.3×6.4-8.8 µm，こん棒形，4 胞子性；偽担子器はこん棒形．縁シスチジアは 16-21.6×5-8.5 µm，不稔帯を形成し，亜こん棒形〜不規則な形状をなし，頂部に 1- 数本の短指状突起（幅2.4-4.2 µm）を持ち，無色または淡黄色，非アミロイド，薄壁．側シスチジアはない．子実層托実質は平列型，菌糸は傘実質と同じ．傘表皮組織は毛状末端細胞に接続する亜円柱形の菌糸が平行菌糸被を成す：菌糸細胞は 30-60×6-12 µm，しばしば膨大し，壁は厚さ1 µm 以下，橙褐色の色素が凝着し，偽アミロイド；毛状末端細胞は 230-1200×4-7 µm，匍匐性または直立し，円柱形，鈍頭，一般に頂部に向かって次第に細くなり，時に屈曲し，平滑，強偽アミロイド（紫褐色）に染まり，壁は無色または淡い蜜色で厚さ1.6-3.2 µm，二次隔壁を欠く．傘実質を構成する菌糸は幅 6-16 µm，並列し，円柱形〜亜紡錘形，しばしば膨大し，平滑，淡い蜜色の細胞内色素を持ち，偽アミロイド，薄壁．柄表皮組織は平行菌糸被を成し，束状に配列した毛状末端細胞に接続する：毛状末端細胞は傘と同様．柄実質の菌糸は幅 5.3-8 µm，縦に沿って配列し，円柱形，平滑，淡い蜜色〜帯褐橙色の細胞内色素を有し，偽アミロイド，薄壁．全ての組織においてクランプが存在する．

生態および発生時期：シシアクチなどの常緑低木の枯れ枝上に孤生または散生，4月〜11月，普通種．
分布：沖縄（八重山諸島：石垣島，西表島），マレーシア（Kerekes and Desjardin 2009）．
供試標本：KPM-NC0005014（正基準標本），常緑低木の枯れ枝上，沖縄県西表島浦内川流域，1999年6月2日，高橋春樹採集；CBM-FB-24120，同上，1996年10月1日，高橋春樹採集；CBM-FB-24121，同上，1997年5月15日，高橋春樹採集；CBM-FB-24122，同上，1998年6月6日，高橋春樹採集；TNS-F-48181，常緑低木の枯れ枝上，石垣市オモト山，2012年6月9日；KPM-NC0023876，シシアクチの落枝上，沖縄県石垣市バンナ岳，2013年6月2日，高橋春樹採集．

主な特徴：傘は放射状の溝線を表し，白色の微毛に被われ，成熟時帯褐黄色を呈する；柄は白色または帯褐色，微毛〜剛毛に被われ，根元は基質を被う圧着した白色の菌糸マットに接続する；縁シスチジアは亜こん棒形〜不規則な形状を成し，1- 数個の短指状付属糸を持つ；傘および柄の表面は偽アミロイドに染まる毛状末端細胞に被われる；シシアクチなどの常緑低木の枯れ枝上に発生．

コメント：明るい色素を欠き，相対的に長い（7 mm 以下）中心性の柄，幅9 µm 以下の楕円形の担子胞子，そして側シスチジアを欠く性質は本種がニセホウライタケ属 *Crinipellis*，ニセホウライタケ節 section *Crinipellis*，ニセホウライタケ亜節 subsection *Crinipellis*（Singer 1942, 1986）に属することを示唆している．

　　亜節内においてシラガニセホウライタケは径10 mm 以下の小形の傘と複数の付属糸を持つ縁

7. *Crinipellis canescens* シラガニセホウライタケ —— 63

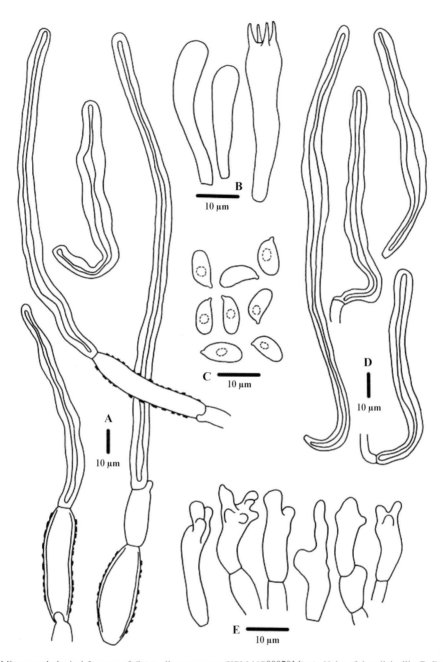

Fig. 46 – Micromorphological features of *Crinipellis canescens* (KPM-NC0005014): **A**. Hairs of the pileipellis. **B**. Basidium and basidioles. **C**. Basidiospores. **D**. Hairs of the stipitipellis. **E**. Cheilocystidia. Illustrated by Takahashi, H.
シラガニセホウライタケの顕鏡図（KPM-NC0005014）：**A**. 傘表皮組織の毛状末端細胞. **B**. 担子器および偽担子器. **C**. 担子胞子. **D**. 柄表皮組織の毛状末端細胞. **E**. 縁シスチジア. 図：高橋春樹.

7. *Crinipellis canescens* シラガニセホウライタケ

シスチジアの特徴において以下の新熱帯産分類群に近縁と思われる：*Crinipellis pseudostipitaria* Singer（Pegler, 1983；Singer, 1942）；*Crinipellis septotricha* Singer（Pegler, 1983；Singer, 1942）；*Crinipellis stupparia*（Berk. & M. A. Curtis）Pat.（Pegler, 1983）．しかしながらこれら新熱帯産分類群は傘が乳頭状中丘を形成し，根元に発達した菌糸体を欠くと言われている．さらに *C. pseudostipitaria* は草本の残骸から発生し，また *C. septotricha* および *C. stupparia* は毛状末端細胞に梯子状二次隔壁を有する．スリランカ産 *Crinipellis omotricha*（Berk.）D. A. Reid（Pegler, 1986）も本種に類似するが，縁シスチジアは付属糸を欠き，草本を基質とする点でシラガニセホウライタケと異なる．マレーシア産 *Crinipellis hepatica* Corner（Corner, 1996）はシラガニセホウライタケに最も近縁と考えられるが，傘は赤褐色を帯びること，柄の根元に発達した菌糸体を持たないこと，亜紡錘形の偽担子器を形成すること，そして付属糸を欠く縁シスチジアを有する性質において本種と区別できる．

　Kerekes and Desjardin（Kerekes and Desjardin 2009）はマレーシア産標本を用いた ITS 領域のシーケンスデータに基づきニセホウライタケ属 *Crinipellis* の無性世代として知られる熱帯産

Fig. 47 – Pileus surface of *Crinipellis canescens*, 18 May 2012, Banna Park, Ishigaki Island. Photo by Takahashi, H.
　シラガニセホウライタケの傘表面．2012年5月18日，石垣島バンナ公園．写真：高橋春樹．

Moniliophthora 属 (Evans et al. 1978; Aime and Phillips-Mora 2005) と本種との組み合わせを提唱している.しかしながら *Moniliophthora* 属の採用にあたっては,以下のような問題点がある.
1. 正基準標本またはそれに準じる標本(topotype)を分子解析に用いていない.
2. *Moniliophthora* 属とニセホウライタケ属の形態学的相違が不明確である.

上記の理由から,ここでは従来の分類概念(Singer 1986)に従い本種をニセホウライタケ属として扱った.

Fig. 48 – Underside view of the basidioma of *Crinipellis canescens*, 18 May 2012, Banna Park, Ishigaki Island. Photo by Takahashi, H.
シラガニセホウライタケの子実体.2012年5月18日.石垣島バンナ公園.写真:高橋春樹.

Fig. 49 – Underside view of the basidiomata of *Crinipellis canescens*, 18 May 2012, Banna Park, Ishigaki Island. Photo by Takahashi, H.
シラガニセホウライタケの子実体，2012年5月18日，石垣島バンナ公園．写真：高橋春樹．

7. *Crinipellis canescens* シラガニセホウライタケ ―― 67

Fig. 50 – Basidiomata of *Crinipellis canescens* on dead twigs of shrubs in an evergreen broad-leaved forest, 18 May 2012, Banna Park, Ishigaki Island. Photo by Takahashi, H.
常緑低木の枯れ枝上に発生したシラガニセホウライタケの子実体，2012年5月18日，石垣島バンナ公園．写真：高橋春樹．

Fig. 51 – Immature basidiomata of *Crinipellis canescens* (KPM-NC0005014), 2 Jun. 1999, Urauchi River, Iriomote Island. Photo by Takahashi, H.
シラガニセホウライタケの幼菌（KPM-NC0005014），1999年6月2日，西表島浦内川流域．写真：高橋春樹．

8. *Crinipellis rhizomorphica* Har. Takah. ミドリニセホウライタケ

Mycoscience 52 (6): 392 (2011) [MB#519031].

Etymology: The specific epithet refers to the white, thread-like rhizomorphs on the substratum.

Macromorphological characteristics (Figs. 54–60): Pileus 5–8 (–12) mm in diameter, at first hemispherical to convex with incurved margin, then broadly convex, sometimes with depressed center, often with a brownish orange (6C7–8 or 7C7–8) or blackish, small umbo at center; surface dry, dull, opaque, radially fibrillose-squamulose with strigose, brownish orange (6C7–8 or 7C7–8) to brown (7D7–8) hairs projecting beyond the margin, often concentrically zoned. Flesh very thin (up to 0.5 mm), white, pliant but easily broken, odor and taste not distinctive. Stipe 8–15 (–20) × 0.7–1.3 mm, subequal or slightly enlarged at the base, central, slender, terete, hollow; surface entirely strigose-fibrillose with brownish orange (6C7–8 or 7C7-8) to brown (7D7–8) hairs; basal mycelium not seen. Lamellae adnexed, 19–23 reach the stipe, with 1–3 series of lamellulae, up to 1.5 mm wide, white then pale brownish in age; edges ciliate to fimbriate, concolorous. Rhizomorphs independent of the formation of basidiomata, growing into the air, scattered on the substratum, 5–15 (–30) × 0.02–0.6 mm, thread-like, gradually tapering toward the apex, tough, not branched, dense, parallel, white.

Micromorphological characteristics (Figs. 52, 53): Basidiospores (9.5–) 11–13 (–15) × (4–) 4.5–5 µm (n = 39, Q = 2.44–2.6), oblong-ellipsoid, smooth, colorless, inamyloid, rarely septate, thin-walled. Basidia 25–27 × 3–8 µm, clavate, 4-spored; basidioles fusiform to subclavate. Cheilocystidia 17–33 × 6–10 µm, gregarious, forming a compact sterile edge, projecting from the hymenium, subclavate, with two or three short cylindrical apical appendages 1–5 × 1–2.5 µm, colorless, inamyloid, hyaline in KOH, thin-walled. Pleurocystidia none. Hyphae of hymenophoral trama 3–8 µm wide, parallel, subcylindrical, not inflated, smooth, colorless, inamyloid, hyaline in KOH, thin-walled. Pileipellis a hypotrichial layer of subcylindrical cells 5–7 µm wide, with smooth, colorless walls 1–2 µm, inamyloid or weakly dextrinoid; hairs of pileus 400–600 × 4–6 µm, arising directly from the hypotrichium, repent or erect, cylindrical, with a rounded apex, sometimes flexuous, with smooth, brownish orange (6C7–8) walls 1–2 µm, strongly dextrinoid, turning olive yellow (3C6–8) in KOH, occasionally with several secondary septa. Hyphae of pileitrama 5–10 µm wide, similar to those of the hymenophoral trama. Hairs of stipe similar to those of the pileus. Stipe trama composed of longitudinally running, cylindrical hyphae 5–13 µm wide, smooth, colorless, inamyloid, hyaline in KOH, thin-walled. Hyphae of rhizomorphs similar to the hairs of basidiomata, 2–5 µm wide, cylindrical, sometimes flexuous, with rounded apex, hyaline in water, inamyloid in the apical portion but otherwise dextrinoid, turning pale olivaceous in KOH, with smooth walls 0.5–2 µm, often with several secondary septa near the apical portion. Clamps present in all tissues.

Habitat and phenology: Gregarious or scattered, on dead twigs (unidentified substrata) of various kinds of low trees in evergreen broad-leaved forests, May to August.

Known distribution: Okinawa (Ishigaki Island).

Specimens examined: KPM-NC0017526 (holotype), on dead twigs of various kinds of low trees in an evergreen broad-leaved forest, Banna Park, Ishigaki-shi, Okinawa Pref., 25 Jul. 2010, coll. Takahashi, H. & Terashima,Y.; KPM-NC0017297, same location, 3 Jul. 2006, coll. Takahashi, H.; KPM-NC0017298, same location, 13 Jun. 2006, coll. Takahashi, H.; KPM-NC0017299, same location, 15 Jun. 2007, coll. Takahashi, H.; KPM-NC0017527, same location, 11 Aug. 2010, coll. Takahashi, H.; TNS-F-48273, same location, 11 Jun. 2012, coll. Taneyama, Y.

Japanese name: Midori-nisehouraitake.

Comments: *Crinipellis rhizomorphica* is characterized by the brownish orange to brown, strigose pileus and stipe with dextrinoid hairs that turn olivaceous in KOH; the oblong-ellipsoid, relatively long basidiospores averaging 12 × 4.75 μm; the subclavate cheilocystidia with two or three short cylindrical apical appendages; the absence of pleurocystidia; and the basidiome formation on dead twigs accompanied by white, thread-like rhizomorphs.

The presence of dextrinoid hairs turning olivaceous in KOH suggests that the present fungus is best accommodated in the genus *Crinipellis*, section *Grisentinae* (Singer) Singer (Singer 1976, 1986).

Within the section *Grisentinae*, *C. rhizomorphica* has macromorphological similarities to the following 4 species: *Crinipellis sapindacearum* Singer from Brazil (Singer 1976); *Crinipellis trichialis* (Lév.) Pat. ex Antonín R. Ryoo & H.D. Shin (Léveillé 1846; Saccardo 1887; Antonín et al. 2009), redescribed by Singer from Venezuela (Singer 1976) and by Kerekes & Desjardin from Indonesia and Malaysia (Kerekes and Desjardin 2009); *Crinipellis tucumanensis* Singer from Argentina (Singer 1976); and *Crinipellis rhizomaticola* Antonín from Republic of Korea (Antonín et al. 2009). These taxa mainly differ from *C. rhizomorphica* in having well developed pleurocystidia and lacking rhizomorphs. Furthermore, *C. sapindacearum* has much smaller basidiospores: 7.5-8.2 × 3-3.5 μm (Singer 1976), and a habitat of dead fallen coriaceous leaves of Sapindaceae. *Crinipellis trichialis* produces shorter and broader basidiospores: (8.5-) 9.6-11.5 × (5.5-) 6-7 (-7.4) μm (Kerekes and Desjardin 2009). *Crinipellis tucumanensis* forms much shorter basidiospores: 5.5-8.5 μm long (Singer 1976). *Crinipellis rhizomaticola* is distinct in having a chestnut-brown, larger pileus: 12-22 mm in diameter (Antonín et al. 2009), and significantly shorter basidiospores: 8.5-10 μm long (Antonín et al. 2009). *Crinipellis rhizomorphica* also shares characteristics such as olivaceous-colored hairs in KOH, copious rhizomorphs, and a radially fibrillose-strigose pileus with a minute blackish papilla in umbilicus in common with the following two species: Southeast Asian *Crinipellis actinophora* (Berk. & Broome) Singer (Berkeley and Broome 1874; Singer, 1955; Pegler, 1986; Corner 1996; Kerekes and Desjardin 2009); and *Crinipellis nigricaulis* Har. Takah. from Japan (Takahashi 2000) and Republic of Korea (Antonín et al. 2009). These two taxa, however, can be discerned from *C. rhizomorphica* by forming a dark brown stipe occasionally associating with the much longer, dark brown hair-like rhizomorphs, significantly shorter basidiospores: 6-10 μm long (Kerekes and Desjardin 2009), and cheilocystidia with numerous apical appendages.

References 引用文献

Antonín V, Ryoo R, Shin HD. 2009. Marasmioid and gymnopoid fungi of the Republic of Korea. 1. Three interesting species of *Crinipellis* (Basidiomycota, Marasmiaceae). Mycotaxon 108: 429-440.
Berkeley MJ, Broome CE. 1874. Enumeration of the fungi of Ceylon. Part II. Journ Linn Soc Bot 14: 29-141.
Corner EJH. 1996. The agaric genera *Marasmius, Chaetocalathus, Crinipellis, Heimiomyces, Resupinatus, Xerula* and *Xerulina* in Malesia. Beih Nova Hedwigia 111: 1-175.
Kerekes J, Desjardin DE. 2009. A monograph of the genera *Crinipellis* and *Moniliophthora* from Southeast Asia including a molecular phylogeny of the nrITS region. Fung Diver 37: 101-152.
Léveillé JH. 1846. Descriptions des champignons de l'herbier du Muséum de Paris. Annls Sci Nat, Bot, sér 35: 111-167.
Pegler DN. 1986. Agaric flora of Sri Lanka. Kew Bull, Addit Ser 12: 1-519.
Saccardo PA. 1887. Sylloge Hymenomycetum, Vol. I. Agaricineae. Syll fung 5: 1-1146.
Singer R. 1955. Type studies on Basidiomycetes VIII. Sydowia 9: 367-431.
Singer R. 1976. Marasmieae (Basidiomycetes - Tricholomataceae). Fl Neotrop Monogr 17: 1-347.
Singer R. 1986. The Agaricales in modern taxonomy, 4th edn. Koeltz, Koenigstein.
Takahashi H. 2000. Three new species of *Crinipellis* found in Iriomote Island, southwestern Japan, and central Honshu, Japan. Mycoscience 41 (2): 171-182.
Takahashi H. 2011. Two new species of Agaricales and a new Japanese record for *Chaetocalathus fragilis* from Ishigaki Island, a southwestern island of Japan. Mycoscience 52 (6): 392-400.

8. ミドリニセホウライタケ *Crinipellis rhizomorphica* Har. Takah.

Mycoscience 52 (6)：392 (2011) [MB#519031].

肉眼的特徴 (Figs. 54-60)：傘は径 5-8 (-12) mm, 最初半球形～饅頭形で縁部は内側に巻き, のち中高偏平になり, 時に中央部が凹み, しばしば橙褐色～黒褐色の小形中丘を具える；表面は乾性, 全体に縁部に向かって放射状に走る帯褐橙色～褐色の毛に被われ, しばしば同心円状の環紋を表す. 肉は非常に薄く (0.5 mm 以下), 白色, 多少柔軟性があるが壊れやすい, 特別な味や臭いはない. 柄は 8-15 (-20)×0.7-1.3 mm, ほぼ上下同大または基部がやや拡大し, 中心生, やせ型, 中空；表面は全体に帯褐橙色～褐色の毛に被われる；根元に発達した菌糸体は見られない. ヒダは上生, 柄に到達するヒダは19-23, 1-3の小ヒダを交え, 幅 1.5 mm 以下, 白色, 老成時は淡褐色を帯びる；縁部は長縁毛があり, 同色. 根状菌糸束は5-15 (-30)×0.02-0.6 mm, 糸状, 頂部に向かって次第に細くなり, 強靱, 白色, 無分岐, 空中に向かって伸長し, 基質上に散生し, 子実体から独立して形成される.

顕微鏡的特徴 (Figs. 52, 53)：担子胞子は (9.5-) 11-13 (-15)×(4-) 4.5-5 µm (n = 39, Q = 2.44-2.6), 長楕円形, 平坦, 無色, 非アミロイド, 稀に隔壁を有し, 薄壁. 担子器は25-27×3-8 µm, こん棒形, 4胞子性；偽担子器は紡錘形または亜こん棒形. 縁シスチジアは17-33×6-10 µm, 群生し, 縁部に不稔帯を形成し, 子実層から突出し, 亜こん棒形, 2-3個の短指状頂生付属糸 (1-5×1-2.5 µm) を持ち, 無色, 非アミロイド, 水酸化カリウム溶液において無色, 薄壁. 側シスチジアはない. 子実層托実質の菌糸は幅 3-8 µm, 平列し, 亜円柱形, 膨大せず, 平滑, 無色, 非アミロイド, 水酸化カリウムによる反応は陰性, 薄壁. 傘表皮組織は毛状細胞に接続する亜円柱形の菌糸 (幅5-7 µm) が層をなし, 壁は無色で厚さ1-2 µm, 非アミロイドまたは弱偽アミロイド；傘の毛状細胞は400-600×4-6 µm, 匍匐性または直立し, 円柱形, 鈍頭, 時に屈曲し, 壁は帯褐橙色で厚さ1-2 µm, 強偽アミロイド, 水酸化カリウム溶液においてオリーブ緑色に染まり, しばしば複数の二次隔壁が梯子状構造を形成する. 傘実質の菌糸は幅5-10 µm, 子実層托実質の菌糸に類似する. 柄の毛状細胞は傘と同様. 柄の実質は縦に沿って配列した円柱形の菌糸 (幅5-13 µm) からなり, 無色, 非アミロイド, 水酸化カリウムによる反応は陰性. 根状菌糸束の菌糸は幅2-5 µm, 子実体の毛状細胞に類似し, 円柱形, 頂部に向かって次第に細くなり, 密集し, 並列し, 鈍頭, 壁は厚さ0.5-2 µm, 無色 (水封), 非アミロイドの頂部以外は偽アミロイドに染まり, 水酸化カリウム溶液において無色または淡オリーブ緑色, 頂部付近においてしばしば二次隔壁が存在する. 全ての組織において菌糸にクランプが見られる.

生態および発生時期：常緑広葉樹林内低木の枯れ枝上に散生または群生, 5月～8月.

分布：沖縄 (石垣島).

供試標本：KPM-NC0017526 (正基準標本), 常緑広葉樹林内低木の枯れ枝上, 沖縄県石垣市バンナ公園, 2010年7月25日, 高橋春樹 & 寺嶋芳江採集；KPM-NC0017297, 同上, 2006年7月3日, 高橋春樹採集；KPM-NC0017298, 同上, 2006年6月13日, 高橋春樹採集；KPM-NC0017299, 同上, 2007年6月15日, 高橋春樹採集；KPM-NC0017527, 同上, 2010年8月11日, 高橋春樹採集；TNS-F-48273, 同上, 2012年6月11日, 種山裕一採集.

主な特徴：1) 傘と柄の表面は帯褐色橙色～褐色を帯び, 偽アミロイドに染まり, アルカリ溶液 (5%水酸化カリウム溶液または水酸化ナトリウム) によりオリーブ緑色に変わる毛状細胞に被われる. 2) 担子胞子はやや長形で, 長さ11-13 µm に達する. 3) 縁シスチジアは亜こん棒形で2-3個の短指状頂生付属糸を具える；4) 側シスチジアを欠く. 5) 子実体は常緑低木の枯れ枝上に発生し, 白色糸状菌糸束を伴う.

コメント：偽アミロイドに染まり且つ水酸化カリウム溶液によりオリーブ緑色に変わる毛状細胞は Singer (1976, 1986) の分類概念によるニセホウライタケ属 *Crinipellis*, ミドリニセホウラ

8. *Crinipellis rhizomorphica* ミドリニセホウライタケ

イタケ節（高橋春樹新称）section *Grisentinae*（Singer）Singer に属すると考えられる．ミドリニセホウライタケ節内において本種は以下の4種と肉眼的類似性が見られる：1）ブラジル産 *Crinipellis sapindacearum* Singer（Singer 1976）；2）新熱帯産 *Crinipellis trichialis*（Lev.）Pat. ex Antonin R. Ryoo ＆ H. D. Shin（Léveillé 1846；Saccardo 1887；Antonín et al. 2009）：3）アルゼンチン産 *Crinipellis tucumanensis* Singer（Singer 1976）；4）韓国産 *Crinipellis rhizomaticola* Antonin（Antonin et al. 2009）．これら4種は主に発達した側シスチジアを有し，より短形の担子胞子を持ち，基質上に菌糸束が存在しない点で本種と区別できる．更に *C. sapindacearum* はムクロジ科（Sapindaceae）の樹木の落ち葉上に発生し，*C. rhizomaticola* はより大型（径12-22 mm）で赤褐色の傘を形成する点で有意差が認められる．ミドリニセホウライタケはまた水酸化カリウム溶液により緑変する毛状細胞，発達した根状菌糸束の存在，そして傘中央部がへそ状に凹み暗色の中丘を具える性質において東南アジア産 *Crinipellis actinophora*（Berk. & Broome）Singer（Berkeley and Broome 1874；Singer, 1955；Pegler, 1986；Corner 1996；Kerekes and Desjardin 2009）並びに東アジア（日本，韓国）産クロカミオチバタケ *Crinipellis nigricaulis* Har. Takah.（Takahashi 2000；Antonín et al. 2009）と共通する．これら2種はしばしば黒褐色の毛状菌糸束につながる暗褐色の柄を持つこと，担子胞子がより短形であること，そして縁シスチジアの頂部に多数の付属糸を具える点でミドリニセホウライタケと異なる．

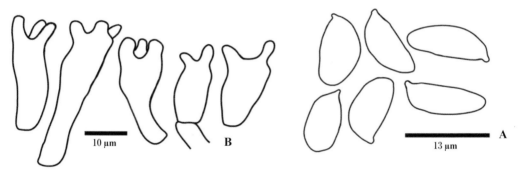

Fig. 52 – Micromorphological features of *Crinipellis rhizomorphica* (KPM-NC0017526): **A**. Basidiospores. **B**. Cheilocystidia. Illustrations by Takahashi, H.
ミドリニセホウライタケの顕微鏡図（KPM-NC0017526）：**A**. 担子胞子．**B**. 縁シスチジア．図：高橋春樹．

8. *Crinipellis rhizomorphica* ミドリニセホウライタケ

Fig. 53 – Hairs of the pileipellis of *Crinipellis rhizomorphica* (in KOH, KPM-NC0017526). Photo by Takahashi, H.
ミドリニセホウライタケの傘表皮組織の毛状細胞（水酸化カリウム溶液でマウント，KPM-NC0017526）．写真：高橋春樹．

Fig. 54 – Basidiomata of *Crinipellis rhizomorphica* on a dead branch of a shrub in an evergreen broad-leaved forest, 18 May 2012, Banna Park, Ishigaki Island. Photo by Takahashi, H.
常緑低木の枯れ枝上に発生したミドリニセホウライタケの子実体．2012年5月18日，石垣島バンナ公園．写真：高橋春樹．

8. *Crinipellis rhizomorphica* ミドリニセホウライタケ —— 73

Fig. 55 – Basidiomata of *Crinipellis rhizomorphica* on a dead branch of a shrub in an evergreen broad-leaved forest, 18 May 2012, Banna Park, Ishigaki Island. Photo by Takahashi, H.
常緑低木の枯れ枝上に発生したミドリニセホウライタケの子実体，2012年5月18日．石垣島バンナ公園．写真：高橋春樹．

74 —— 8. *Crinipellis rhizomorphica* ミドリニセホウライタケ

Fig. 56 – Basidiomata of *Crinipellis rhizomorphica* (KPM-NC0017298) on a dead twig of a shrub in an evergreen broad-leaved forest, 13 Jun. 2006, Banna Park, Ishigaki Island. Photo by Takahashi, H.
常緑低木の枯れ枝上に発生したミドリニセホウライタケの子実体（KPM-NC0017298），2006年6月13日，石垣島バンナ公園．写真：高橋春樹．

8. *Crinipellis rhizomorphica* ミドリニセホウライタケ

Fig. 57 – Basidiomata of *Crinipellis rhizomorphica* (KPM-NC0017527) on a dead twig of a shrub in an evergreen broad-leaved forest, 11 Aug. 2010, Banna Park, Ishigaki Island. Photo by Takahashi, H.
常緑低木の枯れ枝上に発生したミドリニセホウライタケの子実体（KPM-NC0017527），2010年8月11日，石垣島バンナ公園．写真：高橋春樹．

8. *Crinipellis rhizomorphica* ミドリニセホウライタケ

Fig. 58 – Basidiomata of *Crinipellis rhizomorphica* (KPM-NC0017527) on a dead twig of a shrub in an evergreen broad-leaved forest, 11 Aug. 2010, Banna Park, Ishigaki Island. Photo by Takahashi, H.
常緑低木の枯れ枝上に発生したミドリニセホウライタケの子実体（KPM-NC0017527），2010年8月11日，石垣島バンナ公園．写真：高橋春樹．

8. *Crinipellis rhizomorphica* ミドリニセホウライタケ —— 77

Fig. 59 – Basidiomata of *Crinipellis rhizomorphica* (KPM-NC0017527) on a dead twig of a shrub in an evergreen broad-leaved forest, 11 Aug. 2010, Banna Park, Ishigaki Island. Photo by Takahashi, H.
常緑低木の枯れ枝上に発生したミドリニセホウライタケの子実体（KPM-NC0017527）．2010年8月11日，石垣島バンナ公園．写真：高橋春樹．

8. *Crinipellis rhizomorphica* ミドリニセホウライタケ

Fig. 60 – Rhizomorphs of *Crinipellis rhizomorphica* (KPM-NC0017527) on a dead twig of a shrub in an evergreen broad-leaved forest, 11 Aug. 2010, Banna Park, Ishigaki Island. Photo by Takahashi, H.
常緑低木の枯れ枝上に発生したミドリニセホウライタケの菌糸束（KPM-NC0017527）．2010年8月11日．石垣島バンナ公園．写真：高橋春樹．

9. *Cruentomycena orientalis* Har. Takah. & Taneyama, sp. nov. ガーネットオチバタケ
MycoBank no.: MB 809925.
Etymology: The specific epithet means "Eastern".

Distinctive features of this species consist of mycenoid, deep red to dark red basidiomata with a non-gelatinous, sulcate-striate pileus and marginate, decurrent lamellae; amyloid, ellipsoid to oblong-ellipsoid basidiospores; broadly clavate to subcylindrical cheilocystidia and caulocystidia without any outgrowths; absence of pleurocystidia; smooth elements of the pileipellis and stipitipellis; dextrinoid trama in the hymenophore and stipe; thick-walled, highly inflating, fusiform hyphal cells of the lower stipe trama that are similar to sarcohyphae but shorter (up to 150 μm long); and a foliicolous habit on dead fallen hardwood leaves.

Macromorphological characteristics (Figs. 64–68): Pileus 4–8 mm in diameter, at first hemispherical to convex, then broadly convex, sometimes obtusely subumbonate, radially sulcate-striate toward the eroded margin; surface minutely subpruinose, subhygrophanous, dry, red (10A7 to 11B7–8) to vivid red (10A8), darker on the disk and striations. Flesh very thin (up to 0.3 mm), paler concolorous with the pileus, odor not distinctive, taste unknown. Stipe 15–35 × 0.7–1 mm, cylindrical but somewhat thickened at the base, central, slender, terete, fistulose, smooth; surface minutely pruinose, dry, red (10A7 to 11B7–8) or brownish red (10C7) to deep red (10C8); base covered with conspicuous concolorous mycelioid bristles. Lamellae decurrent, subdistant (14–18 reach the stipe), with 1–2 series of lamellulae, not intervenose, narrow (0.5–0.8 mm broad), thin, paler concolorous with the pileus; edges finely fimbriate under a lens, brownish red (10C7) to deep red (10C8).

Micromorphological characteristics (Figs. 62, 63): Basidiospores (7.1–) 7.9–9.0 (–10.5) × (3.5–) 3.8–4.3 (–5.1) μm (n = 30, mean length = 8.48 ± 0.45, mean width = 4.11 ± 0.27, Q = (2.05–) 2.07–2.09 (–2.28), mean Q = 2.08 ± 0.1), ellipsoid to oblong-ellipsoid, smooth, hyaline, weakly amyloid, thin-walled. Basidia (17.6–) 18.6–22.3 (–23.8) × (5.7–) 6.1–6.8 (–7.0) μm (n = 14, mean length = 20.43 ± 1.86, mean width = 6.44 ± 0.38) in the main body, 2.5–3.2 × 0.7–1.5 μm in sterigmata, clavate, 2-3-4-spored; basidioles clavate. Cheilocystidia (17.5–) 21.0–29.4 (–40.0) × (6.4–) 8.6–12.0 (–15.2) μm (n = 50, mean length = 25.22 ± 4.20, mean width = 10.28 ± 1.73), abundant, forming a sterile lamella edge, broadly clavate to subcylindrical, with an obtuse apex, smooth, with reddish orange (7A7–8) vacuolar pigments in water, with slightly thickened walls (0.5–) 0.6–0.7 (–0.8) μm (n = 23, mean thickness = 0.64 ± 0.08). Pleurocystidia absent. Hymenophoral trama subregular; element hyphae (3.5–) 4.2–8.4 (–11.0) μm wide (n = 31, mean width = 6.29 ± 2.11), more or less parallel, cylindrical or occasionally somewhat inflated, smooth, with reddish orange (7A7–8) vacuolar pigments in water, yellowish in KOH, dextrinoid, with slightly thickened walls 0.6–0.8 (–1.1) μm (n = 20, mean thickness = 0.73 ± 0.12). Outermost layer of pileipellis a cutis of somewhat loosely interwoven, repent hyphae (4.2–) 5.2–7.1 (–8.0) μm wide (n = 21, mean width = 6.16 ± 0.93), cylindrical, smooth, with reddish orange (7A7–8) vacuolar pigments in water, thin-walled; pilocystidia none. Innermost layer of pileipellis (subpellis) pseudoparenchymatous, well differentiated from the upper stratum; constituent hyphal cells (21.5–) 30.9–52.9 (–62.6) × (17.3–) 20.1–26.7 (–29.7) μm (n = 24, mean length = 41.92 ± 11.00, mean width = 23.42 ± 3.29), subparallel, highly inflated and constricted at the septa, subcylindrical, smooth, hyaline or pale reddish orange in water, with thickened walls (0.8–) 1.0–1.3 (–1.5) μm (n = 21, mean thickness = 1.14 ± 0.17). Hyphae of pileitrama (2.7–) 4.2–9.5 (–13.5) μm wide (n = 35, mean width = 6.82 ± 2.64) subparallel, subcylindrical, smooth, hyaline or pale reddish orange in water, with slightly thickened walls (0.5–) 0.6–0.8 (–1.0) μm (n = 25, mean thickness = 0.70 ± 0.14). Stipitipellis a cutis of parallel, repent hyphae (3.2–) 3.6–5.0 (–6.2) μm wide (n = 28, mean width = 4.30 ± 0.72), cylindrical, with reddish orange (7A7-8) vacuolar pigments in water, yellowish in KOH, with slightly thickened walls (0.8-) 0.9-1.1 (-1.2) μm (n = 17, mean thickness = 1.00 ± 0.14); caulocystidia (11.5-) 21.9-38.0 (-48.7) × (6.5-) 8.2-14.9 (-20.4) μm (n = 23, mean length = 29.95 ± 8.09, mean width = 11.54 ± 3.34), scattered, broadly clavate to subcylindrical, with an obtuse apex, smooth, with reddish orange (7A7-8) vacuolar pigments in water, with slightly thickened walls 0.6-0.8 (-0.9) μm (n = 20, mean thickness = 0.67 ± 0.10). Stipe trama composed of longitudinally running, unbranched, cylindrical hyphae (7.4-) 9.9-15.7 (-18.7) μm wide (n = 29, mean width = 12.80 ± 2.90) in the upper portion of stipe, smooth, with reddish orange (7A7-8) vacuolar pigments in water, yellowish in KOH,, dextrinoid, with slightly thickened walls (0.6-) 0.7-1.0 (-1.1) μm (n = 18, mean thickness = 0.88 ± 0.15); stipe trama at the lower portion of stipe largely made up of non-inflated, cylindrical hyphae (9.9-) 12.0-19.7 (-28.1) μm wide (n = 26, mean width = 15.84 ± 3.88), with walls up to 1.0 μm thick, intermixed with scattered, fusiform, inflated hyphal cells (30.1-) 56.9-132.6 (-156.6) × (14.1-) 20.9-36.7 (-42.2) μm (n = 17, mean length = 94.76 ± 37.84, mean width = 28.80 ± 7.89), with distinctly thickened walls (1.7-) 1.8-2.3 (-2.5) μm (n = 13, mean thickness = 2.03 ± 0.26). Elements of basal mycelioid bristles (2.6-) 3.4-4.8 (-5.7) μm wide (n = 29, mean width = 4.11 ± 0.72), cylindrical, smooth, hyaline or pale reddish orange vacuolar pigments in water, thin-walled. All tissues inamyloid, except trama of hymenophore and stipe, with clamp connections.

Habitat and phenology: Solitary or scattered, foliicolous on dead fallen hardwood leaves, October.

Known distribution: Okinawa (Ishigaki Island).

Holotype: TNS-F-61362, on dead fallen hardwood leaves, Mt. Banna, Ishigaki-shi, Okinawa Pref., 20 Oct. 2011, coll. Takahashi, H.

Other specimens examined: TNS-F-61363, on dead fallen hardwood leaves, Mt. Banna, Ishigaki-shi, Okinawa Pref., 20 Oct. 2011, coll. Takahashi, H.; KPM-NC0023862 (Isotype), the same location, 20 Oct. 2011, coll. Takahashi, H.

Gene sequenced specimen and GenBank accession number: TNS-F-61363, AB968239 (ITS).

Japanese name: Ganetto-ochibatake (named by H. Takahashi & Y. Taneyama).

Comments: The genus *Cruentomycena* R.H. Petersen Kovalenko & O. Morozova was established by Petersen and coworkers (Petersen et al. 2008) based on phylogenetic analyses and the following combination of the morphological characteristics: small, mycenoid to marasmielloid, rich reddish crimson to blood red basidiomata; a centrally depressed or almost umbilicate, viscid to glutinous pileus; subdistant, marginate lamellae with decurrent teeth; glutinous to viscid stipe; a pileipellis composed of repent, gelatinous, smooth (non-spinulose) elements; a dextrinoid, pseudoparenchymatous pileitrama; amyloid, elongate pip-shaped basidiospores; absence of pleurocystidia; presence of clamp connections; and a foliicolous habitat. Their research using ribosomal LSU and ITS sequences revealed that *Cruentomycena* belongs to panelloid clade comprising the poroid, pleurotoid, bioluminescent genus *Panellus* (= *Dictyopanus*).

Despite the lack of viscidity in the pileipellis and stipitipellis being out of place, we tentatively accept the validity of *Cruentomycena* particularly given the remarkable morphological similarity with the type species of the genus.

The new species has striking morphological characteristics that are shared with Australasian *Cruentomycena viscidocruenta* (Cleland) R.H. Petersen & Kovalenko (Cleland 1919, 1924; Grgurinovic 1997, 2002; Petersen et al. 2008) and Russian *Cruentomycena kedrovayae* R.H. Petersen, Kovalenko & O.V. Morozova (Petersen et al. 2008). *Cruentomycena viscidocruenta* is distinct in forming a typically umbilicate, distinctly gelatinous pileus and dimorphic cheilocystidia (Petersen et al. 2008).

Cruentomycena kedrovayae has a viscid stipe and elongate-pedicellate caulocystidia.

In spite of its outstanding appearance conferred by vivid pigments, the confinement of the genus to the realm of Australasia and Russia has long been considered to be a quintessence of a disjunct distribution. According to the Dr. Yasuaki Murakami's personal communication, an additional taxon of *Cruentomycena* has also been found in Kyushu. A more comprehensive picture of the distribution of this genus is currently being assembled.

References 引用文献

Cleland JB. 1924. Australian fungi: notes and descriptions. No. 5. Trans Roy Soc S Australia 48: 236-252.
Cleland JB, Cheel EC. 1919. Australian fungi: notes and descriptions. No. 2. Trans Roy Soc S Australia 43: 11-22.
Grgurinovic CA. 1997. Larger fungi of South Australia. The Botanic Gardens of Adelaide and State Herbarium and the Flora and Fauna of South Australia Handbooks Committee, Adelaide.
Grgurinovic CA. 2002. The genus *Mycena* in South-Eastern Australia. Fung Diver Press, Hong Kong.
Petersen RH, Hughes KW, Lickey EB, Kovalenko AE, Morozova OV, Psurtseva NV. 2008. A new genus, *Cruentomycena*, with *Mycena viscidocruenta* as type species. Mycotaxon 105: 119-136.

9. ガーネットオチバタケ（新種；高橋春樹 & 種山裕一新称）*Cruentomycena orientalis* Har. Takah. & Taneyama, **sp. nov.**

肉眼的特徴（Figs. 64-68）：傘は径4-8 mm，最初半球形～丸山形，のち中高偏平，鋸歯状の縁部に向かって放射状の溝状条線を表す；表面はやや微細な粉状，やや吸水性，粘性を欠き，深紅色，傘中央部並びに条線はより暗色を帯びる；縁部は全縁で内側に巻かない．肉は厚さ0.3 mm 以下，傘表面より淡色，特別な匂いはなく，味は不明．柄は15-35×0.7-1 mm，円柱形で基部がやや太くなり，中心生，痩せ型，中空；表面は微細な粉状，乾生，暗赤色～深紅色，根元は柄表面と同色の毛状菌糸に被われる．ヒダは垂生，やや疎（柄に到達するヒダは14-18），1-2の小ヒダを伴い，連絡脈を欠き，幅0.5-0.8 mm，傘表面より淡色；縁部は細かい長縁毛状，暗赤色～深紅色の縁取りがある．

顕微鏡的特徴（Figs. 62, 63）：担子胞子は (7.1-) 7.9-9.0 (-10.5) × (3.5-) 3.8-4.3 (-5.1) μm (n = 30, mean length = 8.48 ± 0.45, mean width = 4.11 ± 0.27, Q = (2.05-) 2.07-2.09 (-2.28), mean Q = 2.08 ± 0.1), 楕円形～長楕円形，平坦，無色，弱アミロイド，薄壁．担子器は (17.6-) 18.6-22.3 (-23.8) × (5.7-) 6.1-6.8 (-7.0) μm (本体), 2.5-3.2×0.7-1.5 μm (ステリグマ), こん棒形，2-3-4胞子性．縁シスチジアは (17.5-) 21.0-29.4 (-40.0) × (6.4-) 8.6-12.0 (-15.2) μm, 群生し，広こん棒形～亜円柱形，鈍頭，分岐物や突起を欠き，橙赤色の内容物が液胞内に存在し，壁は厚さ0.6-0.8 (-1.1) μm. 側シスチジアを欠く．子実層托実質を構成する菌糸は幅 (3.5-) 4.2-8.4 (-11.0) μm, ほぼ平列し，円柱形またはやや膨大し，平滑，偽アミロイド，橙赤色の内容物が液胞内に存在し，壁は厚さ (0.5-) 0.6-0.7 (-0.8) μm. 傘の表皮組織上層はやや錯綜した匍匐性の菌糸からなる；菌糸は幅 (4.2-) 5.2-7.1 (-8.0) μm, 円柱形，平滑，橙赤色の内容物が液胞内に存在し，薄壁；傘シスチジアはない．傘の表皮下層は上層から明瞭に分化し，偽柔組織状；菌糸細胞は (21.5-) 30.9-52.9 (-62.6) × (17.3-) 20.1-26.7 (-29.7) μm, ほぼ並列し，著しく膨大し，隔壁の周囲で急に狭くなり，亜円柱形，平滑，無色または淡橙赤色，壁は厚さ (0.8-) 1.0-1.3 (-1.5) μm. 傘実質は 亜円柱形の菌糸からなり幅 (2.7-) 4.2-9.5 (-13.5) μm, ほぼ並列し，平滑，無色または淡橙赤色，壁は厚さ (0.5-) 0.6-0.8 (-1.0) μm. 柄表皮を構成する菌糸は幅 (3.2-) 3.6-5.0 (-6.2) μm, 匍匐性，並列し，円柱形，平滑，橙赤色の内容物が液胞内に存在し，壁は厚さ (0.8-) 0.9-1.1 (-1.2) μm；柄シスチジアは (11.5-) 21.9-38.0 (-48.7) × (6.5-) 8.2-14.9 (-20.4) μm, 散生，広こ

9. *Cruentomycena orientalis* ガーネットオチバタケ

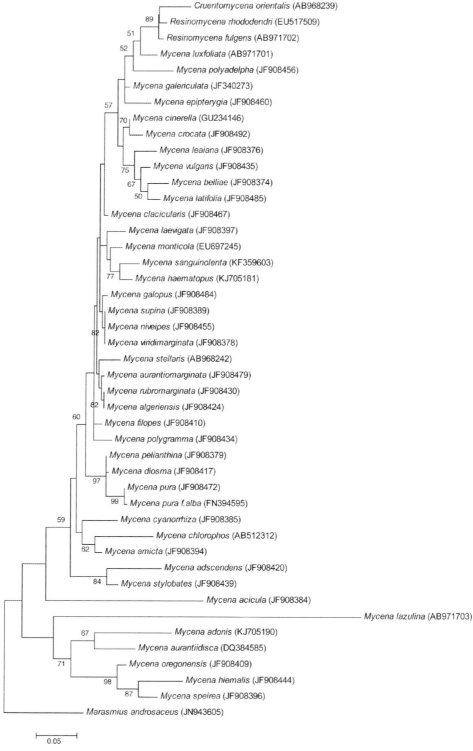

Fig. 61 – A phylogenetic tree of ITS1-5.8S-ITS2 dataset for *Mycena* species, analyzed by maximum likelihood method based on the Tamura-Nei model (Tamura and Nei 1993) in MEGA5 (Tamura et al. 2011). Numbers on the branches represent bootstrap values obtained from 2000 replications (only values greater than 50% are shown).
核リボソーム ITS1-5.8S-ITS2 を用いた最尤法によるクヌギタケ属系統樹.

ん棒形～亜円柱形，鈍頭，分岐物や突起を欠き，橙赤色の内容物が液胞内に存在し，壁は厚さ 0.6-0.8 (-0.9) μm．柄実質の菌糸は縦に沿って並列し，幅 (7.4-) 9.9-15.7 (-18.7) μm (柄の上部)，円柱形，平滑，橙赤色の内容物が液胞内に存在し，偽アミロイド，壁は厚さ (0.6-) 0.7-1.0 (-1.1) μm；柄の下部の菌糸は主に壁の厚さ 1.0 μm 以下の円柱形の菌糸 (幅 (9.9-) 12.0-19.7 (-28.1) μm) からなり，紡錘状に膨大した厚壁 (1.8-2.3 (-2.5) μm) な菌糸細胞 ((30.1-) 56.9-132.6 (-156.6) × (14.1-) 20.9-36.7 (-42.2) μm) が混在する．根元を被う毛状菌糸は幅 (2.6-) 3.4-4.8 (-5.7) μm，円柱形，平滑，無色または淡橙赤色の内容物が液胞内に存在し，薄壁．全ての組織において菌糸は非アミロイド (子実層托実質と柄実質を除く) でクランプを有する．

生態および発生時期：広葉樹の落ち葉上に孤生または散生，10月．

分布：沖縄 (石垣島)．

供試標本：TNS-F-61362 (正基準標本)，広葉樹の落ち葉上，沖縄県石垣市バンナ公園，2011年10月20日，高橋春樹採集；TNS-F-61363，同上，2011年10月20日，高橋春樹採集；KPM-NC0023862 (Isotype)，同上，2011年10月20日，高橋春樹採集．

分子解析に用いた標本並びに GenBank 登録番号：TNS-F-61363，AB968239 (ITS)．

主な特徴：子実体はクヌギタケ型で，深紅色～暗赤色を帯び，粘性を欠き溝状条線を表す傘および

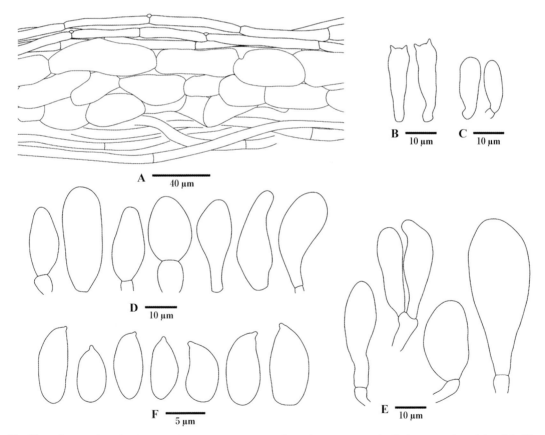

Fig. 62 − Micromorphological features of *Cruentomycena orientalis* (Holotype): **A**. Longitudinal cross section of the pileipellis showing the pseudoparenchymatous layer of the subpellis. **B**. Basidia. **C**. Basidioles. **D**. Cheilocystidea. **E**. Caulocystidia. **F**. Basidiospores. Illustrations by Taneyama, Y.

ガーネットオチバタケの顕微図 (正基準標本)：**A**. 傘の表皮組織および実質の縦断面．**B**. 未熟な担子器．**C**. 偽担子器．**D**. 縁シスチジア．**E**. 柄シスチジア．**F**. 担子胞子．図：種山裕一．

縁取りを持つ垂生のヒダを持つ；担子胞子はアミロイド，楕円形～長楕円形；縁シスチジアおよび柄シスチジアは分岐物や突起を欠き，広こん棒形～亜円柱形；側シスチジアはない；傘および柄の表皮組織の菌糸は平滑；ヒダおよび柄の実質は偽アミロイドで，柄の基部において紡錘状に膨大した厚壁な菌糸が存在するがsarcohyphaより短い（長さ150 μm以下）；広葉樹の落ち葉上に発生．

コメント：ガーネットオチバタケ属（高橋春樹 & 種山裕一新称）*Cruentomycena* R.H. Petersen Kovalenko & O. Morozova は系統解析と以下の特徴の組み合わせに基づきPetersenら（Petersen et al. 2008）によって設立された：子実体は小形のクヌギタケ型～シロホウライタケ型で，深紅色～血石色を帯びる；傘は通常中央部が臍状に凹み，明らかな粘性を表す；ヒダは疎，垂生，縁取りがある；柄は湿時粘性；傘表皮組織を構成する菌糸は匍匐性，ゼラチン質，平滑；傘実質は偽アミロイド，偽柔組織状；胞子はアミロイド，長楕円形～種子形；側シスチジアはない；クランプを有する；葉上性の生態．PetersenらによるITS並びにLSUの領域を用いた系統解析の結果では，ガーネットオチバタケ属はクヌギタケ属 *Mycena* よりもザラメタケ属 *Resinomycena* およびヒラタケ型で管孔を持つスズメタケなどの発光菌を含むワサビタケ属分岐群（panelloid clade）に系統が近いとされている．

表皮組織において粘性を欠く性質は本属として異質であるが，基準種との顕著な形態学的類似性を考慮して暫定的に本種をガーネットオチバタケ属に編入した．

オーストラレーシアに分布する *Cruentomycena viscidocruenta* (Cleland) R.H. Petersen & Kovalenko（Cleland 1919, 1924；Grgurinovic 1997, 2002；Petersen et al. 2008）並びに ロシア産 *Cruentomycena kedrovayae* R.H. Petersen, Kovalenko & O.V. Morozova（Petersen et al. 2008）は際立

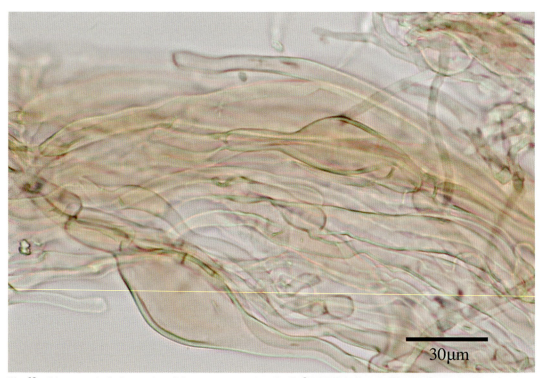

Fig. 63 – Stipe trama in the lower stipe of *Cruentomycena orientalis* (in 3% KOH, holotype). Photo by Taneyama, Y.
ガーネットオチバタケの柄実質（3%水酸化カリウム溶液で封入，正基準標本）．写真：種山裕一．

った形態的特徴をガーネットオチバタケと共有している．しかしながら *C. viscidocruenta* は通常中央部が臍状に凹んだ粘性の傘を形成し，2形性の縁シスチジアを有するとされており（Petersen et al. 2008），一方ロシア産種は湿時粘性の柄を有し，長く伸びた脚部を持つ柄シスチジアを形成すると言われている．

　鮮やかな色彩を帯びた目立つ特徴を持つ分類群であるにも関わらずガーネットオチバタケ属の分布はこれまでオーストラレーシアとロシア以外では報告例がなく，隔離分布の典型と考えられてきたが，最近大分からも近縁種が見つかっており（村上康明氏私信），分布域の実態が解明されつつある．

Fig. 64 – Basidiomata of *Cruentomycena orientalis* (holotype) on dead fallen hardwood leaves, 20 Oct. 2011, Mt. Banna, Ishigaki Island. Photo by Takahashi, H.
常緑広葉樹の落ち葉上に発生するガーネットオチバタケの子実体（正基準標本）．2011年10月20日．石垣島バンナ岳．写真：高橋春樹．

Fig. 65 – Basidiomata of *Cruentomycena orientalis* (holotype) on dead fallen hardwood leaves, 20 Oct. 2011, Mt. Banna, Ishigaki Island. Photo by Takahashi, H.
常緑広葉樹の落ち葉上に発生するガーネットオチバタケの子実体（正基準標本），2011年10月20日，石垣島バンナ岳．写真：高橋春樹．

9. *Cruentomycena orientalis* ガーネットオチバタケ —— 87

Fig. 66 – Basidiomata of *Cruentomycena orientalis* (holotype) on dead fallen hardwood leaves, 20 Oct. 2011, Mt. Banna, Ishigaki Island. Photo by Takahashi, H.
常緑広葉樹の落ち葉上に発生するガーネットオチバタケの子実体（正基準標本），2011年10月20日，石垣島バンナ岳．
写真：高橋春樹．

88 —— 9. *Cruentomycena orientalis* ガーネットオチバタケ

Fig. 67 – Basidioma of *Cruentomycena orientalis* (holotype) on dead fallen hardwood leaves, 20 Oct. 2011, Mt. Banna, Ishigaki Island. Photo by Takahashi, H.
常緑広葉樹の落ち葉上に発生するガーネットオチバタケの子実体（正基準標本），2011年10月20日，石垣島バンナ岳．写真：高橋春樹．

9. *Cruentomycena orientalis* ガーネットオチバタケ —— 89

Fig. 68 – Pileus surface of *Cruentomycena orientalis* (holotype), 20 Oct. 2011, Mt. Banna, Ishigaki Island. Photo by Takahashi, H.
ガーネットオチバタケの傘表面（正基準標本），2011年10月20日，石垣島バンナ岳．写真：高橋春樹．

10. *Gymnopilus iriomotensis* Har. Takah., Taneyama & Wada, **sp. nov.** ミナミホホタケ
MycoBank no.: MB 809924.

Etymology: The specific epithet refers to a toponym of the type locality, Iriomote Island.

Distinctive features of this species include medium-sized fleshy basidiomata forming a fugacious, indistinct, floccose annuliform zone on the stipe apex; a minutely squamulose, orange yellow to light yellow pileus; ovoid-ellipsoid to ellipsoid, verruculose, rusty yellow, small (usually less than 6 μm long) basidiospores; capitulate cheilocystidia; clavate to subcylindrical pleurocystidia; and a lignicolous habit.

Macromorphological characteristics (Figs. 71–75): Pileus 25–60 mm in diameter, at first hemispherical with an involute, smooth margin, then expanding to convex to broadly convex; surface covered overall with minute squamules, dry, subhygrophanous, when young evenly colored brownish orange (7C6–7) to light brown (7D6–7) or at times brownish red (10D6–7) to violet brown (10E6–7), then paler from the margin, finally becoming orange yellow (4A6–7) to light yellow (4A4–5) overall; squamules at first concolorous with the ground, gradually darkening in age. Flesh up to 5 mm thick at the center of pileus, paler concolorous with the pileus, fleshy, soft, unchanging when cut; odor indistinct, taste unknown. Stipe 40–60 × 2.5–5.5 mm, cylindrical, somewhat thickened toward the base, central, terete, solid; surface dry, subhygrophanous, densely covered with orange yellow (4A6–7) to light yellow (4A4–5) floccose to woolly-fibrillose remnants of veil toward the above, longitudinally fibrillose-striate below, darkening in age, often with a fugacious, floccose annuliform zone at the apex; base shrouded by whitish mycelial tomentum. Lamellae adnate, close (80–95 reaching the stipe), up to 4 mm broad, at first light yellow (3A4–5), then becoming orange yellow (4A6–7) to orange (5A6–7) at maturity, unchanging when bruised; edges even, concolorous.

Micromorphological characteristics (Figs. 69, 70): Basidiospores (4.5–) 4.9–5.8 (–6.6) × (3.2–) 3.6–4.2 (–4.8) μm (n = 103, mean length = 5.36 ± 0.45, mean width = 3.86 ± 0.30, Q = (1.22–) 1.31–1.47 (–1.63), mean Q = 1.39 ± 0.08), ovoid-ellipsoid to broadly ellipsoid, rusty yellow in distilled water, verruculose, dextrinoid, orange in KOH, thin-walled, without a germ pore. Basidia (16.2–) 19.3–23.9 (–26.9) × (4.7–) 5.4–6.2 (–6.7) μm (n = 41, mean length = 21.63 ± 2.30, mean width = 5.79 ± 0.42) in main body, (3.6–) 4.2–5.3 (–6.1) × (1.1–) 1.3–1.5 (–1.7) μm (n = 52, mean length = 4.76 ± 0.58, mean width = 1.39 ± 0.11) in sterigmata, clavate, four-spored. Cheilocystidia (22.3–) 23.3–29.0 (–34.7) × (4.2–) 5.2–6.7 (–7.5) μm (n = 34, mean length = 26.17 ± 2.83, mean width = 5.92 ± 0.74), forming a compact sterile edge, subfusiform to cylindrical, often with a capitulate apex, pale yellow in distilled water, smooth, thin-walled. Pleurocystidia (19.4–) 23.2–29.4 (–33.2) × (6.1–) 6.4–7.8 (–9.0) μm (n = 41, mean length = 26.31 ± 3.14, mean width = 7.10 ± 0.72), scattered, subclavate to cylindrical or occasionally subfusiform, hyaline or occasionally with yellowish vacuolar pigment in distilled water, smooth, thin-walled. Trama of lamellae composed of more or less regularly arranged, cylindrical hyphae (5.8–) 8.5–14.0 (–18.8) μm wide (n = 55, mean width = 11.25 ± 2.74), pale yellow in distilled water, smooth, with slightly thickened walls (0.8–) 0.9–1.3 (–1.5) μm (n = 20, mean thickness = 1.11 ± 0.21). Surface of pileus a cutis consisting of compactly arranged, parallel, repent, cylindrical hyphae (5.1–) 7.5–11.1 (–13.8) μm wide (n = 81, mean width = 9.30 ± 1.76) μm, spirally encrusted with yellowish pigment in distilled water, smooth, with slightly thickened walls (0.5–) 0.6–0.9 (–1.0) μm (n = 15, mean thickness = 0.74 ± 0.13), not gelatinized; terminal cells (25.4–) 36.9–66.6 (–95.5) × (6.9–) 8.0–11.5 (–13.7) μm (n = 36, mean length = 51.71 ± 14.86, mean width = 9.73 ± 1.75), cylindrical, undifferentiated. Trama of pileus similar to the hymenophoral trama; constituent hyphae (4.8–) 8.0–14.4 (–20.9) μm wide (n = 54, mean width = 11.24 ± 3.19), with thickened walls (1.0–) 1.2–1.7 (–2.0) μm (n = 23, mean thickness = 1.44 ± 0.29).

Stipitipellis to the pileus surface; constituent hyphae (3.3-) 4.2-5.9 (-7.2) μm wide (n = 57, mean width = 5.01 ± 0.86), with slightly thickened walls (0.5-) 0.5-0.7 (-0.8) μm (n = 14, mean thickness = 0.62 ± 0.11), without caulocystidia. Elements of floccose to woolly-fibrillose annuliform zone at the stipe apex (4.7-) 5.5-7.0 (-8.1) μm wide (n = 47, mean width = 6.25 ± 0.77), similar to those of stipitipellis, with slightly thickened walls (0.6-) 0.7-1.0 (-1.2) μm (n = 17, mean thickness = 0.87 ± 0.17); terminal cells (14.1-) 23.8-42.3 (-56.7) × (5.0-) 5.9-8.0 (-9.0) μm (n = 42, mean length = 33.06 ± 9.24, mean width = 6.97 ± 1.03), cylindrical, undifferentiated. Trama of stipe composed of longitudinally running, cylindrical hyphae (22.3-) 32.9-48.2 (-53.6) × (9.9-) 11.5-16.4 (-20.6) μm (n = 30, mean length = 40.56 ± 7.67, mean width = 13.97 ± 2.45), pale yellow in distilled water, smooth, with thickened walls (1.1-) 1.3-1.7 (-1.8) μm (n = 25, mean thickness = 1.51 ± 0.20). All tissues inamyloid, yellowish in KOH, with clamp connections.

Habitat and phenology: Scattered or gregarious on dead corticated logs of *Pinus luchuensis* Mayr., from March to April.

Known distribution: Okinawa (Iriomote Island).

Holotype: TNS-F-52284, on dead logs of *P. luchuensis* Mayr., Ushiku, Iriomote Island, Taketomi-cho, Yaeyama-gun, Okinawa Pref., 1 Apr. 2012, coll. Wada, S.

Isotype: KPM-NC0023877, on dead logs of *P. luchuensis* Mayr., Ushiku, Iriomote Island, Taketomi-cho, Yaeyama-gun, Okinawa Pref., 1 Apr. 2012, coll. Wada, S.

Gene sequenced specimen and GenBank accession number: TNS-F-52284, AB968238.

Japanese name: Minami-hohotake (named by S. Wada).

Comment: With the basidiospores less than 6 μm long and the poorly developed, fugacious remnants of the veil on the stipe surface, the new species is considered to be a member of the section *Microspori* Hesler in the subgenus *Gymnopilus* of the genus *Gymnopilus* (Hesler 1969), where it appears to be closely related to *Gymnopilus ombrophilus* Miyauchi from Japan (Miyauchi 2004) and *Gymnopilus subtropicus* Hesler from subtropical North America and Hawaii (Hesler 1969). *Gymnopilus ombrophilus*, however, forms lamellae staining brown when bruised and cylindrical narrowly clavate or sublageniform cheilocystidia. *Gymnopilus subtropicus* differs from the present species by having flask-shaped or ventricose, non-capitate cheilocystidia, ventricose, cylindrical or clavate caulocystidia, and interwoven hyphae of pileus trama turning reddish brown in KOH. *Gymnopilus dilepis* (Berk. & Broome) Singer, originally described from Sri Lanka (Berkeley and Broome 1871; Singer 1951; Pegler 1986; Guzmán-Dávalos 2003; Thomas et al. 2003), also comes very close to the present fungus but is distinct in producing markedly larger basidiospores: 6.4-8 (-9.2) × 4.8-5.6 μm (Thomas et al. 2003), utriform, much broader cheilocystidia: 4.5-10.4 μm broad (Thomas et al. 2003) (unlike the slender, often capitate cheilocystidia in *G. iriomotensis*), pseudocystidia with granulose to homogeneous brown contents, and cylindrical to clavate caulocystidia with an obtuse or subcapitate apex.

References 引用文献

Berkeley MJ, Broome CE. 1871. The fungi of Ceylon. (Hymenomycetes, from *Agaricus* to *Cantharellus*). J Linn Soc, Bot 11: 494-567.

Guzmán-Dávalos L, Mueller G, Cifuentes J, Miller AN, Santerre A. 2003. Traditional infrageneric classification of *Gymnopilus* is not supported by ribosomal DNA sequence data. Mycologia 95 (6): 1204-1214.

Hesler LR. 1969. North American species of *Gymnopilus*. Mycol Mem 3: 1-117.

Miyauchi S. 2004. A new species of *Gymnopilus* Sect. *Microspori* from Japan. Mycoscience 45 (1): 76-78.

Pegler DN. 1986. Agaric flora of Sri Lanka. Kew Bull, Addit Ser 12: 1-519.

Singer R. 1949. The Agaricales in modern taxonomy. Lilloa 22: 1-832.

Thomas KA, Guzmán-Dávalos L, Manimohan P. 2003. A new species and new records of *Gymnopilus* from India. Mycotaxon 85: 297-305.

10. ミナミホタケ（新種；和田匠平新称）*Gymnopilus iriomotensis* Har. Takah., Taneyama & Wada, **sp. nov.**

肉眼的特徴（Figs. 71-75）：傘は径25-60 mm，最初半球形で縁部は内側に巻き，のち丸山形〜中高偏平；表面は粘性を欠き，全体に表面と同色の小鱗片に被われ，やや吸水性，幼時均一に褐色または紫褐色を帯び，のち傘周縁部から次第に橙黄色〜淡黄色を呈する；小鱗片は最初地と同色，のち次第に暗色を呈する．肉は傘の中央部において5 mm以下，傘表面より淡色を呈し，柔らかく，肉質，空気に触れても変色せず，特別な匂いはなく，味は不明．柄は40-60×2.5-5.5 mm，円柱形，根元に向かってやや肥大し，中心生，中実；表面は粘性を欠き，やや吸水性，上部に向かって橙黄色〜淡黄色の綿毛に密に被われ，下部は縦に沿って繊維状の筋を表し，老成すると暗色を呈し，しばしば消失性の綿毛状輪状帯を頂部に形成する．根本の菌糸体は白色綿毛状．ヒダは直生，密，L＝80-95，幅4 mm以下，最初淡黄色，のち橙黄色，変色性を欠く；縁部は平坦，同色．

顕微鏡的特徴（Figs. 69, 70）：担子胞子は (4.5-) 4.9-5.8 (-6.6) × (3.2-) 3.6-4.2 (-4.8) μm (n = 103, mean length = 5.36 ± 0.45, mean width = 3.86 ± 0.30, Q = (1.22-) 1.31-1.47 (-1.63), mean Q = 1.39 ± 0.08)，広楕円形〜楕円形，錆黄色，表面は微疣に被われ，偽アミロイド，アルカリ溶液において橙色，薄壁，発芽孔を欠く．担子器は (16.2-) 19.3-23.9 (-26.9) × (4.7-) 5.4-6.2 (-6.7) μm（本体），(3.6-) 4.2-5.3 (-6.1) × (1.1-) 1.3-1.5 (-1.7) μm（ステリグマ），こん棒形，4胞子性．縁シスチジアは (22.3-) 23.3-29.0 (-34.7) × (4.2-) 5.2-6.7 (-7.5) μm，群生し，やや片膨れ状〜亜円柱形，しばしば頂部が頭状に膨れ，淡黄色，平滑，薄壁．側シスチジアは (19.4-) 23.2-29.4 (-33.2) × (6.1-) 6.4-7.8 (-9.0) μm，散生，こん棒形〜亜円柱形または時に亜紡錘形，無色または帯黄色の色素が液胞中に存在し，平滑，薄壁．子実層托実質の菌糸は幅 (5.8-) 8.5-14.0 (-18.8) μm，ほぼ平列し，円柱形，淡黄色，平滑，壁は厚さ (0.8-) 0.9-1.3 (-1.5) μm．傘表皮を構成する菌糸は匍匐性で隙間なく配列し，幅 (5.1-) 7.5-11.1 (-13.8) μm，帯黄色の色素が菌糸の表面をらせん状に取り巻き，平滑，壁は厚さ (0.5-) 0.6-0.9 (-1.0) μm，非ゼラチン質；末端細胞は (25.4-) 36.9-66.6 (-95.5) × (6.9-) 8.0-11.5 (-13.7) μm，円柱形．傘シスチジアはない．傘実質は子実層托実質に類似し，菌糸は幅 (4.8-) 8.0-14.4 (-20.9) μm．柄表皮は傘表皮と同様で，菌糸は幅 (5.1-) 7.5-11.1 (-13.8) μm，壁は厚さ (1.0-) 1.2-1.7 (-2.0) μm，柄シスチジアを欠く．柄上部の綿毛状輪状帯を形成する菌糸は幅 (4.7-) 5.5-7.0 (-8.1) μm，表皮組織の菌糸に類似し，壁は厚さ (0.6-) 0.7-1.0 (-1.2) μm；末端細胞は (14.1-) 23.8-42.3 (-56.7) × (5.0-) 5.9-8.0 (-9.0) μm，円柱形，形態的に未分化．柄実質の菌糸細胞は (22.3-) 32.9-48.2 (-53.6) × (9.9-) 11.5-16.4 (-20.6) μm，平列し，円柱形，淡黄色，平滑，薄壁．全ての組織は非アミロイド，アルカリ溶液において淡黄色を呈し，クランプを持つ．

生態および発生時期：リュウキュウマツの腐木上に散生または群生，3月〜4月．

分布：沖縄（西表島）．

供試標本：TNS-F-52284（正基準標本），リュウキュウマツの朽木上，沖縄県八重山郡竹富町西表島，2012年4月1日，和田匠平採集；KPM-NC0023877（複基準標本），リュウキュウマツの朽木上，沖縄県八重山郡竹富町西表島，2012年4月1日，和田匠平採集．

分子解析に用いた標本並びに GenBank 登録番号：TNS-F-52284，AB968238．

主な特徴：子実体は中型で肉質，消失性の不明瞭な綿毛状輪状帯を柄の頂部に形成する；傘は橙褐

色で小鱗片に被われる；担子胞子は小形（通常長さ6 μm以下），卵状楕円形〜楕円形，疣状微突起に被われ，錆黄色；縁シスチジアは頂部頭状形；側シスチジアはこん棒形〜類円柱形；材上性．

コメント：長さ6 μm以下の担子胞子および発達の悪い消失性の被膜の名残を柄の表面に形成する性質はHesler（Hesler 1969）の分類概念によるチャツムタケ属 *Gymnopilus*，チャツムタケ亜属 subgenus *Gymnopilus*，ホホタケ節（種山裕一 & 和田匠平新称）section *Microspori* Heslerに属することを示唆している．節内において本種は日本から報告されたコガネツムタケ *Gymnopilus ombrophilus* Miyauchi（Miyauchi 2004）並びに北米の亜熱帯地域およびハワイ諸島に分布する *Gymnopilus subtropicus* Hesler（Hesler 1969）に近縁と考えられる．しかしながらコガネツムタケ

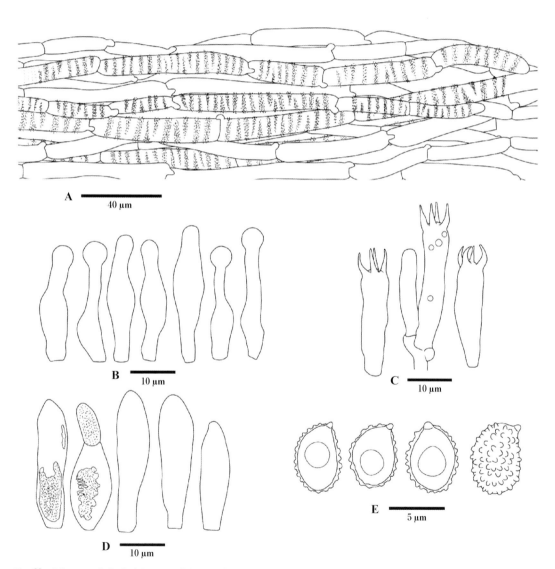

Fig. 69 – Micromorphological features of *Gymnopilus iriomotensis* (holotype): **A**. Longitudinal cross section of the pileipellis. **B**. Cheilocystidia. **C**. Basidia and basidiole. **D**. Pleurocystidia. **E**. Basidiospores. Illustrations by Taneyama, Y.
ミナミホホタケの顕鏡図（正基準標本）：**A**. 傘の表皮組織の縦断面．**B**. 縁シスチジア．**C**. 担子器．**D**. 側シスチジア．**E**. 担子胞子．図：種山裕一．

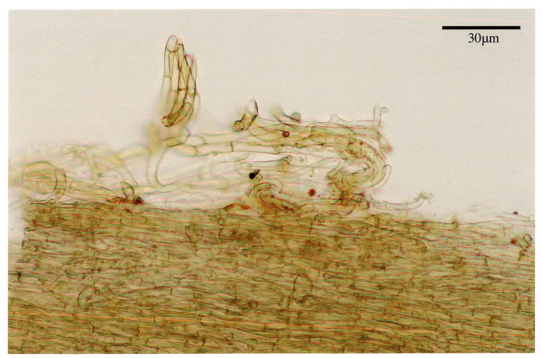

Fig. 70 – Longitudinal cross section of the stipitipellis of *Gymnopilus iriomotensis* (in 3% KOH, holotype) showing the elements of the floccose velar layer. Photo by Taneyama, Y.
ミナミホホタケの柄表皮組織の断面（3％水酸化カリウム溶液で封入，正基準標本）．表皮組織に付着した綿毛状被膜を形成する菌糸を示す．写真：種山裕一．

Fig. 71 – Basidiomata of *Gymnopilus iriomotensis* (holotype) on a dead corticated log of *Pinus luchuensis*, 1 Apr. 2012, Iriomote Island. Photo by Wada, Kimiko.
リュウキュウマツの腐木上に発生したミナミホホタケの子実体（正基準標本），2012年4月1日，西表島．写真：和田貴美子．

は傷を受けると褐変するヒダを持ち，類円柱形～類フラスコ形の縁シスチジアを形成するとされている．*Gymnopilus subtropicus* はフラスコ形または片膨れ状の縁シスチジアを有すること，片膨れ状または円柱形～こん棒形の柄シスチジアを形成すること，そしてアルカリ溶液において赤褐色に染まる錯綜した傘実質の菌糸の特徴においてミナミホホタケと異なる．本種はまたスリランカから記載された *Gymnopilus dilepis*（Berk. & Broome）Singer（Berkeley and Broome 1871；Singer 1951；Pegler 1986；Guzmán-Dávalos 2003；Thomas et al. 2003）に似るが，より大型の担子胞子：6.4-8（-9.2）×4.8-5.6 μm（Thomas et al. 2003），小のう形でより幅の広い縁シスチジア：幅4.5-10.4 μm（Thomas et al. 2003）（ミナミホホタケの縁シスチジアは痩せ型で，しばしば頂部頭状形になる），褐色の内容物を持つ偽シスチジア，そして頂部が頭状に膨れた円柱形～こん棒形の柄シスチジアにおいて *G. dilepis* はミナミホホタケと異なる形質を有する．

Fig. 72 – Basidiomata of *Gymnopilus iriomotensis* (holotype) showing the floccose to woolly-fibrillose annuliform zone at the stipe apex, 1 Apr. 2012, Iriomote Island. Photo by Wada, K.
綿毛状輪状帯を柄の頂部に形成するミナミホホタケの子実体（正基準標本），2012年4月1日，西表島．写真：和田貴美子．

Fig. 73 – Basidiomata of *Gymnopilus iriomotensis* (holotype) on a dead corticated log of *Pinus luchuensis*, 1 Apr. 2012, Iriomote Island. Photo by Wada, K.
　リュウキュウマツの腐木上に発生したミナミホホタケの子実体（正基準標本），2012年4月1日，西表島．写真：和田貴美子．

Fig. 74 – Basidioma of *Gymnopilus iriomotensis* (holotype), 1 Apr. 2012, Iriomote Island. Photo by Wada, K.
　ミナミホホタケの子実体（正基準標本），2012年4月1日，西表島．写真：和田貴美子．

Fig. 75 – Basidiomata of *Gymnopilus iriomotensis* on a dead decorticated log of *Pinus luchuensis*, 1 Apr. 2012, Iriomote Island. Photo by Wada, K.
リュウキュウマツの腐木上に発生したミナミホホタケの子実体，2012年4月1日，西表島．写真：和田貴美子．

11. *Gymnopus albipes* Har. Takah. & Taneyama, **sp. nov.** シロアシホウライタケ
MycoBank no.: MB 809923.

Etymology: The specific epithet refers to the white stipe.

  Distinctive features of this species consist of entirely pure white, collybioid basidiomata with a small, radiating disk of a white, pilose basal mycelium; distinctly intervenose lamellae; inamyloid, ellipsoid to ovoid-ellipsoid basidiospores; cheilocystidia in the form of *Siccus*-typed broom cells; diverticulate terminal elements of the stipitipellis; dextrinoid subhymenium and stipitipellis; and a lignicolous habit.

Macromorphological characteristics (Figs. 78-80): Pileus 9-30 mm in diameter, at first convex with an incurved, entire margin, then broadly convex, radially translucent-striate toward the center; surface hygrophanous, dry, glabrous, pure white. Flesh thin (up to 1.5 mm thick), pure white, tough; odor not distinctive, taste unknown. Stipe $10-20 \times 1-2$ mm, cylindrical, with a small, radiating disk of a white, pilose basal mycelium, central, terete, fistulose, smooth; surface dry, entirely pure white, glabrous. Lamellae adnate to subdecurrent, horizontal to somewhat arched, subdistant (9-13 reach the stipe), with 1-2 series of lamellulae, intervenose in the interstices, up to 3 mm broad, thin, pure white; edges entire, concolorous.

Micromorphological characteristics (Figs. 76, 77): Basidiospores $(7.1-) 8.1-9.5 (-12.1) \times (2.1-) 4.8-5.6 (-6.3)$ μm (n = 146, mean length = 8.80 ± 0.71, mean width = 5.19 ± 0.41, Q = $(1.35-) 1.43-1.99 (-4.66)$, mean Q = 1.71 ± 0.28), ellipsoid to ovoid-ellipsoid, smooth, hyaline, inamyloid, colorless in KOH, thin-walled. Basidia $(26.3-) 28.1-35.7 (-42.7) \times (6.7-) 7.9-9.3 (-10.1)$ μm (n = 34, mean length = 31.91 ± 3.76, mean width = 8.60 ± 0.72) in main body, $(3.5-) 3.8-4.7 (-5.5) \times (1.5-) 1.6-2.0 (-2.2)$ μm (n = 24, mean length = 4.29 ± 0.45, mean width = 1.82 ± 0.18) in sterigmata, clavate, 4-spored; basidioles clavate. Cheilocystidia forming a compact sterile edge, in the form of *Siccus*-typed broom cells with apical setulae, with hyaline walls $(0.6-) 0.8-1.0 (-1.2)$ μm, inamyloid; main cell bodies $(14.6-) 16.8-21.2 (-24.1) \times (7.9-) 8.5-11.4 (-14.2)$ μm (n = 40, mean length = 19.00 ± 2.23, mean width = 9.92 ± 1.45), clavate to broadly clavate; setulae $(3.0-) 3.9-5.5 (-6.8) \times (1.6-) 1.7-2.3 (-2.6)$ μm (n = 25, mean length = 4.70 ± 0.84, mean width = 2.00 ± 0.30), subcylindrical with obtuse apex, somewhat wavy in outline, rarely forked. Pleurocystidia none. Elements of hymenophoral trama $(4.9-) 5.8-9.8 (-13.0)$ μm wide (n = 39, mean width = 7.80 ± 2.01), subparallel, cylindrical, smooth, hyaline, inamyloid but weakly dextrinoid in the subhymenium, with walls $(0.7-) 0.9-1.2 (-1.4)$ μm. Pileipellis a cutis of repent, loosely interwoven hyphae $(11.2-) 18.6-48.7 (-82.3) \times (4.6-) 5.7-7.7 (-9.3)$ μm (n = 47, mean length = 33.63 ± 15.07, mean width = 6.71 ± 1.03), cylindrical, infrequently with nodulose outgrowths or short branches, hyaline, inamyloid, thin-walled. Elements of pileitrama $(2.6-) 3.4-6.2 (-9.1)$ μm wide (n = 40, mean width = 4.80 ± 1.37), loosely interwoven, cylindrical, often branched, smooth, hyaline, inamyloid, with walls 0.6-0.8 μm thick. Stipitipellis a cutis of parallel, repent hyphae $(2.7-) 3.2-4.1 (-4.8)$ μm wide (n = 34, mean width = 3.63 ± 0.47), cylindrical, with occasionally diverticulate terminal elements, hyaline, inamyloid or dextrinoid, with walls 0.6-0.9 μm. Stipe trama composed of longitudinally running, cylindrical hyphae $(6.4-) 8.4-14.7 (-19.0)$ μm wide (n = 38, mean width = 11.55 ± 3.14), cylindrical, with highly inflated, fusiform hyphae $(24.0-) 24.1-37.2 (-43.7)$ μm wide (n = 14, mean width = 30.65 ± 6.56) at the base of stipe, smooth, unbranched, septate, hyaline, inamyloid, with walls $1.7-2.4 (-2.9)$ μm. Elements of pilose basal disk $(1.7-) 2.1-2.8 (-3.3)$ μm wide (n = 47, mean width = 2.48 ± 0.33), filiform, smooth, unbranched, septate, hyaline, inamyloid, with walls 0.7-0.9 μm thick. Clamp connections present in all tissues. All tissues colorless in KOH.

Habitat and phenology: Gregarious or scattered on dead twigs (unidentified substrata) in evergreen broad-leaved forests, May to October.

Known distribution: Okinawa (Ishigaki Island).

Holotype: TNS-F-52276, on dead twigs in an evergreen broad-leaved forest, Mt. Banna, Ishigaki-shi, Okinawa, 18 May 2012, coll. Takahashi, H.

Japanese name: Siroasi-houraitake (named by H. Takahashi & Y. Taneyama).

Comments: The collybioid basidiomata, the inamyloid basidiospores, the non-hymeniform pileipellis elements infrequently with nodulose outgrowths or short branches, and the dextrinoid stipitipellis with occasionally diverticulate terminal elements, suggest that the new species belongs to the genus *Gymnopus*, section *Androsacei* (Kühner) Antonín & Noordel. (Mata et al. 2004; Noordeloos and Antonín 2008; Antonín and Noordeloos 2010).

Marasmiellus candidus (Fr.) Singer, originally described from Europe (Singer 1948; Malençon and Bertault 1970; Antonín and Noordeloos 1997, 2010), macromorphologically resembles the present fungus in the white pileus and the intervenose lamellae. The former has a blackish brown, not discoid base to the stipe, subcylindrical, much longer basidiospores: (10.5-) 11.5-15 (-17.5)×3.5-6.5 μm (Antonín and Noordeloos 2010), filiform to sublageniform cheilocystidia and dermatocystidia with a prolonged neck, and inamyloid hyphae in all tissues.

References 引用文献

Antonín V, Noordeloos ME. 1997. A Monograph of *Marasmius, Collybia* and related genera in Europe. Part 2: *Collybia, Gymnopus, Rhodocollybia, Crinipellis, Chaetocalathus,* and additions to *Marasmiellus*. Lib Bot 17: 1-256.

Antonín V, Noordeloos ME. 2010. A monograph of marasmioid and collybioid fungi in Europe. IHW-Verlag.

Malençon G, Bertault R. 1970. Flore des champignons superieurs du Maroc 1: 1-601.

Mata JL, Hughes KW, Petersen RH. 2004. Phylogenetic placement of *Marasmiellus juniperinus*. Mycoscience 45 (3): 214-221.

Noordeloos ME, Antonín V. 2008. Contribution to a monograph of marasmioid and collybioid fungi in Europe. Czech Mycology 60 (1): 21-27.

Singer R. 1948. New and interesting species of Basidiomycetes. II. Pap Mich Acad Sci 32: 103-150.

11. シロアシホウライタケ（新種；高橋春樹 & 種山裕一新称）*Gymnopus albipes* Har. Takah. & Taneyama, sp. nov.

肉眼的特徴（Figs. 78-80）：傘は径9-30 mm，最初丸山形で全縁な縁部が内側に巻き，のち中高偏平，中央部に向かって放射状に半透明な条線を表す；吸水性，粘性を欠き，平滑，純白色．肉は厚さ1.5 mm 以下，純白色，強靭，特別な匂いはなく，味は不明．柄は10-20×1-2 mm，円柱形，白色軟毛に被われた小形の盤状基部を具え，中心生，痩せ型，中空，平坦；表面は粘性を欠き，全体に純白色を呈し，平滑．ヒダは直生〜やや垂生，水平〜ややアーチ形，疎（柄に到達するヒダは9-13），ヒダの間に連絡脈を表し，幅3 mm 以下，純白色；縁部は同色，全縁．

顕微鏡的特徴（Figs. 76, 77）：担子胞子は (7.1-) 8.1-9.5 (-12.1)×(2.1-) 4.8-5.6 (-6.3) μm (n=146, mean length=8.80±0.71, mean width=5.19±0.41, Q=(1.35-) 1.43-1.99 (-4.66), mean Q=1.71±0.28)，楕円形〜卵状楕円形，平坦，無色，非アミロイド，アルカリ溶液において無色，薄壁．担子器は (26.3-) 28.1-35.7 (-42.7)×(6.7-) 7.9-9.3 (-10.1) μm (本体), (3.5-) 3.8-4.7 (-5.5)×(1.5-) 1.6-2.0 (-2.2) μm (ステリグマ)，こん棒形，4胞子性；偽担子器はこん棒形．縁シスチジアはハリガネオチバタケ型箒状細胞，(14.6-) 16.8-21.2 (-24.1)×(7.9-) 8.5-11.4 (-14.2) μm，群生し，こん棒形〜広こん棒形，頂部に短指状付属糸を具え，無色，非アミロイド，壁の厚さ

は1 μm 以下；短指状付属糸は (3.0-) 3.9-5.5 (-6.8)×(1.6-) 1.7-2.3 (-2.6) μm，亜円柱形，鈍頭，外形はやや不規則に波うち，稀に分岐する．側シスチジアはない．子実層托実質は平列し，菌糸は幅 (4.9-) 5.8-9.8 (-13.0) μm，円柱形，平坦，無色，子実下層は弱偽アミロイド，他は非アミロイド，やや厚壁 (0.9-1.2 μm)．傘の表皮組織は緩く錯綜した匍匐性の菌糸からなる；菌糸細胞は (11.2-) 18.6-48.7 (-82.3)×(4.6-) 5.7-7.7 (-9.3) μm，円柱形，稀に瘤状～短指状分岐物を伴い，無色，非アミロイド，薄壁．傘実質の菌糸は幅 (2.6-) 3.4-6.2 (-9.1) μm，緩く錯綜した並列型，円柱形，しばしば分岐し，平滑，無色，非アミロイド，壁は厚さ0.6-0.8 μm．柄表皮を構成する菌糸は幅 (2.7-) 3.2-4.1 (-4.8) μm，並列し，円柱形，不規則な瘤状分岐物に被われた末端細胞を伴い，無色，非アミロイドまたは偽アミロイド，壁は厚さ0.6-0.9 μm．柄実質の菌糸は幅 (6.4-) 8.4-14.7 (-19.0) μm，平列し，円柱形，柄の基部において紡錘形に著しく膨大した幅 (24.0-) 24.1-37.2 (-43.7) μm の菌糸が混在し，無分岐，隔壁を有し，平滑，無色，非アミロイド，厚壁 (1.7-2.4 (-2.9) μm)．盤状基部の菌糸は幅 (1.7-) 2.1-2.8 (-3.3) μm，糸状，平滑，無分岐，隔壁を持ち，無色，非アミロイド，壁は厚さ0.7-0.9 μm．全ての組織において菌糸はクランプを有し，アルカリ溶液において無色．

生態および発生時期：常緑広葉樹林内枯れ枝上（基質は不明）に群生または散生，5月～10月．

分布：沖縄（石垣島）．

供試標本：TNS-F-52276（正基準標本），常緑広葉樹林内枯れ枝上，沖縄県石垣市バンナ岳，2012年5月18日，高橋春樹採集．

主な特徴：子実体は全体に純白色を呈するモリノカレバタケ型で，白色軟毛に被われた盤状基部を形成する；ヒダは明瞭な連絡脈を表す；胞子は楕円形～卵状楕円形，非アミロイド；縁シスチジアはハリガネオチバタケ型箒状細胞；柄の表皮組織に不規則に分岐するこぶ状の突起に被われた末端細胞が存在する；子実下層および柄表皮組織は偽アミロイド；常緑広葉樹林内の枯れ枝上から発生．

コメント：モリノカレバタケ型の子実体，非アミロイドの担子胞子，稀に瘤状～短指状分岐物を伴い子実層状被を形成しない傘表皮組織，そして時に不規則に分岐するこぶ状突起を持つ末端細胞を伴い偽アミロイドに染まる平行菌糸被からなる柄表皮組織の性質は本種がモリノカレバタケ属 *Gymnopus*，オチバタケ節 section *Androsacei* (Kühner) Antonín & Noordel.（Mata et al. 2004；Noordeloos and Antonín 2008；Antonín and Noordeloos 2010）に属することを示唆している．

　　シロホウライタケ *Marasmiellus candidus* (Fr.) Singer（Singer 1948；Malençon and Bertault 1970；Antonín and Noordeloos 1997, 2010）は外観的特徴において本種に最も似るが，柄の基部は通常盤状に拡大せず，成熟時黒褐色を帯びること，類円柱形でより長い担子胞子：(10.5-) 11.5-15 (-17.5)×3.5-6.5 μm (Antonín and Noordeloos 2010) を持つこと，糸状～フラスコ形の縁シスチジアおよび表皮シスチジアを形成すること，そして全ての組織において非アミロイドの菌糸を有する性質において異なる．

11. *Gymnopus albipes* シロアシホウライタケ —— 101

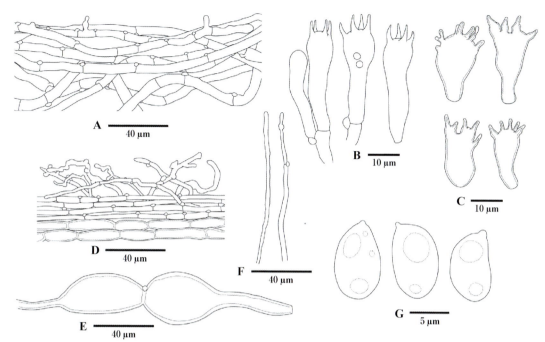

Fig. 76 – Micromorphological features of *Gymnopus albipes* (holotype). **A**: Longitudinal cross section of the pileipellis. **B**: Basidia and basidiole. **C**: Cheilocystidia. **D**: Longitudinal cross section of the stipitipellis and stipe trama. **E**: Inflated hypha of the stipe trama (from the lower portion of stipe). **F**: Elements of the basal mycelium. **G**: Basidiospores. Illustrations by Taneyama, Y.
シロアシホウライタケの顕鏡図（正基準標本）．**A**：傘表皮組織の縦断面．**B**：担子器および偽担子器．**C**：縁シスチジア．**D**：柄表皮組織と柄実質の縦断面．**E**：柄実質の膨大菌糸（柄の下部）．**F**：根元の毛状菌糸体を構成する菌糸．**G**：担子胞子．図：種山裕一．

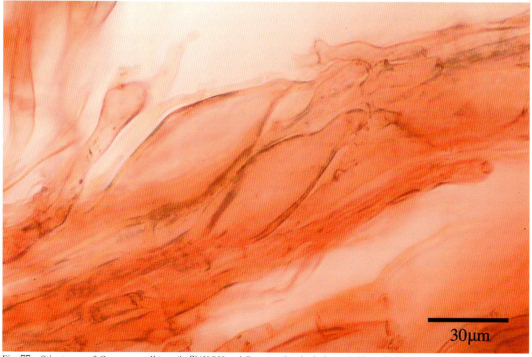

Fig. 77 – Stipetrama of *Gymnopus albipes* (in 3%KOH and Congo red stain, holotype). Photo by Taneyama, Y.
シロアシホウライタケの柄実質（3％水酸化カリウム溶液で封入した後コンゴー赤染色，正基準標本）．写真：種山裕一．

Fig. 78 – Basidiomata of *Gymnopus albipes* on a dead twig (holotype), 18 May 2012, Mt. Banna, Ishigaki Island. Photo by Takahashi, H.
枯れ枝上から発生したシロアシホウライタケの子実体（正基準標本），2012年5月18日，石垣島バンナ岳．写真：高橋春樹．

11. *Gymnopus albipes* シロアシホウライタケ —— 103

Fig. 79 – Basidiomata of *Gymnopus albipes* on a dead twig (holotype), 18 May 2012, Mt. Banna, Ishigaki Island. Photo by Takahashi, H.
枯れ枝上から発生したシロアシホウライタケの子実体（正基準標本），2012年5月18日，石垣島バンナ岳．写真：高橋春樹．

Fig. 80 – Basidiomata of *Gymnopus albipes* on a dead twig (holotype), 18 May 2012, Mt. Banna, Ishigaki Island. Photo by Takahashi, H.
枯れ枝上から発生したシロアシホウライタケの子実体（正基準標本），2012年5月18日，石垣島バンナ岳．写真：高橋春樹．

12. *Gymnopus oncospermatis* (Corner) Har. Takah. ヤシモリノカレバタケ
Mycoscience 43 (5): 397 (2002) [MB#488792].
≡ *Marasmius oncospermatis* Corner, Beih Nova Hedwigia 111: 81 (1996) [MB#442746].

Etymology: The specific epithet derives from the palm genus *Oncosperma* on which the basidiomata grow.

Macromorphological characteristics (Figs. 82–93): Pileus 8–30 mm in diameter, at first convex with incurved margin, then expanding to nearly plane, finally with depressed center and upturned margin, smooth; surface dry, dull, when young covered overall with white, fine pubescence, then pruinose, at times almost glabrescent in age, at first light brown (7D4–5) to reddish brown (8D4–5) overall, then paler toward the margin, finally whitish except the pale brownish center. Flesh very thin (up to 1 mm), whitish, pliant, tough, odor and taste mild. Stipe 20–70 × 1.5–6 mm, subcylindrical, central or somewhat eccentric, terete or compressed, at times longitudinally rugulose, at first entirely white villous then pruinose, at times glabrescent in age, whitish at the apex, light brown (7D4–5) to reddish brown (8D4–5) elsewhere, darkening from the base upwards; base covered with whitish mycelial tomentum attached to a whitish extensive mycelial mat with pale yellowish to pale brownish mycelial veins on the dead leaf bases and root of the palm. Lamellae adnate to subdecurrent, very close (55–70 reach the stipe), with 1–2 series of lamellulae, up to 1.5 mm broad, white; edges even, concolorous. Spore print pure white.

Micromorphological characteristics (Fig. 81): Basidiospores 8–10 × 3.5–4.5 μm (n = 20 spores per two specimens, Q = 2.2–2.3), ellipsoid, smooth, colorless, inamyloid, thin-walled. Basidia 24–30 × 6.5–8 μm, clavate, four-spored; basidioles clavate. Cheilocystidia 20–55 × 4–7 μm, forming a compact sterile edge, subcylindrical to irregularly shaped, with or without one or several digitate projections, colorless, inamyloid, thin-walled. Pleurocystidia none. Hymenophoral trama irregular; element hyphae similar to those of the pileitrama. Pileipellis a cutis of cylindrical cells 4–8 μm wide, thinly encrusted with light brown pigment, inamyloid, thin-walled; pilocystidia 30–60 × 4–8 μm, subcylindrical or irregularly shaped, with or without one or several digitate projections, colorless, thin-walled. Hyphae of pileitrama 5–12 μm wide, parallel to the pileipellis elements, cylindrical, smooth, monomitic, colorless, inamyloid, thin-walled. Stipitipellis a cutis of parallel, repent hyphae 4–7 μm wide, cylindrical, with light brown intercellular pigment, inamyloid, thin-walled; caulocystidia 20–50 × 4–7 μm, subcylindrical to irregularly shaped, with or without one or several digitate projections, colorless, inamyloid, thin-walled. Stipe trama composed of longitudinally running, cylindrical hyphae 5–13 μm wide, monomitic, smooth, colorless, inamyloid, thin-walled. All tissues with clamp connections.

Habitat and phenology: Solitary to scattered, often gregarious, on dead leaf bases and roots of the palm *Satakentia liukiuensis* H.E.Moore, April to November.

Known distribution: Okinawa (Ishigaki Island), Singapore (Corner 1996).

Specimens examined: KPM-NC0013150, on a dead leaf base and root of the palm *S. liukiuensis*, Yonehara, Ishigaki-shi, Okinawa Pref., 22 May 2004, coll. Takahashi, H.; KPM-NC0008777, same location, 18 Oct. 2001, coll. Takahashi, H.; TNS-F-48219, same location, 10 Jun. 2012, coll. Takahashi, H.; TNS-F-48221, same location, 10 Jun. 2012, coll. Takahashi, H.; TNS-F-61377, same location, 3 Jun. 2013, coll. Takahashi, H.; KPM-NC0023870, same location, 10 Jun. 2014, coll. Takahashi, H.

Japanese name: Yashi-morino-karebatake.

Comments: Distinctive features of this species are found in the collybioid basidiomata producing a pale brownish to whitish pileus and a light brown to reddish brown stipe; the very close lamellae; the subcylindrical to irregularly shaped cheilocystidia, pilocystidia and caulocystidia with one or several digitate projections; and the whitish basal tomentum attached to the distinct, whitish extensive mycelial mat with pale yellowish to pale brownish mycelial veins on the dead leaf bases and roots of *S. liukiuensis*.

Its distinct basal tomentum, the cutis with more or less diverticulate terminal elements in the pileipellis, and the presence of well-differentiated cheilocystidia suggest that this species is a member of the subsection *Vestipedes* in the section *Vestipedes* Antonin, Halling & Noordel. of the genus *Gymnopus* (Antonin and Noordeloos 1997; Mata et al. 2004; Noordeloos and Antonín 2008; Antonín and Noordeloos 2010). This species was first described by Corner (Corner 1996) from Singapore as *Marasmius oncospermatis* Corner. It is said to grow on the dead leaf bases of the palm *Oncosperma* in the type locality, and seems to be widely distributed in tropical and subtropical palm forests.

References 引用文献

Antonín V. 2007. Monograph of *Marasmius, Gloiocephala, Palaeocephala* and *Setulipes* in Tropical Africa, Fungus Flora of Tropical Africa Vol. 1. National Botanic Garden of Belgium.

Antonín V. Noordeloos ME. 2010. A monograph of marasmioid and collybioid fungi in Europe. IHW-Verlag.

Corner EJH. 1996. The agaric genera *Marasmius, Chaetocalathus, Crinipellis, Heimiomyces, Resupinatus, Xerula* and *Xerulina* in Malesia. Beih Nova Hedwigia 111: 1-175.

Mata JL, Hughes KW, Petersen RH. 2004. Phylogenetic placement of *Marasmiellus juniperinus*. Mycoscience 45 (3): 214-221.

Noordeloos ME, Antonín V. 2008. Contribution to a monograph of marasmioid and collybioid fungi in Europe. Czech Mycology 60 (1): 21-27.

Takahashi H. 2002. Two new species and one new combination of Agaricales from Japan. Mycoscience 43 (5): 397-403.

12. ヤシモリノカレバタケ *Gymnopus oncospermatis* (Corner) Har. Takah.

Mycoscience 43 (5): 397 (2002) [MB#488792].

≡ *Marasmius oncospermatis* Corner, Beih Nova Hedwigia 111: 81 (1996) [MB#442746].

肉眼的特徴（Figs. 82-93）：傘は径8-30 mm，最初縁部は内側に巻き，饅頭形，のちほぼ平らに開き，老成すると中央部が凹み，縁部が反り返り，平坦で条線や溝線を欠く；表面は乾性，光沢を欠き，最初全体に白色の微細な軟毛に被われ，後粉状，老成するとしばしばほぼ平滑になり，幼時全体に淡褐色〜赤褐色を帯びるがのち周縁から次第に退色し，成熟すると淡褐色の中央部を除きほぼ類白色を呈する．肉は非常に薄く（1.5 mm以下），類白色，弾力性があり，強靱，臭いおよび味は温和．柄は20-70×1.5-6 mm，類円柱形，中心生またはやや偏在生，しばしば偏圧し，時に縦皺を表す；表面は最初全体に白色の微細な軟毛に被われ，後粉状，老成するとしばしばほぼ平滑になり，頂部（類白色）以外は淡褐色〜赤褐色を帯び，根元は暗色になる；根本及び基質（ヤシの根と落ち葉）は淡黄褐色〜淡褐色を帯びた脈状隆起を持つ類白色の綿毛状菌糸体に被われる．ヒダは直生〜僅かに垂生，極めて密（柄に到達するヒダは55-70），1-2の小ヒダを伴い，幅1 mm以下，白色；縁部は平坦で同色．胞子紋は白色．

顕微鏡的特徴（Fig. 81）：担子胞子は8-10×3.5-4.5 µm（n = 20 spores per two specimens，Q = 2.2-2.3），楕円形，平坦，無色，非アミロイド，薄壁．担子器は24-30×6.5-8 µm，こん棒形，4胞子性；偽担子器はこん棒形．縁シスチジアは20-55×4-7 µm，不稔帯を形成し，亜円柱形〜不規則な形状をなし，しばしば複数の短指状突起を有し，無色，非アミロイド，薄壁．側シスチジアはない．子実層托実質の菌糸は不規則に配列し，傘実質の菌糸に類似する．傘表皮の菌糸は平行菌糸被を形成し，幅4-8 µm，円柱形，淡褐色の色素がうすく凝着し，非アミロイド，薄壁；末端細胞は30-60×4-8 µm，類円柱形，しばしば複数の短指状突起を持ち，無色，薄壁．傘実質の菌糸は幅5-12 µm，並列し，円柱形，一菌糸型，平滑，無色，非アミロイド，薄壁．柄表皮組織は

平行菌糸被を成し，菌糸は幅4-7 μm，匍匐性，円柱形，淡褐色の色素が細胞内に存在し，非アミロイド，薄壁；柄シスチジアは20-50×4-7 μm，類円柱形～不規則な形状をなし，しばしば複数の短指状突起を有し，無色，非アミロイド，薄壁．柄実質の菌糸は幅5-13 μm，縦に沿って配列し，円柱形，一菌糸型，平滑，無色，非アミロイド，薄壁．全ての組織はクランプを持つ．

生態および発生時期：ヤエヤマヤシの枯れ葉（特に葉の根もと付近）または枯れた根上に孤生または散生し，しばしば群生，4月～11月．

分布：沖縄（石垣島），シンガポール（Corner 1996）．

供試標本：KPM-NC0013150，ヤエヤマヤシの枯れた根もしくは枯れ葉上，沖縄県石垣島米原，2004年5月22日，高橋春樹採集；KPM-NC0008777，同上，2001年10月18日，高橋春樹採集；TNS-F-48219，同上，2012年6月10日；TNS-F-48221，同上，2012年6月10日；TNS-F-61377，同上，2013年6月3日；KPM-NC0023870，同上，2014年6月10日，高橋春樹採集．

主な特徴：子実体はモリノカレバタケ型で，淡褐色～類白色の傘と淡褐色～赤褐色の柄を持つ；ヒダは著しく密；縁シスチジア，傘シスチジア，柄シスチジアは亜円柱形～不規則な形状を成し，短指状分岐物を持つ；根元の白色菌糸体につながる淡黄褐色～淡褐色の脈状隆起を持つ発達した類白色の菌糸体マットが基質全体（ヤエヤマヤシの枯れた根もしくは枯れ葉の根もと付近）を被う．

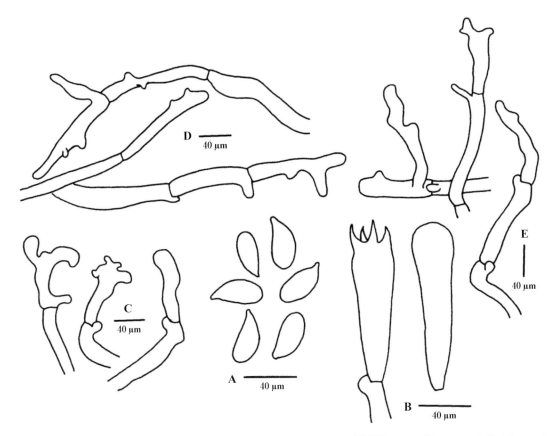

Fig. 81 − Micromorphological features of *Gymnopus oncospermatis* (KPM-NC0008777): **A**. Basidiospores. **B**. Basidium and basidiole. **C**. Cheilocystidia. **D**. Terminal cells in the pileipellis. **E**. Caulocystidia. Illustrations by Takahashi, H.
ヤシモリノカレバタケの顕微図（KPM-NC0008777）：**A**. 担子胞子．**B**. 担子器および偽担子器．**C**. 縁シスチジア．**D**. 傘表皮組織の末端細胞．**E**. 柄シスチジア．図：高橋春樹．

コメント：子実体の根元および基質を被う発達した綿毛状菌糸体，不規則に分岐する短指状またはこぶ状の分岐物を持つ平行菌糸被型の傘表皮組織，そして分化した縁シスチジアの存在は本種がモリノカレバタケ属 *Gymnopus*，アマタケ節 section *Vestipedes* Antonín, Halling & Noordel., アマタケ亜節 subsection *Vestipedes*（Antonín and Noordeloos 1997；Mata et al. 2004；Noordeloos and Antonín 2008；Antonín and Noordeloos 2010）に属することを示唆している．本種は Corner（Corner 1996）によってホウライタケ属 *Marasmius* の所属種としてシンガポールから最初に報告された．タイプ産地ではコブダネヤシ属 *Oncosperma* の枯れ葉の根元に発生すると言われており，熱帯～亜熱帯のヤシ林におそらく広く分布するものと思われる．

Fig. 82 – Mature basidiomata of *Gymnopus oncospermatis* (KPM-NC0023870) on a dead leaf base of the palm *S. liukiuensis*, 10 Jun. 2014, Yonehara, Ishigaki Island. Photo by Wada, K.
ヤエヤマヤシの枯れ葉の根元から発生したヤシモリノカレバタケの成熟した子実体（KPM-NS0023870），2014年6月10日，石垣島米原．撮影：和田貴美子．

12. *Gymnopus oncospermatis* ヤシモリノカレバタケ —— 109

Fig. 83 – Mature basidiomata of *Gymnopus oncospermatis* (TNS-F-48219) on a dead leaf base of the palm *S. liukiuensis*, 10 Jun. 2012, Yonehara, Ishigaki Island. Photo by Taneyama, Y.
ヤエヤマヤシの枯れ葉の根元から発生したヤシモリノカレバタケの成熟した子実体（TNS-F-48219），2012年6月10日，石垣島米原．撮影：種山裕一．

110 —— 12. *Gymnopus oncospermatis* ヤシモリノカレバタケ

Fig. 84 – Mature basidioma of *Gymnopus oncospermatis* on a dead leaf base of the palm *S. liukiuensis*, 15 Dec. 2001, Yonehara, Ishigaki Island. Photo by Takahashi, H.
ヤエヤマヤシの枯れ葉の根元から発生したヤシモリノカレバタケの成熟した子実体，2001年12月15日，石垣島米原．
撮影：高橋春樹．

Fig. 85 – Mature basidioma of *Gymnopus oncospermatis* on a dead leaf base of the palm *S. liukiuensis*, 15 Dec. 2001, Yonehara, Ishigaki Island. Photo by Takahashi, H.
ヤエヤマヤシの枯れ葉の根元から発生したヤシモリノカレバタケの成熟した子実体，2001年12月15日，石垣島米原．
撮影：高橋春樹．

12. *Gymnopus oncospermatis* ヤシモリノカレバタケ —— 111

Fig. 86 – Mature basidioma of *Gymnopus oncospermatis*, showing the pileus surface, 15 Dec. 2001, Yonehara, Ishigaki Island. Photo by Takahashi, H.
ヤシモリノカレバタケの傘表面，2001年12月15日，石垣島米原．撮影：高橋春樹．

Fig. 87 – Mature basidioma of *Gymnopus oncospermatis*, showing the stipe and the lamellae, 15 Dec. 2001, Yonehara, Ishigaki Island. Photo by Takahashi, H.
ヤシモリノカレバタケのヒダおよび柄，2001年12月15日，石垣島米原．撮影：高橋春樹．

12. *Gymnopus oncospermatis* ヤシモリノカレバタケ

Fig. 88 – Immature basidiomata of *Gymnopus oncospermatis* (KPM-NC0023870), attached to an extensive mycelial mat with pale brownish, distinct mycelial ribs similar to veins of a leaf, 10 Jun. 2014, Yonehara, Ishigaki Island. Photo by Takahashi, H.
淡褐色の脈状隆起を持つ菌糸体マットから発生するヤシモリノカレバタケの幼菌（KPM-NC0023870），2014年6月10日，石垣島米原．撮影：高橋春樹．

12. *Gymnopus oncospermatis* ヤシモリノカレバタケ ―― 113

Fig. 89 – Immature basidiomata of *Gymnopus oncospermatis* (KPM-NC0023870), attached to an extensive mycelial mat on a dead leaf base of the palm *S. liukiuensis*, 10 Jun. 2014, Yonehara, Ishigaki Island. Photo by Takahashi, H.
ヤエヤマヤシの枯れ葉の根元から発生したヤシモリノカレバタケの幼菌（KPM-NC0023870），2014年6月10日，石垣島米原．撮影：高橋春樹．

Fig. 90 – Immature basidiomata of *Gymnopus oncospermatis* (KPM-NC0023870), attached to an extensive mycelial mat on a dead leaf base of the palm *S. liukiuensis*, 10 Jun. 2014, Yonehara, Ishigaki Island. Photo by Takahashi, H.
ヤエヤマヤシの枯れ葉の根元から発生したヤシモリノカレバタケの幼菌（KPM-NC0023870），2014年6月10日，石垣島米原．撮影：高橋春樹．

Fig. 91 – Immature basidiomata of *Gymnopus oncospermatis* (KPM-NC0023870), on a dead leaf base of the palm *S. liukiuensis*, 10 Jun. 2014, Yonehara, Ishigaki Island. Photo by Takahashi, H.
ヤエヤマヤシの枯れ葉の根元から発生したヤシモリノカレバタケの幼菌（KPM-NC0023870），2014年6月10日，石垣島米原．撮影：高橋春樹．

12. *Gymnopus oncospermatis* ヤシモリノカレバタケ —— 115

Fig. 92 – Extensive mycelial mat of *Gymnopus oncospermatis* on a dead leaf base of the palm *S. liukiuensis,* showing the distinct, pale brownish mycelial ribs similar to leaf veins, 10 Jun. 2012, Yonehara, Ishigaki Island. Photo by Takahashi, H.
ヤエヤマヤシの枯れ葉上に広がるヤシモリノカレバタケの菌糸体マット，淡褐色～淡黄褐色を帯びた脈状隆起を形成する菌糸体を示す，2012年6月10日，石垣島米原．撮影：高橋春樹．

Fig. 93 – Extensive mycelial mat of *Gymnopus oncospermatis* on a dead leaf base of the palm *S. liukiuensis,* showing the distinct, pale brownish mycelial ribs similar to leaf veins, 10 Jun. 2012, Yonehara, Ishigaki Island. Photo by Taneyama, Y.
ヤエヤマヤシの枯れ葉上に広がるヤシモリノカレバタケの菌糸体マット，淡褐色を帯びた脈状隆起を形成する菌糸体を示す，2012年6月10日，石垣島米原．撮影：種山裕一．

13. *Gymnopus phyllogenus* Har. Takah., Taneyama & Terashima, **sp. nov.** アシグロカレハタケ

MycoBank no.: MB 809922.

Etymology: The specific epithet means "born on leaves", referring to the foliicolous habit.

Distinctive features of this species are found in minute marasmioid basidiomata with a pure white, subpruinose pileus and a blackish, pruinose stipe without a basal mycelium; inamyloid, oblong-ellipsoid to obscurely amygdaliform basidiospores; clavate to fusiform basidioles; *Rotalis*-typed broom cells in the pileipellis and in the edge of lamellae; subcylindrical to irregularly shaped caulocystidia; dextrinoid stipitipellis and stipe trama; and an invariably foliicolous habit.

Macromorphological characteristics (Figs. 96–98): Pileus 1–1.5 mm in diameter, at first hemispherical then convex, at first almost smooth but soon radially grooved at a minutely crenulate to eroded margin; surface non-hygrophanous, dry, subpruinose when young then glabrescent in age, pure white. Flesh thin (up to 0.1 mm thick), pure white in the pileus, dark brown in the lower stipe, soft; odor not distinctive, taste unknown. Stipe 1–2.5 × 0.1–0.2 mm, cylindrical, at times somewhat discoid at the insititious base, central, terete, smooth; surface dry, at first entirely white, soon brownish, finally becoming dark brown to blackish toward the base, consistently whitish at the apex, when young white pruinose, denser toward the base, glabrescent in age. Lamellae adnexed, convex to horizontal, distant (5–8 reach the stipe), not intervenose, up to 0.4 mm broad, thin, pure white; edges entire, concolorous.

Micromorphological characteristics (Figs. 94, 95): Basidiospores (6.4–) 7.7–9.3 (–10.3) × (3.1–) 3.9–4.8 (–5.6) μm (n = 163, mean length = 8.47 ± 0.80, mean width = 4.32 ± 0.46, Q = (1.56–) 1.81–2.13 (–2.40), mean Q = 1.97 ± 0.16), oblong-ellipsoid to obscurely amygdaliform, smooth, hyaline, inamyloid, colorless in KOH, thin-walled. Basidia (16.1–) 17.3–20.9 (–22.5) × (5.8–) 6.1–6.9 (–7.5) μm (n = 31, mean length = 19.11 ± 1.82, mean width = 6.49 ± 0.37) in main body, (2.7–) 3.0–4.1 (–4.6) × (1.1–) 1.3–1.6 (–1.8) μm (n = 20, mean length = 3.52 ± 0.54, mean width = 1.47 ± 0.17) in sterigmata, clavate, 4-spored; basidioles (15.7–) 18.9–25.6 (–27.1) × (4.9–) 5.5–6.8 (–7.4) μm (n = 26, mean length = 22.25 ± 3.31, mean width = 6.12 ± 0.64), clavate to fusiform. Cheilocystidia in the form of *Rotalis*-typed broom cells (15.9–) 18.2–23.8 (–26.6) × (8.9–) 10.2–13.7 (–15.8) μm (n = 26, mean length = 21.02 ± 2.83, mean width = 11.95 ± 1.71), abundant, broadly clavate, usually covered with numerous, subcylindrical to subconical, evenly spaced apical verrucae (1.1–) 1.5–2.4 (–2.9) × (0.8–) 0.8–1.2 (–1.6) μm (n = 29, mean length = 1.98 ± 0.44, mean width = 0.97 ± 0.18), rarely smooth, hyaline, inamyloid, thin-walled. Pleurocystidia none. Pileipellis a cutis of repent, somewhat irregularly arranged hyphae, occasionally with broom cells of the *Rotalis*-type; constituent hyphae (35.5–) 37.5–54.3 (–67.9) × (6.9–) 7.7–11.2 (–13.0) μm (n = 18, mean length = 45.91 ± 8.37, mean width = 9.43 ± 1.74), cylindrical, unbranched, smooth or densely verrucose, hyaline, inamyloid, thin-walled; pilocystidia in the form of *Rotalis*-typed broom cells (15.1–) 14.5–24.4 (–30.6) × (10.0–) 10.2–13.7 (–14.5) μm (n = 10, mean length = 19.46 ± 4.95, mean width = 11.98 ± 1.76), similar to the cheilocystidia. Elements of pileitrama (2.1–) 2.3–3.4 (–4.5) μm wide (n = 28, mean width = 2.83 ± 0.58), parallel, cylindrical, smooth, hyaline, inamyloid, thin-walled. Stipitipellis a cutis of parallel, repent hyphae (2.5–) 3.0–3.9 (–4.9) μm wide (n = 26, mean width = 3.46 ± 0.46), cylindrical, smooth, brownish, dextrinoid, thin-walled. Caulocystidia (6.3–) 9.6–18.3 (–21.6) × (6.7–) 8.1–11.1 (–13.2) μm (n = 25, mean length = 13.95 ± 4.32, mean width = 9.60 ± 1.50), scattered, subcylindrical to irregularly shaped, smooth, hyaline, inamyloid, with walls 0.5–0.8 μm thick. Stipe trama composed of longitudinally running, cylindrical hyphae (4.0–) 4.4–5.8 (–6.5) μm wide (n = 27, mean width = 5.08 ± 0.68), smooth, unbranched, hyaline or pale brownish, weakly dextrinoid, with walls (0.9-) 1.0–1.6 (-2.0) μm thick. Clamp connections present in the base of basidia and the hymenophoral and stipe trama. All tissues colorless in KOH.

Habitat and phenology: Scattered on dead fallen leaves of *Castanopsis sieboldii* (Makino) Hatus. ex T. Yamaz. et Mashiba, October.

Known distribution: Okinawa (Iriomote Island).

Holotype: TNS-F-52270, on dead fallen leaves of *C. sieboldii,* Oomija, Iriomote Island, Taketomi-cho, Yaeyama-gun, Okinawa Pref., 14 Oct. 2011, coll. Terashima, Y.

Japanese name: Asigro-karehatake (named by H. Takahashi, Y. Taneyama & Y. Terashima).

Comments: The marasmioid basidiomata, the inamyloid basidiospores, the cheilocystidia in the form of *Rotalis*-typed broom cells, the non-hymeniform pileipellis made up of irregularly arranged broom cells, and the dextrinoid stipitipellis suggest placement of the new species in the genus *Gymnopus*, section *Androsacei* (Kühner) Antonín & Noordel. (Mata et al. 2004; Noordeloos and Antonín 2008; Antonín and Noordeloos 2010). Within the section, *Marasmius kisangensis* Singer from Congo (Singer 1964; Pegler 1977; Antonín 2003, 2007) appears to be taxonomically aligned with *G. phyllogenus* due to the small, white pileus and a dark brown, pruinose stipe, but the former possesses dark brown rhizomorphs, irregularly shaped, diverticulate cheilocystidia, and setiform caulocystidia.

Although *G. phyllogenus* shares a insititious stipe, inamyloid basidiospores, and fusiform basidioles with the genus *Marasmiellus* Murrill as defined by Singer (Singer 1986), its evenly verrucose (not irregularly diverticulate) pileipellis elements and dextrinoid stipitipellis and stipe trama are absolutely foreign to the genus.

Mycena parsimonia Corner from Malaysia (Corner 1994) has a superficial similarity to *G. phyllogenus* in the minute, white basidiomata and the inamyloid basidiospores. The Malaysian taxon, however, has adnate-decurrent lamellae and a lignicolous habitat, and lacks cheilocystidia and clamp connections.

Gymnopus phyllogenus is also identical to the genus *Mycena,* section *Hiemales* Konrad & Maubl., subsection *Hiemales* Maas Geest. (Maas Geesteranus, 1980, 1991) in the adnexed, convex to horizontal lamellae, and the inamyloid basidiospores associated with the 4-spored, clamped basidia. Unlike the genus *Mycena*, the new species has a somewhat heterogeneous combination of characteristics, such as a marasmioid habit, cheilocystidia and pileipellis elements similar in morphology to *Rotalis*-typed broom cells, fusiform basidioles, and dextrinoid stipitipellis.

References 引用文献

Antonín V. 2003. New species of marasmioid genera (Basidiomycetes, Tricholomataceae) from tropical Africa - II. *Gloiocephala, Marasmius, Setulipes* and two new combinations. Mycotaxon 88: 53–78.

Antonín V. 2007. Monograph of *Marasmius, Gloiocephala, Palaeocephala* and *Setulipes* in Tropical Africa, Fungus Flora of Tropical Africa Vol. 1. National Botanic Garden of Belgium.

Antonín V. Noordeloos ME. 2010. A monograph of marasmioid and collybioid fungi in Europe. IHW-Verlag.

Corner EJH. 1994. Agarics in Malesia. II Mycenoid. Beih Nova Hedwigia. 109: 165–271.

Maas Geesteranus RA. 1980. Studies in Mycenas. 15. A tentative subdivision of the genus *Mycena* in the northern Hemisphere. Persoonia 11: 93–120.

Maas Geesteranus RA. 1991. Conspectus of the Mycenas of the Northern Hemisphere.15. Sections *Hiemales* and *Exornatae*. Proc K Ned Akad Wet 94: 81–102.

Mata JL, Hughes KW, Petersen RH. 2004. Phylogenetic placement of *Marasmiellus juniperinus*. Mycoscience 45 (3): 214–221.

Noordeloos ME, Antonín V. 2008. Contribution to a monograph of marasmioid and collybioid fungi in Europe. Czech Mycology 60 (1): 21–27.

Pegler DN. 1977. A preliminary agaric flora of East Africa. Kew Bull, Addit Ser 6: 1–615.
Singer, R. 1964. *Marasmius* congolais recuillis par Mme. Goosens-Fontana et d'autrescollecteurs Belges. Bull Jard bot État Brux 34: 317–388.
Singer R. 1986. The Agaricales in modern taxonomy, 4th edn. Koeltz, Koenigstein.

13. アシグロカレハタケ（新種；高橋春樹，種山裕一 & 寺嶋芳江新称）*Gymnopus phyllogenus* Har. Takah., Taneyama & Terashima, **sp. nov.**

肉眼的特徴（Figs. 96-98）：傘1-1.5 mm，最初半球形，のち饅頭形，最初ほぼ平坦まもなく周縁部において放射状の溝線を表す；非吸水性，粘性を欠き，表面はやや粉状，老成すると平滑になり，純白色；縁部は細鋸歯状で内側に巻かない．肉は厚さ0.1 mm以下，傘において純白色，柄の下部において暗褐色を帯び，軟質，特別な匂いはなく，味は不明．柄は1-2.5×0.1-0.2 mm，円柱形，時に根元がやや盤状に拡大し，中心生，痩せ型；表面は粘性を欠き，最初全体に白色まもなく下方に向かって帯褐色〜黒褐色を帯び，頂部類白色，若いとき下半部を中心に白粉状を呈し，老成時平滑になる；根元に発達した菌糸体は見られない．ヒダは上生，丸山形〜水平，疎（柄に到達するヒダは5-8），連絡脈を欠き，幅0.4 mm以下，純白色；縁部は同色，全縁．

顕微鏡的特徴（Figs. 94, 95）：担子胞子は (6.4-) 7.7-9.3 (-10.3) × (3.1-) 3.9-4.8 (-5.6) μm (n = 163, mean length = 8.47±0.80, mean width = 4.32±0.46, Q = 1.81-2.13)，長楕円形〜類アーモンド形，平坦，無色，非アミロイド，アルカリ溶液において無色，薄壁．担子器は (16.1-) 17.3-20.9 (-22.5) × (5.8-) 6.1-6.9 (-7.5) μm（本体），(2.7-) 3.0-4.1 (-4.6) × (1.1-) 1.3-1.6 (-1.8) μm（ステリグマ），こん棒形，4胞子性；偽担子器は (15.7-) 18.9-25.6 (-27.1) × (4.9-) 5.5-6.8 (-7.4) μm，こん棒形〜紡錘形．縁シスチジアはシロヒメホウライタケ型箒状細胞，(15.9-) 18.2-23.8 (-26.6) × (8.9-) 10.2-13.7 (-15.8) μm，群生し，広こん棒形，通常上半分に疣状微突起が密生し，稀に平滑，無色，非アミロイド，壁は厚さ0.8-1.2 (-1.4) μm；疣状微突起は (1.1-) 1.5-2.4 (-2.9) × (0.8-) 0.8-1.2 (-1.6) μm，亜円柱形〜亜円錐形．側シスチジアはない．傘の表皮組織はやや不規則に配列した匍匐性の菌糸からなり，シロヒメホウライタケ型箒状細胞が混在する；菌糸細胞は (35.5-) 37.5-54.3 (-67.9) × (6.9-) 7.7-11.2 (-13.0) μm，円柱形，無分岐，平滑またはイボ状突起が密生し，非アミロイド，壁は厚さ0.9-1.3 μm；傘シスチジアはシロヒメホウライタケ型箒状細胞，(15.1-) 14.5-24.4 (-30.6) × (10.0-) 10.2-13.7 (-14.5) μm，縁シスチジアに類似する．傘実質の菌糸は幅 (2.1-) 2.3-3.4 (-4.5) μm，並列し，円柱形，平滑，無色，非アミロイド，薄壁．柄表皮を構成する菌糸は幅 (2.5-) 3.0-3.9 (-4.9) μm，並列し，円柱形，平滑，帯褐色，偽アミロイド，薄壁．柄シスチジアは (6.3-) 9.6-18.3 (-21.6) × (6.7-) 8.1-11.1 (-13.2) μm，散在し，類円柱形〜不規則な形状を成し，平滑，無色，非アミロイド，壁の厚さは0.5-0.8 μm．柄実質の菌糸は幅 (4.0-) 4.4-5.8 (-6.5) μm，平列し，円柱形，無分岐，平滑，無色または淡褐色，弱偽アミロイド，壁は厚さ (0.9-) 1.0-1.6 (-2.0) μm．担子器の根元，ヒダ実質および柄実質にクランプを有する．全ての組織はアルカリ溶液において無色．

生態および発生時期：スダジイの枯れ葉上に散生，10月．

分布：沖縄（西表島）．

供試標本：TNS-F-52270（正基準標本），スダジイの枯れ葉上，沖縄県八重山郡竹富町西表島大見謝，2011年10月14日，寺嶋芳江採集．

主な特徴：子実体は微小なホウライタケ型；傘はやや粉状で純白色；柄は黒褐色で，白色粉状物に被われ，根元に発達した菌糸体はない；胞子は非アミロイド，長楕円形〜類アーモンド形；偽担子器はこん棒形または紡錘形；ヒダ縁部と傘表皮組織にシロヒメホウライタケ型箒状細胞が存在する；柄シスチジアは類円柱形〜不規則な形状を成す；柄表皮組織と柄実質は偽アミロイ

ド：スダジイの枯れ葉上に発生．

コメント：ホウライタケ型の子実体，非アミロイドの担子胞子，シロヒメホウライタケ型箒状細胞の縁シスチジア，箒状細胞が混在した平行菌糸被からなる傘表皮組織，そして偽アミロイドに染まる柄表皮組織の性質は本種がモリノカレバタケ属 *Gymnopus*，オチバタケ節 section *Androsacei* (Kühner) Antonín & Noordel. (Mata et al. 2004；Noordeloos and Antonín 2008；Antonín and Noordeloos 2010) に近縁であることを示唆している．節内においてコンゴ産 *Marasmius kisangensis* Singer は小形で白色の傘と粉状暗褐色の柄を有する点で本種に似ているが，暗褐色の根状菌糸束を伴うこと，不規則な形状の縁シスチジアを持つこと，そして剛毛体型柄シスチジアの存在において異なる．

根元に発達した菌糸体を欠き，非アミロイドの担子胞子および紡錘形の偽担子器を形成する性質は Singer の分類概念 (Singer 1986) によるシロホウライタケ属 *Marasmiellus* Murrill と共通するが，箒状細胞を持つ傘表皮組織および偽アミロイドに染まる柄表皮組織の性質は明らかに異質である．

マレーシア産 *Mycena parsimonia* Corner (Corner 1994) は微小で白色の子実体と非アミロイドの担子胞子を形成する点で本種に類似するが，マレーシア産種は材上生で，直生〜垂生のヒダを持ち，縁シスチジア並びにクランプを欠くと言われている．

本種はまた上生する丸山形〜水平のヒダ，非アミロイドの担子胞子，4胞子性でクランプを持つ担子器において Maas の分類概念 (Maas Geesteranus 1980, 1991) によるクヌギタケ属 *Mycena*，ウスズミコジワタケ節 section *Hiemales* Konrad & Maubl., ウブゲオチエダタケ亜節 subsection *Hiemales* Maas Geest. と一致する．しかしながらホウライタケ型の子実体の類型，シロヒメホウライタケ型箒状細胞，紡錘形の偽担子器，そして偽アミロイドに染まる柄表皮組織はクヌギタケ属に見られない特徴の組み合わせである．

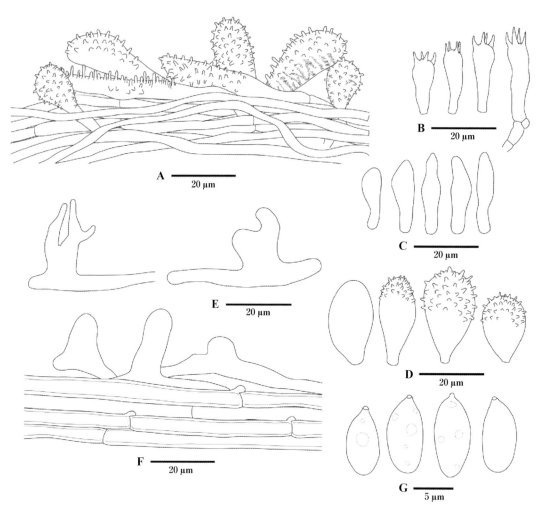

Fig. 94 – Micromorphological features of *Gymnopus phyllogenus* (holotype): **A**. Longitudinal cross section of the pileipellis. **B**. Basidia. **C**. Basidioles. **D**. Cheilocystidia. **E**. Caulocystidia. **F**. Longitudinal section of the stipitipellis. **G**. Basidiospores. Illustrations by Taneyama, Y.

アシグロカレハタケの顕鏡図（正基準標本）：**A**. 傘表皮組織の縦断面. **B**. 担子器. **C**. 偽担子器. **D**. 縁シスチジア. **E**. 柄シスチジア. **F**. 柄表皮組織の縦断面. **G**. 担子胞子. 図：種山裕一.

13. *Gymnopus phyllogenus* アシグロカレハタケ —— 121

Fig. 95 – Longitudinal cross section of the sterile edge of a lamella of *Gymnopus phyllogenus* (in 3%KOH and Congo red stain, holotype), showing gregarious cheilocystidia. Photo by Taneyama, Y.
アシグロカレハタケのヒダ縁部の縦断面（3％水酸化カリウム溶液で封入した後コンゴー赤染色，正基準標本）．ヒダ縁部に群生した縁シスチジアが不稔帯を形成する．図：種山裕一．

Fig. 96 – Basidiomata of *Gymnopus phyllogenus* occurring on a dead fallen leaf of *C. sieboldii* (holotype), 14 Oct. 2011, Oomija, Iriomote Island. Photo by Takahashi, H.
スダジイの枯れ葉上から発生したアシグロカレハタケの子実体（正基準標本）．2011年10月14日，西表島大見謝．写真：高橋春樹．

13. *Gymnopus phyllogenus* アシグロカレハタケ

Fig. 97– Basidiomata of *Gymnopus phyllogenus* on a dead fallen leaf of *C. sieboldii* (hotype), 14 Oct. 2011, Oomija, Iriomote Island. Photo by Takahashi, H.
スダジイの枯れ葉上から発生したアシグロカレハタケの子実体（正基準標本），2011年10月14日，西表島大見謝．写真：高橋春樹．

13. *Gymnopus phyllogenus* アシグロカレハタケ —— 123

Fig. 98 − Basidiomata of *Gymnopus phyllogenus* on a dead fallen leaf of *C. sieboldii* (holotype), 14 Oct. 2011, Oomija, Iriomote Island. Photo by Takahashi, H.
スダジイの枯れ葉上から発生したアシグロカレハタケの子実体（正基準標本）．2011年10月14日，西表島大見謝．写真：高橋春樹．

14. *Inocybe fuscomarginata* Kühner フチドリトマヤタケ

Bull Soc Myc France 71 (3): 169 (1956) [MB#298910].

Macromorphological characteristics: Basidiocarp small. Pileus convex to plane, brown. Lamellae adnate to adnexed, with brownish fimbriate edges. Stipe equal, surface tomentose, dark brown.

Micromorphological characteristics (Fig. 99): Basidiospores 10.5-13.3 × 7.0-8.5 μm (N = 20, average: 12.1 × 7.7 μm), Q = 1.5-1.7 (average 1.6), amygdaliform, occasionally oblong, usually with a subconical apex in side view. Basidia 34-42 × 10.0-12.0 μm, 4-spored, narrowly clavate, thin-walled, pale yellow. Pleurocystidia absent. Cheilocystidia often catenate, thin-walled, with terminal cells 24-48 × 15.0-27.5 μm, cylindrical to broadly ellipsoid, occasionally with a pedicel, filled with pale greyish brown to pale rusty brown contents. Hymenophoral trama subregular; hyphae 5.0-8.3 μm in diameter, occasionally swollen (< 15.0 μm), almost colorless. Caulocystidia absent. Pileipellis of interwoven hyphae, yellow to yellowish brown, with fusiform to clavate terminals, < 18.8 μm in diameter, walls somewhat thickened (< 2.0 μm). Clamp connections abundant in all tissues, but not at all septa.

Habitat and phenology: On ground in a *Casuarina stricta* Ait. coastal forest, May.

Known distribution: Japan (Okinawa, Honshu), Europe.

Specimen examined: OSA-MY 3892, on ground in a *C. stricta* forest, Urauchi, Iriomote Island, Yaeyama-gun, Okinawa Pref., 9 May 1999, leg. D. Sakuma.

Japanese name: Fuchidori-tomayatake (named by T. Kobayashi).

Comments: This species belongs to the subgenus *Inosperma* Kühner, section *Dulcamarae* R. Heim, because it possesses catenate cheilocystidia and smooth basidiospores.

The morphological characteristics of the present collection coincides well with *I. fuscomarginata* reported by Kühner (1956).

References 引用文献

Kühner R. 1955. Complements a la "Flore analytique" VI) Inocybes goniospores et Inocybes acystidies. - Especes nouvelles ou critiques. Bull Soc Myc France 71 (3): 169-201 (published in 1956).

14. フチドリトマヤタケ（小林孝人新称）*Inocybe fuscomarginata* Kühner

Bull Soc Myc France 71 (3): 169 (1956) [MB#298910].

肉眼的特徴：子実体は小型．傘はまんじゅう型～平ら，褐色．ヒダは直生～上生，縁は褐色，粉状．柄は上下同大，表面は毛状，暗褐色．

微鏡的特徴 (Fig. 99)：担子胞子は 10.5-13.3 × 7.0-8.5 μm (N = 20, 平均値：12.1 × 7.7 μm), Q = 1.5-1.7（平均値：1.6），アーモンド形，時々長楕円形，通常円錐形の頂部を持つ．担子器は 34-42 × 10.0-12.0 μm, 4胞子性，狭棍棒状，薄壁，淡黄色．側シスチジアを欠く．縁シスチジアはしばしば連鎖し，薄壁，頂部の細胞は 24-48 × 15.0-27.5 μm, 円柱状～広楕円体，時々柄があり，淡灰褐色～淡さび褐色の内容物で充満する．ヒダの実質は平行状菌糸より構成され，菌糸の幅は5.0-8.3 μm, 時々膨張し（幅15.0 μm まで），ほぼ無色．柄シスチジアを欠く．傘の表皮は錯綜型菌糸より構成され，黄色～黄褐色，紡錘形～棍棒状の頂端細胞を持ち，幅は18.8 μm まで，細胞壁は幾分肥厚する（厚さ2.0 μm まで）．クランプ結合は豊富だが，全ての隔壁で観察されるわけではない．

生態および発生時期：モクマオウ海岸林内地上，5月．

分布：日本（沖縄，本州），欧州．

14. *Inocybe fuscomarginata* フチドリトマヤタケ ―― 125

供試標本：OSA-MY 3892，八重山郡西表島浦内モクマオウ海岸林，1999年5月9日，佐久間大輔採集.
コメント：本種は縁シスチジアがしばしば連鎖し，担子胞子が平滑なため，アオアシアセタケ亜属 subgenus *Inosperma* Kühner，マレンソントマヤタケ節 section *Dulcamarae* R. Heim に所属する．本試料の特徴は Kühner（1956）による *I. fuscomarginata* の原記載に一致する．

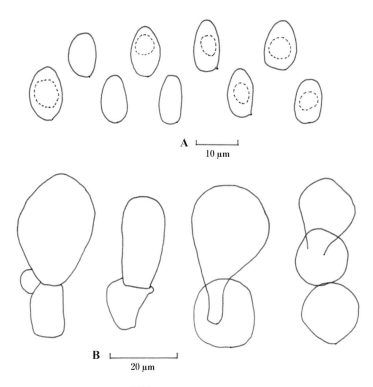

Fig. 99 – *Inocybe fuscomarginata* (OSA-MY 3892): **A**. Basidiospores. **B**. Cheilocystidia. Illustrations by Kobayashi, T.
フチドリトマヤタケ（OSA-MY 3892）：**A**. 担子胞子．**B**. 縁シスチジア．図：小林孝人．

15. *Inocybe humilis* (J. Favre & E. Horak) Esteve-Rav. & Vila コカブラアセタケ

Revta Catal Micol 21: 192 (1998) [MB#483713].
≡ *Astrosporina humilis* J. Favre & E. Horak, Arctic and Alpine Mycology 2: 231 (1987) [MB#131370].
≡ *Inocybe humilis* J. Favre, Ergebn wiss Unters schweiz NatnParks 6 (42): 480, 587 (1960) nom. inval.

Macromorphological characteristics: Basidiocarp small. Pileus convex, umbonate, surface smooth, rimulose, dark reddish brown. Lamellae adnate to sinuate, yellowish brown, with fimbriate white edges. Stipe equal with a marginately bulbous base, surface pruinose overall, reddish brown.

Micromorphological characteristics (Fig. 100): Basidiospores $9.0\text{-}11.3 \times 6.5\text{-}8.3$ μm (N = 20, average: 10.1×7.3 μm), Q = 1.1-1.5 (average: 1.4), nodulose, yellowish brown to greyish brown. Basidia $28\text{-}33 \times 10.0\text{-}12.3$ μm, 4-spored, clavate, thin-walled, slightly lemon-colored. Pleurocystidia $53\text{-}64 \times 17.5\text{-}22.0$ μm, ventricose to fusiform, with a short pedicel, thick-walled (< 4.3 μm), almost colorless. Cheilocystidia similar to pleurocystidia, thick-walled. Paracystidia present along with cheilocystidia, often catenate with terminal cells clavate to cylindrical, thin-walled, almost colorless. Hymenophoral trama subregular; hyphae 3.3-7.5 μm in diameter, occasionally swollen (< 17.5 μm), almost colorless. Caulocystidia descending to base, similar to pleurocystidia. Cauloparacystidia present along with caulocystidia, similar to paracystidia. Pileipellis a subregularly arrayed cutis, duplex; upper layer < 98 μm thick, composed of hyphae 6.3-20.0 μm, walls somewhat thickened (< 2.5 μm), almost colorless to slightly grey; subtending layer < 73 μm thick, composed of hyphae 5.8-15.0 μm, walls somewhat thickened (< 3.3 μm), brown to dark brown. Clamp connections abundant in all tissues, but not at all septa.

Habitat and phenology: On ground in evergreen broad-leaved forests, June.

Known distribution: Japan (Okinawa, Honshu), Europe.

Specimen examined: TNS-F-48269, on ground in an evergreen broad-leaved forest, Banna Park, Ishigaki-shi, Okinawa Pref., 11 Jun. 2012, leg. K. Hosaka.

Japanese name: Ko-kabura-asetake (named by T. Kobayashi)

Comments: This species belongs to the subgenus *Inocybe*, section *Marginatae* Kühner, because it possesses caulocystidia throughout and has nodulose basidiospores.

The morphological features of the present collection accord with *I. humilis* from Europe (Favre 1960; Horak 1987; Esteve-Raventós and Vila 1998).

References 引用文献

Favre J. 1960. Catalogue descriptif des Champignons supérieurs de la zone subalpine du Parc National Suisse. Ergebn wiss Unters schweiz NatnParks 6 (42): 323-610.

Horak E. 1987. *Astrosporina* in the alpine zone of the Swiss National Park (SNP) and adjacent regions. "Arctic and alpine mycology II: Proceeding of the Second International Symposium of Arctic and alpine mycology", pp 205-234. Plenum Publishing Corp., New York & London.

Esteve-Raventós F, Vila J. 1998. Algunos *Inocybe* de la zona alpina de los Pirineos de Cataluña. II. Revta Catal Micol 21: 185-195.

15. コカブラアセタケ（小林孝人新称）*Inocybe humilis* (J. Favre & E. Horak) Esteve-Rav. & Vila

Revta Catal Micol 21: 192 (1998) [MB#483713].
≡ *Astrosporina humilis* J. Favre & E. Horak, Arctic and Alpine Mycology 2: 231 (1987) [MB#131370].
≡ *Inocybe humilis* J. Favre, Ergebn wiss Unters schweiz NatnParks 6 (42): 480, 587 (1960) nom. inval.

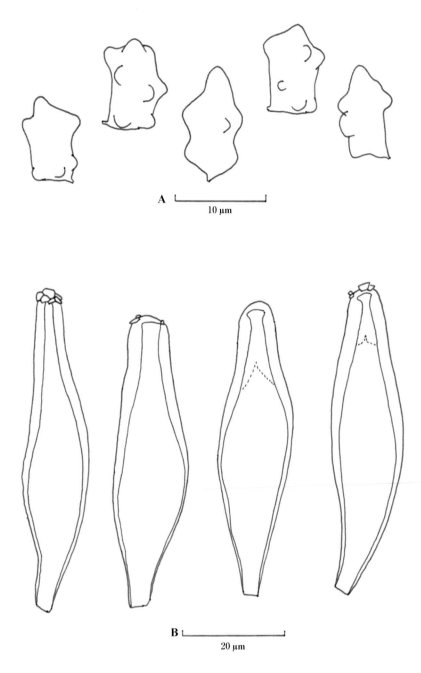

Fig. 100. *Inocybe humilis* (TNS-F-48269)：**A**. Basidiospores. **B**. Cheilocystidia. Illustrations by Kobayashi, T.
コカブラアセタケ（TNS-F-48269）：**A**. 担子胞子．**B**. 縁シスチジア．図：小林孝人．

15. *Inocybe humilis* コカブラアセタケ

肉眼的特徴：子実体は小型．傘はまんじゅう型で中高の平，表面は平滑で放射状に細かく裂け，暗紫褐色．ヒダは直生から湾生，黄褐色，縁は白色で粉状．柄は凹頭の球根状膨大部を基部に有し，表面は全面が粉状，赤褐色．

微鏡的特徴：担子胞子は 9.0-11.3×6.5-8.3 µm（N = 20，平均値：10.1×7.3 µm），Q = 1.1-1.5（平均値：1.4），コブを有し，黄褐色～灰褐色．担子器は 28-33×10.0-12.3 µm，4胞子性，棍棒状，薄壁，わずかにレモン色．側シスチジアは53-64×17.5-22.0 µm，便腹状～紡錘形，常に短い柄があり，厚壁（4.3 µmまで），内容物はほぼ無色．縁シスチジアは側シスチジアと同様，薄壁．パラシスチジアは縁シスチジアと混在し，しばしば連鎖し頂部の細胞は棍棒状～円柱形，薄壁，ほぼ無色．ヒダの実質は平行状菌糸より構成され，菌糸の幅は3.3-7.5 µm，時々膨張し（幅17.5 µmまで），ほぼ無色．柄シスチジアは柄の全面に存在し，側シスチジアと同様，厚壁．柄パラシスチジアは柄シスチジアと混在し，パラシスチジアと同様，薄壁．傘の表皮は平行状菌糸より構成される2層よりなり，外層は厚さ98 µmまで，菌糸の幅は6.3-20.0 µm，細胞壁は幾分肥厚し（2.5 µmまで），ほぼ無色～わずかに灰色；第2層は厚さ73 µmまで，菌糸の幅は5.8-15.0 µm，細胞壁は幾分肥厚し（3.3 µmまで），褐色～暗褐色．クランプ結合は豊富だが，全ての隔壁で観察されるわけではない．

生態および発生時期：常緑広葉樹林内地上，6月．

分布：日本（沖縄，本州），欧州．

供試標本：TNS-F-48269，常緑広葉樹林内地上，八重山郡石垣市字登野城バンナ公園，2012年6月11日，保坂健太郎採集．

コメント：本種は柄シスチジアが柄の全面にあり，コブを持った胞子を有するためクロニセトマヤタケ亜属 subgenus *Inocybe*，カブラアセタケ節 section *Marginatae* Kühner に所属する．
　本試料の形態的特徴は欧州産 *I. humilis*（Favre 1960；Horak 1987；Esteve-Raventós and Vila 1998）に一致する．

16. *Leccinellum rhodoporosum* (Har. Takah.) Har. Takah., **comb. nov**. ウラベニヤマイグチ

MycoBank no.: MB 809942.

Basionym: *Leccinum rhodoporosum* Har. Takah., Mycoscience 48 (2): 92 (2007) [MB#529790].

Etymology: The specific epithet means "*rhodo-* (rose-) + *porosum* (porous)", referring to the deep-red pores.

Macromorphological characteristics (Figs. 102-109): Pileus 40-70 mm in diameter, at first hemispherical, expanding to convex to broadly convex, even or occasionally rugulose-pitted; margin slightly appendiculate; surface dry, subtomentose to nearly glabrous, at first evenly colored reddish-brown (8E7-8) to dark reddish brown (8F6-8), then somewhat paler from the margin, unchanging when bruised. Flesh up to 10 mm thick, firm, whitish in the pileus, whitish to light yellow in the stipe, slowly changing to blue when cut; odor and taste indistinct. Stipe 30-60 × 10-20 mm, subequal or somewhat enlarged toward the base, central, terete, solid, not reticulate; surface dry, distinctly covered overall with violet-brown (11E8-11F8, when young) to blackish-brown (in age), furfuraceous scales on a whitish ground color; base covered with whitish mycelial tomentum. Tubes up to 12 mm deep, deeply depressed around the stipe, pale yellow (3A4) to pastel yellow (3A4), slowly staining blue when cut; pores small (2-3 per millimeter), subcircular, at first concolorous, soon becoming deep red (9A8-10A8) to reddish-orange (7A8), then brownish-orange (7C8) in age, slowly staining blue where handled.

Micromorphological characteristics (Fig. 101): Basidiospores 15-16.5 × 4.5-5 μm (n = 87 spores of 6 basidiocarps, Q = 3.3), inequilateral with a suprahilar depression in profile, elongate-fusiform in face view, smooth, brown (7E6-7) in water, thick-walled (up to 1 μm). Basidia 18-27 × 8-11 μm, clavate, four-spored. Cheilocystidia gregarious, 22-37 × 4-7 μm, subclavate, smooth, brownish-red in water, thin-walled. Pleurocystidia scattered, 37-55 × 7-11 μm, fusoid-ventricose, smooth, colorless, thin-walled. Hymenophoral trama composed of elements 6-20 μm wide, cylindrical, divergent, smooth, pale yellow, thin-walled. Pileipellis composed of a trichodermial palisade formed by vertically and compactly arranged hyphae 5-12 μm wide, cylindrical, encrusted with fine, granular, brown pigment, thin-walled. Pileitrama of cylindrical, loosely interwoven hyphae 6-30 μm wide, smooth, colorless, thin-walled. Stipitipellis a disrupted hymeniform, consisting of caulocystidia and caulobasidia. Caulocystidia of two types: 1) broadly clavate, 25-35 × 6-13 μm, smooth, brownish-red in water, thin-walled; 2) narrowly fusoid-ventricose, 40-65 × 7-11 μm, smooth, brownish-red in water, thin-walled. Stipe trama composed of longitudinally arranged, cylindrical hyphae 7-12 μm wide, unbranched, smooth, colorless, thin-walled. Clamps absent in all tissues.

Habitat and phenology: Solitary to scattered on ground in subtropical evergreen broad-leaved forests dominated by *Quercus miyagii* Koidz. and *Castanopsis sieboldii* (Makino) Hatus. ex T. Yamaz. et Mashiba and warm temperate *Quercus-Pinus* forests dominated by *Q. acuta* Thunb. ex Murray, *Q. myrsinaefolia* Blume, and *Pinus densiflora* Sieb. et Zucc., May to September.

Known distribution: Okinawa (Ishigaki Island), Hyogo.

Specimens examined: KPM-NC0010093 (holotype), on ground in a evergreen broad-leaved forest dominated by *Q. miyagii* and *C. sieboldii,* Banna Park, Ishigaki-shi, Okinawa Pref., 30 May 2002, coll. Takahashi, H.; KPM-NC0013136, same location, 4 Jun. 2004, coll. Takahashi, H.; KPM-NC0023878, same location, 11 Aug. 2010, coll. Takahashi, H.

Extralimital specimen examined: KPM-NC0014100, on ground in mixed forests dominated by *Q. acuta*, *Q. myrsinaefolia* and *P. densiflora,* Kita-ku, Kobe-shi, Hyogo Pref., 11 Sep. 2006, coll. Koutoku, S.

Japanese name: Urabeni-yamaiguchi.

Comments: This species is characterized by the reddish-brown to dark reddish brown pileus; the whitish stipe covered overall with violet-brown to blackish-brown, furfuraceous scales; the deep-red to reddish-orange pores; the large, fusoid-cylindrical, brown basidiospores; the pileipellis composed of a trichodermial palisade; the dimorphic, broadly clavate and narrowly fusoid-ventricose caulocystidia; and the habitat in subtropical evergreen broad-leaved forests and warm temperate *Quercus-Pinus* forests. Except for the discolored red pores, which are characteristic feature of the genus *Boletus*, section *Luridi* Fr. as defined in Singer (Singer 1986), the combination of the violet-brown, coarsely furfuraceous scales on the stipe surface, the rather large (15-16.5 μm long), elongate-fusiform, brown basidiospores, the yellow hymenophore, the pileipellis of a trichodermial palisade, and the lowland evergreen broad-leaved forest habitat, all strongly suggest that the present species belongs to the genus *Leccinellum* (Bresinsky & Besl 2003), and we therefore make the formal transfer herein.

Sutorius australiensis (Bougher & Thiers) Halling & Fechner (≡ *Leccinum australiense* Bougher & Thiers), reported from tropical northern Australia (Bougher and Thiers 1991; Halling et al. 2012), appears to be closely related to *L. rhodoporosum* in having discolorous pores, a reddish-brown pileus, an appressed-squamulose stipe, and a palisade-trichoderm structure in the pileipellis. The former species, however, has a white context soon turning pale grey, unchanging tubes, pores discolored dark reddish brown, and a habitat in *Eucalyptus* forests.

The North American *Boletus morrisii* Peck (Bessette et al. 2000; Singer 1947), belonging to the genus *Boletus*, section *Luridi*, is closely similar in discolorous red pores and brownish-red scales on the stipe surface; however, it also has striking differences such as, it yields rufescent flesh, does not exhibit darkening of the scales of the stipe, and has yellowish basidiospores.

References 引用文献

Bessette AE, Roody WC, Bessette AR. 2000. North American Boletes. A color guide to the fleshy pored mushrooms. Syracuse University Press, New York.

Bougher NL, Thiers HD. 1991. An indigenous species of *Leccinum* (Boletaceae) from Australia. Mycotaxon 42: 225-262.

Bresinsky A, Besl H. 2003. Beiträge zu einer Mykoflora Deutschlands - Schlüssel zur Gattungsbestimmung der Blätter-, Leisten- und Röhrenpilze mit Literaturhinweisen zur Artbestimmung. Regensb Mykol Schr 11: 1-236.

Halling RE, Nuhn M, Fechner NA, Osmundson TW, Soytong K, Hibbett DS, Binder M. 2012. *Sutorius*: a new genus for *Boletus eximius*. Mycologia 104 (4): 951-961.

Singer R. 1947. The Boletineae of Florida with notes on extralimital species. III. Am Midl Nat 37: 1-135.

Singer R. 1986. Agaricales in modern taxonomy, 4th edn. Koeltz, Koenigstein.

Takahashi H. 2007. Five new species of the Boletaceae from Japan. Mycoscience 48 (2): 90-99.

16. ウラベニヤマイグチ *Leccinellum rhodoporosum* (Har. Takah.) Har. Takah., **comb. nov.**

Basionym: *Leccinum rhodoporosum* Har. Takah., Mycoscience 48 (2): 92 (2007) [MB#529790].

肉眼的特徴（Figs. 102-109）：傘は径40-70 mm，最初半球形，のち丸山形～中高偏平，平坦もしくは時に皺状凹凸を表す；縁部は管孔側に僅かに突出する；表面は粘性を欠き，密綿毛状～ほぼ平滑，最初赤褐色～暗赤褐色のち周縁部からやや褪色し，傷を受けても変色しない．肉は厚さ10 mm 以下，堅く，傘は類白色，柄は淡黄色，空気に触れると徐々に青変し，特別な味や臭いはない．柄は30-60×10-20 mm，ほぼ上下同大または下方に向かってやや太くなり，中心生，中実，網目を欠く；表面は乾性，全体に類白色の地に暗紫褐色（幼時）～黒褐色（老成時）の粒状細鱗片に密に被われる；根元は類白色の綿毛状菌糸体に被われる．管孔は長さ12 mm 以下，柄の周囲において深く嵌入し，淡黄色，空気に触れると徐々に青変する；孔口は小型（2-3 per mm），類

円形，最初管孔と同色，まもなく赤色〜橙赤色を帯び，老成時帯褐橙色になり，触れた部分は徐々に青変する．

顕微鏡的特徴（Fig. 01）：担子胞子は 15-16.5×4.5-5 μm（n = 87 spores of 6 basidiocarps，Q = 3.3），紡錘状円柱形，下部側面になだらかな凹みがあり（イグチ型），平坦，褐色，厚壁．担子器は 18-27×8-11 μm，こん棒形，4胞子性．縁シスチジアは群生し，22-37×4-7 μm，亜こん棒形，平滑，帯褐赤色（水封），薄壁．側シスチジアは散在し，37-55×7-11 μm，片脹れ状紡錘形，平滑，無色，薄壁．傘表皮組織は柵状毛状被を形成し，菌糸は幅 5-12 μm，円柱形，褐色の細かい粒状色素が細胞壁の外側に沈着し，薄壁．傘実質を構成する菌糸は幅 6-30 μm，緩く錯綜し，円柱形，平滑，無色，薄壁．柄表皮組織は柄シスチジアと柄担子器が不連続な子実層状被を形成する．柄シスチジアは2形性；1）広こん棒形，25-35×6-13 μm，平滑，帯褐赤色（水封），薄壁；2）幅の狭い片脹れ状紡錘形，40-65×7-11 μm，平滑，帯褐赤色（水封），薄壁．柄実質の菌糸は縦に沿って配列し，幅 7-12 μm，円柱形，無分岐，平滑，無色，薄壁．全ての組織において菌糸はクランプを欠く．

生態および発生時期：スダジイ，オキナワウラジロガシを主体とする亜熱帯性常緑広葉樹林内およびシラカシ，アカガシを主体とする温帯性常緑広葉樹林内地上に孤生または散生，5月〜9月．

分布：沖縄（石垣島），兵庫．

供試標本：KPM-NC0010093（正基準標本），スダジイ，オキナワウラジロガシを主体とする亜熱帯性常緑広葉樹林内地上，沖縄県石垣市バンナ公園，2002年5月30日，高橋春樹採集；KPM-NC0013136，同上，2004年6月4日，高橋春樹採集；KPM-NC0023878，同上，2010年8月11日，高橋春樹採集．

地域外供試標本：KPM-NC0014100，アカマツが混在したシラカシ，アカガシを主体とする常緑広葉樹林内地上，兵庫県神戸市北区，2006年9月11日，幸徳伸也採集．

主な特徴：傘は赤褐色〜暗赤褐色；柄は全体に暗紫褐色を帯びた粒状細鱗片に密に被われる；孔口は赤色〜橙赤色；担子胞子は大型，紡錘状円柱形，褐色；傘表皮組織は柵状毛状被からなる；柄表皮組織は広こん棒形と幅の狭い片脹れ状紡錘形の2形性柄シスチジアを持つ；スダジイ，オキナワウラジロガシを主体とする亜熱帯性常緑広葉樹林およびシラカシ，アカガシを主体とする温帯性常緑広葉樹林内地上に発生．

コメント：孔口が管孔の色と異なり，赤色を帯びる点は Singer（Singer 1986）の分類概念によるヤマドリタケ属 *Boletus*，ウラベニイロガワリ節 section *Luridi* Fr. と共通するが，暗色の粒状細鱗片に被われた柄，長い紡錘状円柱形（長さ 15 μm 以上）で褐色の担子胞子，黄色の子実層托，柵状毛状被を形成する傘表皮組織，そして低地の常緑広葉樹林内に発生する生態はクロヤマイグチ属 *Leccinellum*（Bresinsky & Besl 2003）により近縁であることを強く示唆している．

　　熱帯オーストラリア産 *Sutorius australiensis*（Bougher & Thiers）Halling & Fechner（Bougher and Thiers 1991；Halling et al. 2012）は，管孔と色の異なる孔口，赤褐色の傘，圧着した小鱗片に被われた柄，柵状毛状被を形成する傘表皮組織の特徴において本種と共通性が認められるが，肉が空気に触れると灰色に変色すること，管孔は変色しないこと，孔口は暗赤褐色を帯びること，そしてユーカリ属の樹下に発生する性質においてウラベニヤマイグチと異なる．

　　ヤマドリタケ属，ウラベニイロガワリ節に所属する北米産 *Boletus morrisii* Peck（Bessette et al. 2000；Singer 1947）は赤色の孔口と柄の表面を被う帯褐赤色の明瞭な粒状細鱗片を有する点でウラベニヤマイグチに類似するが，肉に赤変性があり，柄の鱗片は暗色にならず，担子胞子は黄色を帯びる．

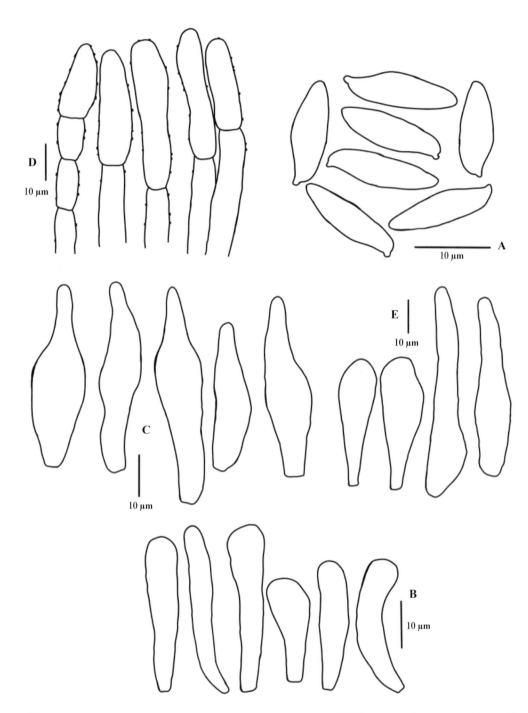

Fig. 101 – Micromorphological features of *Leccinellum rhodoporosum* (KPM-NC0010093): **A**. Basidiospores. **B**. Cheilocystidia. **C**. Pleurocystidia. **D**. Elements of the pileipellis. **E**. Caulocystidia. Illustrations by Takahashi, H.
ウラベニヤマイグチの顕微鏡図 (KPM-NC0010093)：**A**. 担子胞子．**B**. 縁シスチジア．**C**. 側シスチジア．**D**. 傘表皮組織．**E**. 柄シスチジア．図：高橋春樹．

Fig. 102 – Mature basidioma of *Leccinellum rhodoporosum* (KPM-NC0013136), 4 Jun. 2004, Banna Park, Ishigaki Island. Photo by Takahashi, H.
ウラベニヤマイグチの成熟した子実体（KPM-NC0013136），2004年6月4日，石垣島バンナ公園．写真：高橋春樹．

134 —— 16. *Leccinellum rhodoporosum* ウラベニヤマイグチ

Fig. 103 − Stipe surface of *Leccinellum rhodoporosum* (KPM-NC0013136), 4 Jun. 2004, Banna Park, Ishigaki Island. Photo by Takahashi, H.
ウラベニヤマイグチの柄表面（KPM-NC0013136），2004年6月4日，石垣島バンナ公園．写真：高橋春樹．

Fig. 104 − Underside view of the basidioma of *Leccinellum rhodoporosum* (KPM-NC0013136), 4 Jun. 2004, Banna Park, Ishigaki Island. Photo by Takahashi, H.
ウラベニヤマイグチの管孔側（KPM-NC0013136），2004年6月4日，石垣島バンナ公園．写真：高橋春樹．

16. *Leccinellum rhodoporosum* ウラベニヤマイグチ

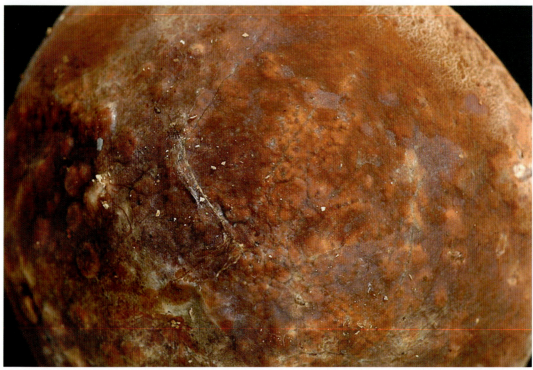

Fig. 105 – Pileus surface of *Leccinellum rhodoporosum* (KPM-NC0013136), 4 Jun. 2004, Banna Park, Ishigaki Island. Photo by Takahashi, H.
ウラベニヤマイグチの傘表面 (KPM-NC0013136), 2004年6月4日, 石垣島バンナ公園. 写真：高橋春樹.

16. *Leccinellum rhodoporosum* ウラベニヤマイグチ — 137

Fig. 106 – Immature basidioma of *Leccinellum rhodoporosum* (KPM-NC0010093), 30 May 2002, Banna Park, Ishigaki Island. Photo by Takahashi, H.
ウラベニヤマイグチの幼菌 (KPM-NC0010093), 2002年5月30日, 石垣島バンナ公園. 写真：高橋春樹.

Fig. 107 – Mature basidioma of *Leccinellum rhodoporosum* (KPM-NC0010093), 30 May 2002, Banna Park, Ishigaki Island. Photo by Takahashi, H.
ウラベニヤマイグチの成熟した子実体（KPM-NC0010093），2002年5月30日，石垣島バンナ公園．写真：高橋春樹．

16. *Leccinellum rhodoporosum* ウラベニヤマイグチ —— 139

Fig. 108 – Stipe surface of *Leccinellum rhodoporosum* (KPM-NC0010093), 30 May 2002, Banna Park, Ishigaki Island. Photo by Takahashi, H.
ウラベニヤマイグチの柄表面（KPM-NC0010093），2002年5月30日，石垣島バンナ公園．写真：高橋春樹．

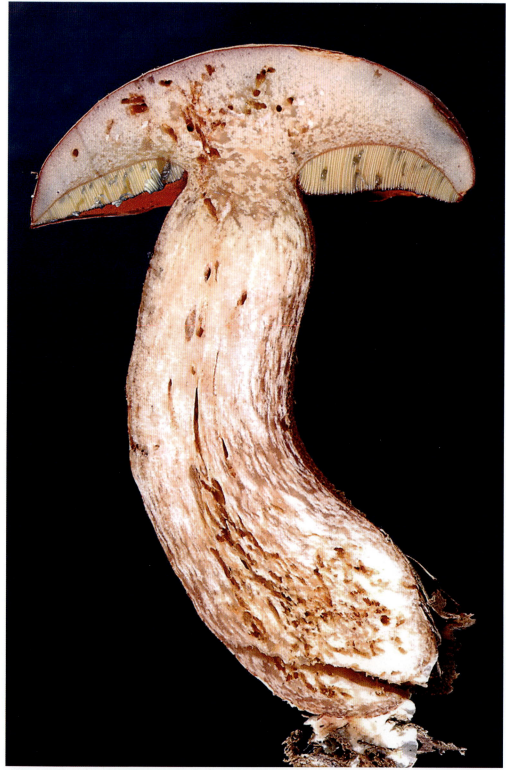

Fig. 109 – Longitudinal cross section of the basidioma of *Leccinellum rhodoporosum* (KPM-NC0010093), 30 May 2002, Banna Park, Ishigaki Island. Photo by Takahashi, H.
ウラベニヤマイグチの子実体縦断面（KPM-NC0010093），2002年5月30日，石垣島バンナ公園．写真：高橋春樹．

17. *Marasmiellus arenaceus* Har. Takah., Taneyama & Wada, **sp. nov.** シロスナホウライタケ

MycoBank no.: MB 809929.

Etymology: The specific epithet means "arenaceous, sandy".

Distinctive features of this species are found in a pale brownish to almost whitish, small, collybioid basidiomata; broadly ellipsoid, inamyloid basidiospores; subcylindrical to subclavate cheilocystidia with apical nodose-coralloid excrescences; usually non-diverticulate terminal elements in the pileipellis which concentrate around the pileus margin; a stipitipellis of densely entangled hyphae; lack of caulocystidia; inamyloid trama; and a habitat of living roots and stems of various kinds of beach grasses in arenaceous soil of coastal dunes.

Macromorphological characteristics (Figs. 113–122): Pileus 9–18 mm in diameter, at first hemispherical, then broadly convex to applanate, often with slightly depressed center, smooth or at times more or less uneven; surface non-hygrophanous, dry, minutely appressed fibrillose to felted-tomentose, brownish orange (6C4–5) to light brown (6D4–5) at the center, paler toward the crenulate margin, at times almost whitish overall. Flesh thin (up to 1.5 mm thick), whitish, tough; odor not distinctive, taste mild. Stipe 15–20 × 1.7–2 mm, cylindrical, central, terete, solid, smooth; surface dry, concolorous with the pileus, felted-tomentose; basal mycelioid tomentum whitish. Lamellae subdecurrent, subdistant to subclose (29–34 reach the stipe), with 1–3 series of lamellulae, not intervenose, up to 3 mm broad, thin, whitish; edges minutely fimbriate under a lens, concolorous.

Micromorphological characteristics (Figs. 110–112): Basidiospores (7.5–) 9.5–10.8 (–11.9) × (5.6–) 6.4–7.3 (–8.3) μm (n = 146, mean length = 10.16 ± 0.66, mean width = 6.87 ± 0.47, Q = (1.27–) 1.40–1.57 (–1.69), mean Q = 1.48 ± 0.08), broadly ellipsoid, smooth, hyaline, inamyloid, colorless in KOH, thin-walled. Basidia (26.0–) 31.6–39.6 (–44.1) × (6.5–) 8.2–10.2 (–11.1) μm (n = 32, mean length = 35.60 ± 4.03, mean width = 9.20 ± 0.98) in main body, (4.3–) 4.5–5.7 (–6.2) × (2.0–) 2.1–2.5 (–2.6) μm (n = 22, mean length = 5.08 ± 0.57, mean width = 2.29 ± 0.18) in sterigmata, clavate, 4-spored; basidioles clavate. Cheilocystidia (22.5–) 28.5–40.7 (–46.2) × (4.9–) 6.2–8.1 (–9.0) μm (n = 30, mean length = 34.60 ± 6.06, mean width = 7.18 ± 0.96), abundant, subcylindrical to subclavate or irregularly shaped, often apically nodose-coralloid or bifurcate, thin-walled; outgrowths (3.8–) 6.8–14.5 (–17.8) × (2.1–) 2.9–4.7 (–5.6) μm (n = 38, mean length = 10.64 ± 3.85, mean width = 3.79 ± 0.92). Pleurocystidia absent. Hymenophoral trama subregular; element hyphae (2.9–) 4.6–7.0 (–8.6) μm wide (n = 57, mean width = 5.78 ± 1.20), cylindrical, smooth, thin-walled. Pileipellis a cutis of loosely interwoven, repent, cylindrical hyphae (2.7–) 3.8–5.4 (–6.8) μm wide (n = 56, mean width = 4.63 ± 0.78), with pale brownish, slightly thickened walls 0.4–0.5 μm (n = 13, mean thickness = 0.47 ± 0.04), often encrusted with fine granules concolorous with the walls, smooth, usually non-diverticulate, without *Rameales*-structure; terminal cells (25.5–) 25.7–58.4 (–73.5) × (4.9–) 4.4–7.1 (–8.8) μm (n = 8, mean length = 42.03 ± 16.36, mean width = 5.75 ± 1.38), subcylindrical, smooth, concentrated around the pileus margin. Elements of pileitrama (3.7–) 4.6–7.1 (–8.0) μm wide (n = 46, mean width = 5.84 ± 1.27), cylindrical, parallel, smooth, thin-walled. Stipitipellis composed of densely entangled hyphae (2.8–) 3.1–4.9 (–7.5) μm wide (n = 41, mean width = 3.97 ± 0.91), cylindrical, occasionally branched, smooth, thin-walled; caulocystidia undifferentiated. Stipe trama composed of longitudinally running, cylindrical hyphae (3.7–) 5.4–9.5 (–13.7) μm wide (n = 52, mean width = 7.43 ± 2.07), monomitic, unbranched, smooth, with thickened walls (1.0–) 1.1–1.4 (–1.5) μm (n = 20, mean thickness = 1.29 ± 0.15). Elements of basal mycelioid tomentum (2.8–) 3.5–4.8 (–5.3) μm wide (n = 36, mean width = 4.17 ± 0.62), cylindrical, occasionally branched, smooth, with thickened walls (0.6–) 0.7–1.0 (–1.1) μm (n = 23, mean thickness = 0.85 ± 0.13). All tissues inamyloid, colorless in water and KOH except for pileipellis elements, with clamp connections.

Habitat and phenology: Solitary or scattered on living roots and stems of various kinds of beach grasses in arenaceous soil of coastal dunes, June.

Known distribution: Okinawa (Yaeyama Islands: Ishigaki Island and Iriomote Island), Hyogo.

Holotype: TNS-F-48339, on living roots and stems of various kinds of beach grasses, Ishigaki-shi, Okinawa Pref., Japan, 12 Jun. 2012, coll. Hosaka, K.

Other specimens examined: TNS-F-61369, on living stems and roots of beach grasses (unidentified substrata), Iriomote Island, Taketomi-cho, Yaeyama-gun, Okinawa Pref., 13 Jun. 2014, coll. Wada, S.; TNS-F-61370, same location, 16 Jun. 2014, coll. Wada, S.; TNS-F-61371, same location, 16 Jun. 2014, coll. Wada, S; KPM-NC0023867, same location, 16 Jun. 2014, coll. Wada, S.

Extralimital specimens examined: TNS-F-61372, on living stems and roots of *Carex kobomugi* Ohwi, Suma-ku, Kobe-shi, Hyogo Pref., 22 Jun. 2014, coll. Wada, S; KPM-NS0023868, same location, 22 Jun. 2014, coll. Wada, S.

Japanese name: Sirosuna-houraitake (named by S. Wada).

Comments: The pale brownish to almost whitish, small, collybioid basidiomata with a whitish base to the stipe, the inamyloid basidiospores, the apically nodose-coralloid cheilocystidia, the inamyloid trama, and the lack of *Rameales*-structure in the pileipellis suggest placement of the new species in the genus *Marasmiellus* Murrill, section *Dealbati* (Bat.) Singer (Singer 1973, 1986; Noordeloos 1995, 2010).

Marasmiellus mesosporus Singer from North America (Singer et al. 1973; Robich et al. 1991; Desjardin et al. 1992; Antonín and Noordeloos 1993, 2010; Takehashi et al. 2007), commonly known as a *Marasmius*-blight fungus (Lucas et al. 1971; Warren 1972; Warren and Lucas 1973, 1975), and *Marasmiellus carneopallidus* (Pouzar) Singer reported from Czech Republic (Pouzar 1966; Singer 1973; Antonín and Noordeloos 1993, 2010), share striking morphological and ecological similarities with the new species. The former taxon clearly differs from *M. arenaceus* by a pale pinkish brown pileus and stipe, ellipsoid to oblong, slightly narrower basidiospores: (7.5–) 9–11 (–11.8)×(4–) 5.5–6.5 (–7) μm (Singer et al. 1973), more or less evenly distributed ventricose terminal elements in the pileipellis, and sparsely diverticulate caulocystidia. *Marasmiellus carneopallidus* is distinct from *M. arenaceus* in possessing non-diverticulate, broadly clavate cheilocystidia, a weak *Rameales*-structure of the pileipellis which consists of distinctly coralloid terminal elements and a habitat preference for old decaying roots and stems of various kinds of grasses in xerophytic grassland.

References 引用文献

Antonín V, Noordeloos ME. 1993. A monograph of *Marasmius, Collybia* and related genera in Europe. Part 1: *Marasmius, Setulipes,* and *Marasmiellus*. Lib Bot 8: 157–163.

Antonín V, Noordeloos ME. 2010. A monograph of marasmioid and collybioid fungi in Europe. IHW-Verlag.

Desjardin DE, Wong GJ, Hemmes DE. 1992. Agaricales of the Hawaiian Island. 1. Marasmioid fungi: new species, new distributional records, and poorly known taxa. Can J Bot 70: 530–542.

Lucas LT, Warren TB, Woodhouse Jr WW, Seneca ED. 1971. *Marasmius* Blight, a new disease of American beachgrass. Plant Dis Rep 55 (7): 582–585.

Pouzar Z. 1966. *Micromphale carneo-pallidum* spec. nov., a new steppe fungus similar to *Marasmius oreades*. Česká Mykologie 20 (1): 18–24.

Robich G, Moreno G, Pöder R. 1991. *Marasmiellus dunensis* (Marasmiaceae, Agaricales), a new species from the European Mediterranean. Mycotaxon 42: 181–186.

Singer R. 1973. The genera *Marasmiellus, Crepidotus,* and *Simocybe* in the Neotropics. Beih Nova Hedwigia 44: 1–517.

Singer R. 1986. The Agaricales in modern taxonomy, 4th edn. Koeltz, Koenigstein.
Singer R, Lucas LT, Warren TB. 1973. The *Marasmius*-blight fungus. Mycologia 65 (2): 468–473.
Takehashi S, Kasuya T, Kakishima M. 2007. *Marasmiellus mesosporus*, a *Marasmius*-blight fungus newly recorded from sand dunes of the Japanese coast. Mycoscience 48 (6): 407–410.
Warren TB. 1972. *Marasmius* Blight, a New Disease of American Beachgrass on the Outer Banks of North Carolina (M.S. Thesis). North Carolina State University at Raleigh.
Warren TB, Lucas LT. 1973. Histopathology of *Marasmius* Blight of American Beachgrass. Phytopathology 63: 725–728.
Warren TB, Lucas LT. 1975. Susceptibility of American beachgrass and other dune plants to *Marasmiellus mesosporus*. Phytopathology 65: 690–692.

17. シロスナホウライタケ（新種；和田匠平新称）*Marasmiellus arenaceus* Har. Takah., Taneyama & Wada, **sp. nov.**

肉眼的特徴（Figs. 113-122）：傘は径9-18 mm，最初半球形，のち饅頭形～ほぼ平らに開き，しばしば中央部がやや凹み，平坦または多少凹凸を生じる；非吸水性，粘性を欠き，圧着した繊維状または羊毛状～密綿毛状，中央部淡褐色，周縁部に向かって色が薄くなる；縁部は細円鋸歯状．肉は厚さ1.5 mm以下，類白色，強靱，特別な匂いはなく，味は温和．柄は15-20×1.7-2 mm，円柱形，中心生，痩せ型，中実，平坦；表面は粘性を欠き，傘とほぼ同色，羊毛状～密綿毛状；根元の綿毛状菌糸体は類白色．ヒダはやや垂生～やや密，やや疎（柄に到達するヒダが29-34），小ヒダは1-3，ヒダの間に連絡脈を欠き，幅3 mm以下，類白色；縁部は同色，微細な長縁毛状．

顕微鏡的特徴（Figs. 110-112）：担子胞子は (7.5-) 9.5-10.8 (-11.9) × (5.6-) 6.4-7.3 (-8.3) μm (n = 146, mean length = 10.16±0.66, mean width = 6.87±0.47, Q = (1.27-) 1.40-1.57 (-1.69), mean Q = 1.48±0.08)，広楕円形，平坦，無色，非アミロイド，アルカリ溶液において無色，薄壁．担子器は (26.0-) 31.6-39.6 (-44.1) × (6.5-) 8.2-10.2 (-11.1) μm（本体），(4.3-) 4.5-5.7 (-6.2) × (2.0-) 2.1-2.5 (-2.6) μm（ステリグマ），こん棒形，4胞子性；偽担子器はこん棒形．縁シスチジアは (22.5-) 28.5-40.7 (-46.2) × (4.9-) 6.2-8.1 (-9.0) μm，群生し，亜円柱形～亜こん棒形または不規則な形状を成し，頂部が短指状に分岐し，薄壁；短指状付属糸は (3.8-) 6.8-14.5 (-17.8) × (2.1-) 2.9-4.7 (-5.6) μm．側シスチジアはない．子実層托実質は多少平列し，菌糸は幅 (2.9-) 4.6-7.0 (-8.6) μm，円柱形，平滑，薄壁．傘の表皮組織は緩く配列した匍匐性の菌糸からなり，通常分岐物を欠き，ラメアレス構造はない；菌糸は幅 (2.7-) 3.8-5.4 (-6.8) μm，円柱形，しばしば淡褐色の微細な凝着物に被われ，壁は淡褐色で厚さ 0.4-0.5 μm；末端細胞は (25.5-) 25.7-58.4 (-73.5) × (4.9-) 4.4-7.1 (-8.8) μm，亜円柱形，平滑，傘周縁部に集中して見られる．傘実質の菌糸は幅 (3.7-) 4.6-7.1 (-8.0) μm，並列し，円柱形，平滑，薄壁．柄表皮を構成する菌糸は幅 (2.8-) 3.1-4.9 (-7.5) μm，密に絡み合い，円柱形，平滑，時に分岐し，無色，薄壁；柄シスチジアはない．柄実質の菌糸は幅 (3.7-) 5.4-9.5 (-13.7) μm，平列し，円柱形，一菌糸型，無分岐，平滑，壁は厚さ (1.0-) 1.1-1.4 (-1.5) μm．柄の根元の綿毛状菌糸は幅 (2.8-) 3.5-4.8 (-5.3) μm，円柱形，時に分岐し，平滑，壁は厚さ (0.6-) 0.7-1.0 (-1.1) μm．全ての組織において菌糸はクランプを有し，非アミロイド，傘表皮組織を除き水封およびアルカリ溶液において無色．

生態および発生時期：海岸砂浜の草本植物の柄または根から孤生または散生，6月．

分布：沖縄（石垣島，西表島），兵庫．

供試標本：TNS-F-48339（正基準標本），海岸砂浜の草本植物の柄または根から発生，沖縄県石垣市，2012年6月12日，保坂健太郎採集；TNS-F-61369，海岸砂浜の草本植物の柄または根から発生，沖縄県八重山郡竹富町西表島，2014年6月13日，和田匠平採集；TNS-F-61370，同上，2014年6月16日，和田匠平採集；TNS-F-61371，同上，2014年6月16日，和田匠平採集；KPM-NC0023867，

144 —— 17. *Marasmiellus arenaceus* シロスナホウライタケ

　　同上，2014年6月16日，和田匠平採集.
地域外供試標本：TNS-F-61372, 海岸砂浜のコウボウムギの柄または根から発生, 兵庫県神戸市須磨区，2014年6月22日，和田匠平採集；KPM-NS0023868, 同上，2014年6月22日，和田匠平採集.
主な特徴：子実体は全体に淡褐色～類白色を呈する小形のモリノカレバタケ型；担子胞子は広楕円形，非アミロイド；縁シスチジアは亜円柱形～亜こん棒形または不規則な形状を成し，しばしば頂部が短指状に分岐する；傘表皮組織の菌糸は通常平滑で，末端細胞は傘周縁部に集中する傾向が見られる；柄表皮を構成する菌糸は密に絡み合い，柄シスチジアを欠く；実質は非アミロイド；海岸砂浜の草本植物の柄または根から発生.
コメント：類白色の柄を持つ淡褐色～類白色を呈する小形のモリノカレバタケ型の子実体，非アミロイドの担子胞子，頂部が短指状に分岐する縁シスチジア，非アミロイドの実質，そしてラメアレス構造を形成しない性質は本種がシロホウライタケ属 *Marasmiellus* Murrill，ハルノキノボリホウライタケ節 section *Dealbati*（Bat.）Singer（Singer 1973, 1986；Noordeloos 1995, 2010）に属

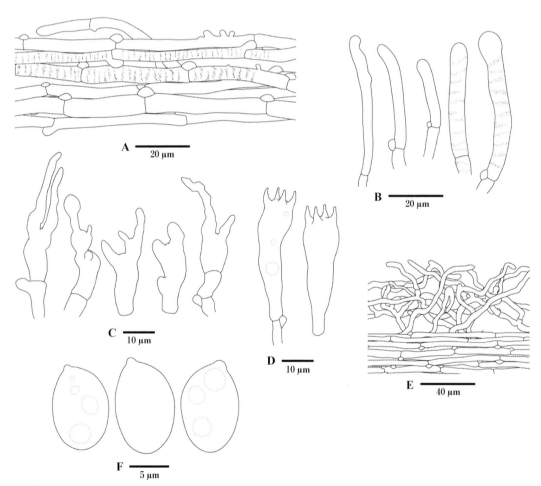

Fig. 110 - Micromorphological features of *Marasmiellus arenaceus* (holotype): **A**. Longitudinal cross section of the pileipellis. **B**. Terminal cells in the pileipellis. **C**. Cheilocystidia. **D**. Basidia. **E**. Longitudinal section of the stipitipellis. **F**. Basidiospores. Illustrations by Taneyama, Y.
　　シロスナホウライタケの顕鏡図（正基準標本）：**A**. 傘表皮組織の縦断面．**B**. 傘表皮組織の末端細胞．**C**. 縁シスチジア．**D**. 担子器．**E**. 柄表皮組織の縦断面．**F**. 担子胞子．図：種山裕一．

17. *Marasmiellus arenaceus* シロスナホウライタケ —— 145

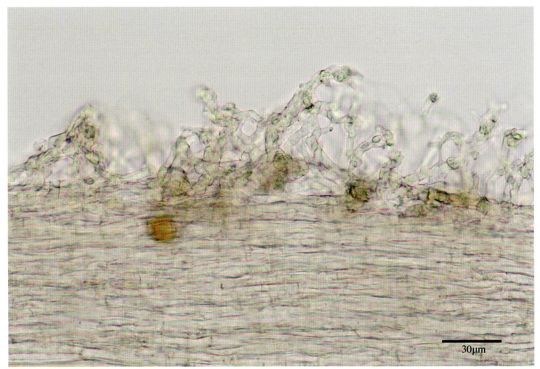

Fig. 111 – Longitudinal cross section of the stipitipellis of *Marasmiellus arenaceus* (in 3% KOH, holotype). Photo by Taneyama, Y.
シロスナホウライタケの柄表皮組織の縦断面（3%水酸化カリウム溶液で封入，正基準標本）．写真：種山裕一．

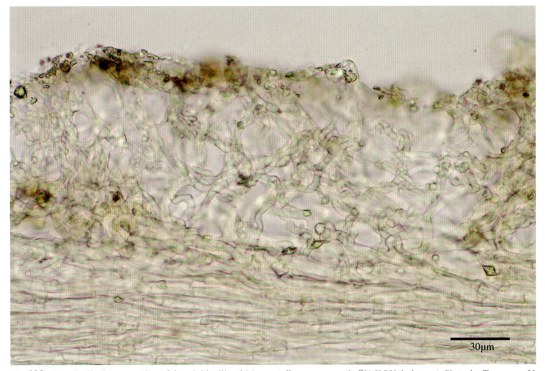

Fig. 112 – Longitudinal cross section of the stipitipellis of *Marasmiellus arenaceus* (in 3% KOH, holotype). Photo by Taneyama, Y.
シロスナホウライタケの柄表皮組織の縦断面（3%水酸化カリウム溶液で封入，正基準標本）．写真：種山裕一．

することを示唆している.

　海浜の草本植物の病原菌（Lucas et al. 1971；Warren 1972；Warren and Lucas 1973, 1975）として広く知られている北米から記載されたスナジホウライタケ Marasmiellus mesosporus Singer（Singer et al. 1973；Robich et al. 1991；Desjardin et al. 1992；Antonín and Noordeloos 1993, 2010；Takehashi et al. 2007）並びにチェコ共和国から報告された Marasmiellus carneopallidus（Pouzar）Singer（Pouzar 1966；Singer 1973；Antonín and Noordeloos 1993, 2010）は，形態学的にも生態学的にも際立った特徴をシロスナホウライタケと共有している．しかしながら M. mesosporus は，傘と柄が淡紅褐色を帯びること，楕円形～長楕円形のやや幅の狭い担子胞子：(7.5-) 9-11 (-11.8)×(4-) 5.5-6.5 (-7) μm（Singer et al. 1973）を持つこと，傘表皮組織の末端細胞は片膨れ状で均等に分布すること，そして瘤状突起にまばらに被われた柄シスチジアが存在する特徴においてシロスナホウライタケと区別される．Marasmiellus carneopallidus は，広こん棒形で平滑な縁シスチジアを有し，珊瑚状に分岐した末端細胞を形成するラメアレス構造が傘表皮組織に存在し，そして乾燥地に生育する枯れた草本植物の柄および根から発生すると言われている．

Fig. 113 – Basidiomata of *Marasmiellus arenaceus* (TNS-F-61370) on a root of beach grass (unidentified substrata), 13 Jun. 2014, Iriomote Island. Photo by Wada, Kimiko.
草本植物の根元から発生したシロスナホウライタケの子実体（TNS-F-61370），2014年6月13日，西表島．写真：和田貴美子．

17. *Marasmiellus arenaceus* シロスナホウライタケ —— 147

Fig. 114 – Basidiomata of *Marasmiellus arenaceus* (TNS-F-61369) on a root of beach grass (unidentified substrata), 13 Jun. 2014, Iriomote Island. Photo by Wada, K.
草本植物の根元から発生したシロスナホウライタケの子実体（TNS-F-61369），2014年6月13日，西表島．写真：和田貴美子．

17. *Marasmiellus arenaceus* シロスナホウライタケ

Fig. 115 – Basidiomata of *Marasmiellus arenaceus* (TNS-F-61369) on stem of beach grasses, 13 Jun. 2014, Iriomote Island. Photo by Wada, K.
草本植物の茎から発生したシロスナホウライタケの子実体（TNS-F-61369），2014年6月13日，西表島．写真：和田貴美子．

Fig. 116 – Immature basidiomata of *Marasmiellus arenaceus* (TNS-F-61369) on a stem of beach grass, 13 Jun. 2014, Iriomote Island. Photo by Wada, K.
草本植物の茎から発生したシロスナホウライタケの未熟な子実体（TNS-F-61369），2014年6月13日，西表島．写真：和田貴美子．

Fig. 117 – Basidiomata of *Marasmiellus arenaceus* (TNS-F-61371) on a root of beach grass, 16 Jun. 2014, Iriomote Island. Photo by Wada, K.
草本植物の根元から発生したシロスナホウライタケの子実体（TNS-F-61371），2014年6月16日，西表島．写真：和田貴美子．

17. *Marasmiellus arenaceus* シロスナホウライタケ

Fig. 118 – Immature basidioma of *Marasmiellus arenaceus* (TNS-F-61371) occurring on a stem of beach grass, 16 Jun. 2014, Iriomote Island. Photo by Wada, K.
草本植物の茎から発生したシロスナホウライタケの未熟な子実体（TNS-F-61371），2014年6月16日，西表島．写真：和田貴美子．

Fig. 119 – Immature basidioma of *Marasmiellus arenaceus* (TNS-F-61371) on a stem of beach grass, 16 Jun. 2014, Iriomote Island. Photo by Wada, K.
草本植物の茎から発生したシロスナホウライタケの未熟な子実体（TNS-F-61371）．2014年6月16日．西表島．写真：和田貴美子．

Fig. 120 – Basidiomata of *Marasmiellus arenaceus* (TNS-F-61372) on a root of *Carex kobomugi* Ohwi, 22 Jun. 2014, Suma-ku, Kobe-shi, Hyogo Pref. Photo by Wada, S.
コウボウムギの根元から発生したシロスナホウライタケの子実体（TNS-F-61372）．2014年6月22日．兵庫県神戸市須磨区．写真：和田匠平．

Fig. 121 – Basidiomata of *Marasmiellus arenaceus* (TNS-F-61372) on roots of *Carex kobomugi* Ohwi, 22 Jun. 2014, Suma-ku, Kobe-shi , Hyogo Pref. Photo by Wada, S.
コウボウムギの根元から発生したシロスナホウライタケの子実体 (TNS-F-61372). 2014年6月22日. 兵庫県神戸市須磨区. 写真：和田匠平.

17. *Marasmiellus arenaceus* シロスナホウライタケ

Fig. 122 – Basidiomata of *Marasmiellus arenaceus* (TNS-F-61372) on a root of *Carex kobomugi* Ohwi, 22 Jun. 2014, Suma-ku, Kobe-shi, Hyogo Pref. Photo by Wada, K.
コウボウムギの根元から発生したシロスナホウライタケの子実体（TNS-F-61372），2014年6月22日，兵庫県神戸市須磨区．写真：和田貴美子．

18. *Marasmiellus lucidus* Har. Takah., Taneyama & S. Kurogi, **sp. nov.** ヒメホタルタケ
MycoBank no.: MB 809927.

Etymology: The specific epithet comes from the Latin word for "lucid, shining", referring to the luminescence of the basidiomata.

The distinctive features of the present species are luminescent, small, whitish to light brown, pleurotoid basidiomata with or without a lateral, extremely short (up to 1.5 mm long) pseudostipe, a reduced, distant, shallowly lamellate to vein-like hymenophore, and the basidiome formation on dead twigs or branches of broad-leaved trees or dead leaves of *Abies firma* Sieb. & Zucc.; microscopically, inamyloid, ellipsoid to oblong-ellipsoid basidiospores, lack of cheilocystidia, distinctly encrusted pleurocystidia (pseudocystidia), and diverticulate terminal elements around the marginal region of pileus forming a discontinuous *Rameales*-structure.

Macromorphological characteristics (Figs. 125–129): Basidiomata small (4–8 mm in diameter), tough, sessile or with a reduced pseudostipe, lateral or sometimes dorsal, reniform to semiorbicular. Pileus at first convex with an incurved, entire margin, then broadly convex to almost applanate, smooth; surface dry, non-hygrophanous, subpruinose to silky-fibrillose toward the marginal region, glabrescent in age, when young and fresh brownish orange (7C4–7C5) to light brown (7D4–7D5) at the center, paler toward the margin, becoming entirely whitish with age. Flesh up to 1 mm thick, whitish, tough, odor not distinctive, taste undetermined. Pseudostipe 1–1.5 × 0.5–1 mm, almost lateral from the first, reduced to an extremely short expansion of tissue which merely attaches the pileus to the substrate; surface silky-fibrillose, whitish; basal mycelium none. Hymenophore reduced to shallow lamellae or almost vein-like ridges, adnate to subdecurrent, distant (3–9 reach the stipe), at times bifurcate, not intervenose, up to 1.3 mm broad, whitish; edges finely fimbriate under a lens, concolorous. Spore print pure white.

Luminescence: The whole basidiomata emit yellowish green light; mycelium luminescence unknown.

Micromorphological characteristics (Figs. 123, 124): Basidiospores (8.9–) 10.0–11.8 (–12.8) × (5.5–) 6.0–7.2 (–8.1) μm (n = 43, mean length = 10.90 ± 0.88, mean width = 6.57 ± 0.59, Q = (1.33–) 1.53–1.80(–1.94), mean Q = 1.66 ± 0.13), ellipsoid to oblong-ellipsoid, smooth, hyaline, colorless in KOH, inamyloid, thin-walled. Basidia (33.1–) 37.7–49.1 (–52.9) × (8.1–) 9.5–11.0 (–11.7) μm (n = 26, mean length = 43.38 ± 5.67, mean width = 10.28 ± 0.76) in the main body, (4.6–) 5.1–6.7 (–8.0) × (2.0–) 2.2–2.7 (–2.9) μm (n = 23, mean length = 5.90 ± 0.76, mean width = 2.45 ± 0.24) in the sterigmata, clavate, 4-spored; basidioles clavate. Cheilocystidia none. Pleurocystidia (32.9–) 37.0–46.3 (–50.1) × (7.6–) 8.0–9.6 (–10.7) μm (n = 11, mean length = 41.67 ± 4.66, mean width = 8.82 ± 0.82), subcylindrical to slenderly fusoid, covered with hyaline, granular crystals in the upper half, thin-walled. Hymenophoral trama subregular; element hyphae (2.3–) 2.7–3.6 (–4.9) μm wide (n = 46, mean width = 3.16 ± 0.46), cylindrical, smooth, with walls (0.3–) 0.4–0.6 μm (n = 16, mean thickness = 0.51 ± 0.08). Pileipellis a cutis with a poorly developed and discontinuous *Rameales*-structure of loosely interwoven, filiform hyphae (1.8–) 2.0–2.9 (–4.2) μm wide (n = 37, mean width = 2.43 ± 0.47), smooth or infrequently with scattered rod-like or knob-like outgrowths, densely packed with intermittently diverticulate terminal elements in the marginal regions, occasionally thinly spirally encrusted, thin-walled; diverticulate terminal elements at the marginal regions (24.5–) 25.9–45.3 (–48.1) × (2.4–) 2.5–3.9 (–4.5) μm (n = 6, mean length = 35.61 ± 9.69, mean width = 3.23 ± 0.71), with scattered rod-like or knob-like diverticula. Hyphae of pileitrama (1.9–) 2.6–3.8 (–4.4) μm wide (n = 37, mean width = 3.15 ± 0.60), monomitic, subparallel, cylindrical, not inflated, smooth, with slightly thickened walls (0.5–) 0.6–0.9 (–1.0) μm (n = 18, mean thickness = 0.73 ± 0.13). Cortical layer of pseudostipe a cutis of loosely interwoven hyphae

(1.8–) 2.3–3.1 (–3.4) μm wide (n = 30, mean width = 2.71 ± 0.40), cylindrical, smooth or occasionally diverticulate on the hymenophore-side of pseudostipe, hyaline in water, brownish or yellowish in KOH on the pileipellis-side of pseudostipe, with somewhat thickened walls (0.7–) 0.8–1.0 μm (n = 16, mean thickness = 0.88 ± 0.09). Pseudostipe trama composed of longitudinally running, cylindrical hyphae (2.1–) 2.9–3.9 (–4.3) μm wide (n = 37, mean width = 3.40 ± 0.54), monomitic, smooth, with thickened walls (0.8–) 0.9–1.2 (–1.5) μm (n = 26, mean thickness = 1.06 ± 0.17). All tissues colorless in water and KOH, except for the cortical layer of pseudostipe, inamyloid, with clamp connections.

Habitat and phenology: Gregarious or scattered, on dead twigs or branches of *Lonicera hypoglauca* Miq., *Quercus sessilifolia* Blume and *Lonicera affinis* Hooker & Arnott or dead leaves of *A. firma*, March to June.

Known distribution: Kyushu (Miyazaki).

Holotype: KPM-NS0017309, on dead branches of *L. hypoglauca*, Miyazaki-shi, Miyazaki Pref. Japan, 8 May 2006, coll. Kurogi, S.

Other specimens examined: KPM-NC0017310, on dead twigs of *L. hypoglauca*, Miyazaki-shi, Miyazaki Pref., Japan, 7 May 2006, coll. Kurogi, S.; KPM-NC0017311, same location, 6 May 2006, coll. Kurogi, S.; KPM-NC0017312, same location, 13 May 2006, coll. Kurogi, S.; KPM-NC0017313, same location, 13 May 2006, coll. Kurogi, S.; KPM-NC0017314, same location, 7 May 2007, coll. Kurogi, S.; KPM-NC0017315, same location, 29 Apr. 2007, coll. Kurogi, S.; KPM-NC0017316, same location, 27 Jun. 2009, coll. Kurogi, S.; KPM-NC0017317, on dead leaves of *A. firma*, Kobayashi-shi, Miyazaki pref., Japan, 30 Jun. 2009, coll. Kurogi, S.; KPM-NC0017318, on dead twigs of *L. affinis*, Kushima-shi, Miyazaki pref., Japan, 22 Mar. 2010, coll. Kurogi, S.

GenBank accession number: AB968237.

Japanese name: Hime-hotarutake (named by S. Kurogi).

Comments: The combination of the features including the small, pleurotoid basidiomata with a reduced, distant, shallowly lamellate to vein-like hymenophore, the inamyloid reaction in all tissues, the discontinuous *Rameales*-structure in the pileipellis, the lack of basal mycelium, and the presence of clamp connections suggests that *M. lucidus* is closely related to the genus *Marasmiellus*, section *Distantifolii* Singer as defined by Singer (Singer 1961, 1973, 1986). However, unambiguous taxonomic placement remains elusive owing to the poorly developed hymenophore and the distinctly encrusted pleurocystidia (pseudocystidia).

Within the genus, *Marasmiellus afer* Pegler from East Africa (Pegler 1968, 1977) is somewhat similar to *M. lucidus* that it has small, pleurotoid basidiomata with a reduced hymenophore of arcuate, distant, 5–7 lamellae. The former, however, does not correspond with the present species because of a pure white pileus, cheilocystidia with irregular, short diverticulae, and lack of pleurocystidia.

The pleurotoid basidiomata of the present species with a poorly developed hymenophore are also reminiscent of the reduced allies of *Marasmiellus*, such as the genus *Cymatella* (Patouillard 1899; Singer 1986), which can be distinguished from *M. lucidus* by having a developed stipe and completely lacking a hymenophore and hymenial cystidia.

References 引用文献

Patouillard NT. 1899. Champignons de la Guadeloupe. Bull Soc mycol Fr 15: 191–209.
Pegler DN. 1968. Studies on African Agaricales: I. Kew Bull 21 (3): 499–533.
Pegler DN. 1977. A preliminary agaric flora of East Africa. Kew Bull, Addit Ser 6: 1–615.

Singer R. 1961. Diagnoses fungorum novorum Agaricalium II. Sydowia 15 (1-6): 45-83.
Singer R. 1973. The genera *Marasmiellus*, *Crepidotus* and *Simocybe* in the Neotropics. Beih Nova Hedwigia 44: 1-517.
Singer R. 1986. The Agaricales in modern taxonomy, 4th edn. Koeltz, Koenigstein.

18. ヒメホタルタケ（新種；黒木秀一新称）*Marasmiellus lucidus* Har. Takah., Taneyama & S. Kurogi, sp. nov.

肉眼的特徴（Figs. 125-129）：子実体は小形（4-8 mm）で, 強靱, 無柄または退化した偽柄を持ち, 側生もしくは時に背着生, 半球形〜腎臓形. 傘は最初丸山形で縁部が内側に巻き, のち饅頭形〜ほぼ平らに開き, ほぼ平坦, 縁部は全縁；表面は粘性を欠き, 非吸水性, 中央部はほぼ平滑, 周縁部に向かって粉状〜繊維状を呈し, 老成すると全体に平滑, 中央部は淡褐色, 周縁部に向かって褪色し, しばしば全体に類白色を呈する. 肉は厚さ1 mm以下, 類白色, 強靱, 特別な匂いはなく, 味は不明. 偽柄は1-1.5×0.5-1 mm（傘の子実層側）, 最初からほぼ側生し, 極めて短いため傘の表面側では柄がほとんど退化して傘が直接基質に付着して見える；表面は粘性を欠き, 粉状〜繊維状を呈し, 類白色；根本に発達した菌糸体はない. 子実層托は浅いヒダ状〜脈状隆起に退化し, 直生〜やや垂生, 疎（柄に到達するヒダは3-9）, 連絡脈を欠き, 幅1.3 mm以下, 類白色；縁部は同色, 長縁毛状. 胞子紋は純白色.

発光性：子実体全体が黄緑色に発光；菌糸体の発光性は未確認.

顕微鏡的特徴（Figs. 123, 124）：担子胞子は (8.9-) 10.0-11.8 (-12.8) × (5.5-) 6.0-7.2 (-8.1) μm（n = 43, mean length = 10.90 ± 0.88, mean width = 6.57 ± 0.59, Q = (1.33-) 1.53-1.80 (-1.94), mean Q = 1.66 ± 0.13）, 楕円形〜長楕円形, 平坦, 無色, 非アミロイド, アルカリ溶液において無色, 薄壁. 担子器は (33.1-) 37.7-49.1 (-52.9) × (8.1-) 9.5-11.0 (-11.7) μm（本体）, (4.6-) 5.1-6.7 (-8.0) × (2.0-) 2.2-2.7 (-2.9) μm（ステリグマ）, こん棒形, 4胞子性；偽担子器はこん棒形. 縁シスチジアはない. 側シスチジアは (32.9-) 37.0-46.3 (-50.1) × (7.6-) 8.0-9.6 (-10.7) μm, 類円柱形〜幅の狭い紡錘形, 上部は無色の結晶物に被われ, 薄壁. 子実層托実質を構成する菌糸細胞は幅 (2.3-) 2.7-3.6 (-4.9) μm, ほぼ平列し, 円柱形, 平滑, 壁は厚さ (0.3-) 0.4-0.6 μm. 傘の表皮組織は匍匐性の菌糸からなり, 傘の周縁部を中心に断続的なラメアレス構造が存在する；菌糸は径 (1.8-) 2.0-2.9 (-4.2) μm, 円柱形, 平滑または稀に瘤状〜短指状突起に被われ, 時に凝着物が薄く付着し, 薄壁；傘周縁部の末端細胞は (24.5-) 25.9-45.3 (-48.1) × (2.4-) 2.5-3.9 (-4.5) μm, 部分的に複数の瘤状〜短指状突起に被われ, 断続的なラメアレス構造を形成する. 傘実質の菌糸は幅 (1.9-) 2.6-3.8 (-4.4) μm, 円柱形, 平滑, やや厚い壁 (0.5-) 0.6-0.9 (-1.0) μmを形成する. 偽柄の表皮組織を構成する菌糸は緩く並列した平行菌糸被をなし, 幅 (1.8-) 2.3-3.1 (-3.4) μm, 円柱形, 平滑または子実層托側の偽柄表皮組織においてしばしば不規則に分岐する樹枝状, 短指状またはこぶ状の突起に被われ, 無色, 傘側の偽柄表皮組織はアルカリ溶液において淡褐色〜淡黄色を呈し, やや厚壁 (0.8-1.0 μm)；柄シスチジアはない. 偽柄実質の菌糸は幅 (2.1-) 2.9-3.9 (-4.3) μm, 平列し, 円柱形, 平滑, 隔壁を有し, 壁は厚さ (0.8-) 0.9-1.2 (-1.5) μm. 全ての組織はクランプを有し, 非アミロイド, 偽柄の表皮組織を除き水封並びにアルカリ溶液において無色.

生態および発生時期：キダチニンドウ, ツクバネガシ, ハマニンドウの枯れ枝上またはモミの枯れ葉上に群生または散生, 3月〜6月.

分布：九州（宮崎）.

供試標本：KPM-NC0017309（正基準標本）, キダチニンドウの枯れ枝上, 宮崎県宮崎市, 2006年5月8日, 黒木秀一採集；KPM-NC0017310, 同上, 2006年5月7日, 黒木秀一採集；KPM-NC0017311,

同上，2006年5月6日，黒木秀一採集；KPM-NC0017312，同上，2006年5月13日，黒木秀一採集；KPM-NC0017313，同上，2006年5月13日，黒木秀一採集；KPM-NC0017314，同上，2007年5月7日，黒木秀一採集；KPM-NC0017315，同上，2007年4月29日，黒木秀一採集；KPM-NC0017316，同上，2009年6月27日，黒木秀一採集；KPM-NC0017317，モミの枯れ葉上，宮崎県小林市，2009年6月30日，黒木秀一採集；KPM-NC0017318，ハマニンドウの枯れ枝上，串間市，2010年3月22日，黒木秀一採集．

GenBank 登録番号：AB968237．

主な特徴：子実体は発光性を持つ小形（4-8 mm）のヒラタケ型で，淡褐色〜類白色，無柄または退化した微小な偽柄を持ち，側生もしくは時に背着生；子実層托は浅いヒダ状〜脈状に退化する；担子胞子は楕円形〜長楕円形，非アミロイド；縁シスチジアを欠く；側シスチジアは上部に結晶物が付着する偽シスチジア形；傘周縁部の末端細胞は部分的に複数の瘤状〜短指状突起に被われた断続的なラメアレス構造を形成する；広葉樹の枯れ枝上またはモミの枯れ葉上に発生．

コメント：根元に綿毛状菌糸体を欠き，ヒラタケ型で非アミロイドの子実体，脈状に退化した子実層托，不連続なラメアレス構造を形成する傘表皮組織，そしてクランプの存在はツキヨタケ科 Omphalotaceae に分類されるシロホウライタケ属 *Marasmiellus*，ディスタンティフォリイ節

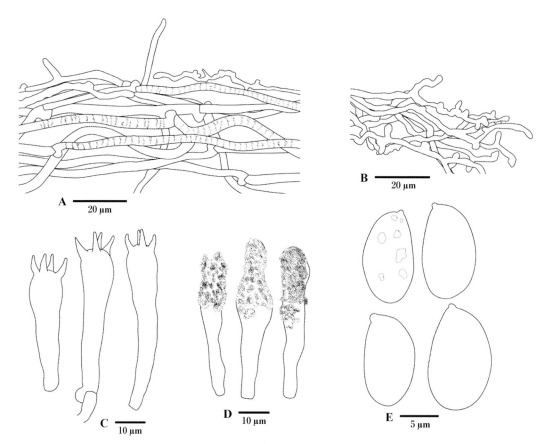

Fig. 123 – Micromorphological features of *Marasmiellus lucidus* (holotype): **A**. Longitudinal cross section of the pileipellis (from the center of pileus). **B**. Intermittently diverticulate terminal elements in the marginal region of pileipellis. **C**. Basidia. **D**. Pleurocystidia, with the upper half covered with hyaline crystals. **E**. Basidiospores. Illustrations by Taneyama, Y.
ヒメホタルタケの顕微鏡図（正基準標本）：**A**．傘表皮組織の縦断面（傘中央部）．**B**．部分的に短指状突起に被われた傘周縁部の表皮組織．**C**．担子器．**D**．無色の結晶物に被覆された側シスチジア．**E**．担子胞子．図：種山裕一．

section *Distantifolii*（Singer 1961, 1973, 1986）との類縁関係を示唆している．しかしながら，退化した子実層托および結晶物に被われる側シスチジア（偽シスチジア）の特徴はシロホウライタケ属として異質であり，分類学的位置については疑問が残る．

　属内において本種は退化した子実層托と小形のヒラタケ型子実体を形成する東アフリカ産 *Marasmiellus afer* Pegler（Pegler 1968, 1977）に類似するが，後者は傘が純白色を呈すること，不規則な短指状突起に被われた縁シスチジアを持つこと，そして側シスチジアを欠く性質において本種と異なる．

　またヒメホタルタケの発達の悪い子実層托を形成するヒラタケ型の子実体はシロホウライタケ属に近縁とされる退化系の *Cymatella* 属（Patouillard 1899；Singer 1986）を想起させるが，*Cymatella* 属は発達した柄を有し，子実層托並びにシスチジアを欠くとされている．

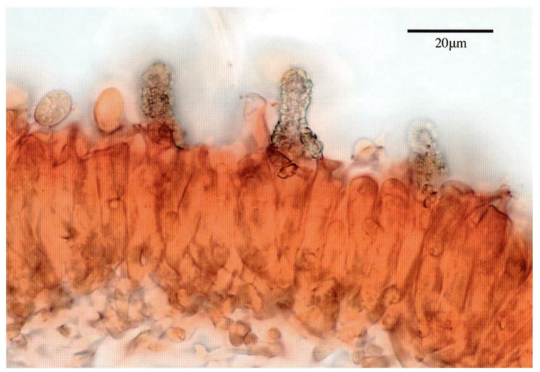

Fig. 124 – Hymenium of *Marasmiellus lucidus* (in 3%KOH and Congo red stain, holotype), showing pleurocystidia with the upper half covered with hyaline crystals. Photo by Taneyama, Y.
ヒメホタルタケの子実層（3％水酸化カリウム溶液で封入した後コンゴー赤染色，正基準標本）．無色の結晶物に被覆された側シスチジアを示す．写真：種山裕一．

Fig. 125 – Basidiomata of *Marasmiellus lucidus* (holotype), on a dead branch of *L. hypoglauca*, 8 May 2006, Miyazaki-shi, Miyazaki Pref. Photo by Takahashi, H.
キダチニンドウの枯れ枝上に発生したヒメホタルタケ（正基準標本），2006年5月8日，宮崎県宮崎市．写真：高橋春樹．

18. *Marasmiellus lucidus* ヒメホタルタケ —— 161

Fig. 126 – Basidiomata of *Marasmiellus lucidus* (holotype) on a dead branch of *L. hypoglauca*, 8 May 2006, Miyazaki-shi, Miyazaki Pref. Photo by Takahashi, H.
キダチニンドウの枯れ枝上に発生したヒメホタルタケ（正基準標本），2006年5月8日，宮崎県宮崎市．写真：高橋春樹．

Fig. 127 – Immature basidiomata of *Marasmiellus lucidus* (KPM-NC0017311) on a dead branch of *L. hypoglauca*, 6 May 2006, Miyazaki-shi, Miyazaki Pref. Photo by Takahashi, H.
キダチニンドウの枯れ枝上に発生したヒメホタルタケの幼菌（KPM-NC0017311），2006年5月6日，宮崎県宮崎市．写真：高橋春樹．

18. *Marasmiellus lucidus* ヒメホタルタケ —— 163

Fig. 128 - Immature basidiomata of *Marasmiellus lucidus* (KPM-NC0017311) on a dead branch of *L. hypoglauca*, 6 May 2006, Miyazaki-shi, Miyazaki Pref. Photo by Takahashi, H.
キダチニンドウの枯れ枝上に発生したヒメホタルタケの幼菌（KPM-NC0017311），2006年5月6日，宮崎県宮崎市．写真：高橋春樹．

164 —— 18. *Marasmiellus lucidus* ヒメホタルタケ

Fig. 129 – Basidiomata of *Marasmiellus lucidus* (KPM-NC0017316) emitting yellowish green light on a dead branch of *L. hypoglauca*, 27 Jun. 2009, Miyazaki-shi, Miyazaki Pref. Photo by Takahashi, H.
キダチニンドウの枯れ枝上において黄緑色に発光するヒメホタルタケ（KPM-NC0017316），2009年6月27日，宮崎県宮崎市．写真：高橋春樹．

19. *Marasmiellus venosus* Har. Takah., Taneyama & A. Hadano, **sp. nov.** ヒメヒカリタケ
MycoBank no.: MB809928.

Etymology: The specific epithet comes from the Latin word for "having veins", referring to the conspicuously veined lamellae.

This species is characterized by a light brown, pruinose to minutely pubescent, semiorbicular to reniform pileus; sessile or with an almost lateral, reduced, extremely short (up to 2 mm long) pseudostipe; a well-developed, white basal mycelium attached to an extensive mycelial mat on the substrate; distinctly transvenose lamellae; inamyloid, ellipsoid to ovoid-ellipsoid basidiospores; subcylindrical to subclavate cheilocystidia with irregularly branched, apical diverticulae; a distinct *Rameales*-structure in the pileipellis; luminescence in the basidiomata and mycelia; and basidiome formation on dead logs.

Macromorphological characteristics (Figs. 133-140): Basidiomata small, pleurotoid, sessile or with a reduced pseudostipe. Pileus 2-5 mm in diameter, semiorbicular to reniform, at first convex with an incurved, entire margin, soon applanate and depressed at the center, sulcate-striate toward the margin; surface dry, non-hygrophanous, pruinose to minutely pubescent overall, sometimes glabrescent in age, when young brownish orange (6C3-6C4) to light brown (6D4-6D6), becoming paler or almost whitish with age. Flesh up to 0.5 mm thick, whitish, tough; odor not distinctive, taste unknown. Pseudostipe 1-1.5×0.5-1.5 mm, lateral from the first, reduced to an extremely short expansion of tissue which merely attaches the pileus to the substrate; surface entirely pruinose to pubescent, whitish or paler concolorous with the pileus; base covered with whitish or brownish orange (6C3-6C4), strigose mycelial hairs or tomentum attached to an extensive mycelial mat on the substratum. Lamellae adnate to subdecurrent, subdistant (8-13 reach the stipe), with 1-3 series of lamellulae, up to 1 mm broad, distinctly transvenose, whitish; edges finely fimbriate under a lens, concolorous. Spore print pure white.

Luminescence: The entire basidiomata and the mycelia on the substratum emit yellowish green light.

Micromorphological characteristics (Figs. 130-132): Basidiospores (5.9-) 7.0-8.0 (-8.7)×(2.4-) 4.3-5.1 (-5.7) μm (n = 120, mean length = 7.52±0.53, mean width = 4.68±0.43, Q = (1.35-) 1.43-1.81 (-3.18), mean Q = 1.62±0.19), ellipsoid to ovoid-ellipsoid, smooth, hyaline, colorless in KOH, inamyloid, thin-walled. Basidia (25.3-) 28.7-34.9 (-38.6)×(6.7-) 6.9-7.6 (-8.0) μm (n = 28, mean length = 31.80±3.06, mean width = 7.28±0.34) in the main body, (4.0-) 4.3-6.1 (-8.8)×(1.6-) 1.7-2.0 (-2.2) μm (n = 27, mean length = 5.19±0.92, mean width = 1.85±0.14) in the sterigmata, clavate, four-spored; basidioles clavate. Cheilocystidia (13.1-) 15.4-19.9 (-22.6)×(3.5-) 4.2-5.6 (-6.2) μm (n = 29, mean length = 17.65±2.29, mean width = 4.87±0.71), gregarious, subcylindrical to subclavate, with irregularly branched, apical diverticulae (3.5-) 5.6-13.2 (-20.0)×(1.5-) 1.9-2.7 (-3.5) μm, colorless, thin-walled. Pleurocystidia none. Hymenophoral trama subregular, monomitic; element hyphae (1.9-) 2.6-3.4 (-4.1) μm wide (n = 42, mean width = 3.03±0.42), cylindrical, smooth, hyaline, with walls 0.5-0.6 μm (n = 13, mean thickness = 0.56±0.05); subhymenium weakly dextrinoid. Pileipellis a cutis with a well-developed *Rameales*-structure; constituent hyphae (1.7-) 2.0-2.5 (-2.9) μm wide (n = 43, mean width = 2.24±0.26), parallel, cylindrical, with abundant warty or finger-like protuberances, colorless, thin-walled. Hyphae of pileitrama (1.8-) 2.5-3.4 (-4.0) μm wide (n = 44, mean width = 2.96±0.45), subparallel, cylindrical, not inflated, smooth, colorless, with walls 0.5-0.7 (-0.8) μm (n = 25, mean thickness = 0.63±0.09). Hyphae of pseudostipe trama similar to those of pileitrama, (2.4-) 2.9-3.9 (-4.6) μm wide (n = 50, mean width = 3.42±0.52), with walls (0.5-) 0.6-0.8 (-1.0) μm (n = 28, mean thickness = 0.70±0.12). Elements of basal mycelium (1.7-) 1.8-2.2 (-2.4) μm wide (n = 15, mean width = 2.01±0.22), filiform, not branched, smooth, hyaline, thin-walled. All tissues colorless in KOH, inamyloid (except weakly dextrinoid subhymenium), with abundant clamp connections.

Habitat and phenology: Gregarious or scattered, lignicolous on dead fallen logs (unidentified substrata), May to June.

Known distribution: Kyushu (Oita).

Holotype: TNS-F-52281, on dead logs in broad-leaved forests, Mt. Takahaze, hasama, Yufu-shi, Oita Pref., 21 Jun. 2012, coll. Hadano, E.

Other specimens examined: TNS-F-61373, on dead logs in a broad-leaved forest, Mt. Takahaze, Hasama, Yufu-shi, Oita Pref., 29 May 2012, coll. Hadano, E.; TNS-F-61374, same location, 23 Jun. 2012, coll. Hadano, E.

Gene sequenced specimen and GenBank accession number: TNS-F-52281, AB968236.

Japanese name: Hime-hikaritake (named by A. Hadano).

Comments: The combination of features, including the pleurotoid habit, the pale colored pileus, the inamyloid reaction in all tissues, the distinct *Rameales*-structure in the pileipellis, and the presence of clamp connections, suggests placement in the genus *Marasmiellus* Murrill, section *Marasmiellus*, subsection *Inodermini* Singer, as defined by Singer (Singer 1973, 1986).

Marasmiellus purpureoalbus (Petch) Singer, originally described from Sri Lanka (Petch 1947; Singer 1961; Pegler 1977, 1986), most closely approximates the present species in appearance. The former differs from *M. venosus* in having a pale purple, glabrous pileus, not transvenose lamellae, an insititious stipe, significantly longer basidiospores: 10-14 μm in length (Petch 1947), and cheilocystidia with or without unbranched, short apical outgrowths.

Marasmiellus goossensiae (Beeli) Pegler, originally described from East Africa (Beeli 1928; Petch 1947; Pegler 1977, 1986), macromorphologically resembles *M. venosus* in having small, pleurotoid basidiomata with a reduced, extremely short stipe. However, the former species clearly differs from the present species in having dark purplish-brown basidiomata with lamellae that are not transvenose, an insititious stipe, subfusoid, markedly narrower basidiospores: 2.7-3.7 μm wide (Pegler 1986), a poorly developed *Rameales*-structure, and cheilocystidia with fewer branched, short apical diverticulae.

References 引用文献

Beeli M. 1928. Contribution a l'étude de la flore mycologique du Congo. VI Fungi Goossensiani. Agaricacées rhodosporées. Bull Soc R Bot Belg 61 (1): 78-103.

Pegler DN. 1977. A preliminary agaric flora of East Africa. Kew Bull, Addit Ser 6: 1-615.

Pegler DN. 1983. Agaric flora of the Lesser Antilles. Kew Bull, Addit Ser 9: 1-668.

Pegler DN. 1986. Agaric flora of Sri Lanka. Kew Bull, Addit Ser 12: 1-519.

Petch T. 1947. A revision of Ceylon Marasmii. Trans Brit Mycol Soc 31 (1): 19-44.

Singer R. 1961. Diagnoses fungorum novorum Agaricalium II. Sydowia 15 (1-6): 45-83.

Singer R. 1973. The genera *Marasmiellus*, *Crepidotus* and *Simocybe* in the Neotropics. Beih Nova Hedwigia 44: 1-517.

Singer R. 1986. The Agaricales in modern taxonomy, 4th edn. Koeltz, Koenigstein.

19. ヒメヒカリタケ（新種；波多野敦子新称）*Marasmiellus venosus* Har. Takah., Taneyama & A. Hadano, **sp. nov.**

肉眼的特徴（Figs. 133-140）：子実体は小形のヒラタケ型で，半球形〜腎臓形の傘と退化した側生の偽柄（pseudostipe）からなる．傘は径2-5 mm，最初丸山形で縁部が内側に巻き，まもなく平らに開いて中央部がへこみ，周縁部に向かって放射状の溝線を表す；縁部は全縁；表面は粘性を

19. *Marasmiellus venosus* ヒメヒカリタケ —— 167

欠き，非吸水性，粉状〜微毛を帯び，最初淡褐色，のち次第に褪色し，老成するとしばしば類白色を呈する．肉は厚さ0.5 mm以下，類白色，強靱，特別な匂いはなく，味は不明．偽柄は1-1.5×0.5-1.5 mm（傘の子実層側），側生，極めて短いため傘の表面側では柄がほとんど退化して傘が直接基質に付着して見える；表面は粘性を欠き，粉状〜微毛を帯び，傘より淡色または類白色；根本は発達した剛毛状〜綿毛状の白色菌糸体に被われ，基質上の広範囲に菌糸マットを形成する．ヒダは直生〜やや垂生，やや疎（柄に到達するヒダは 8-13)，1-3の小ヒダを伴い，ヒダの表面に著しい側脈を表し，幅1 mm以下，類白色；縁部は同色，長縁毛状．胞子紋は純白色．

発光性：子実体全体および基質上の菌糸体が黄緑色に発光．

顕微鏡的特徴（Figs. 130-132）：担子胞子は (5.9-) 7.0-8.0 (-8.7) × (2.4-) 4.3-5.1 (-5.7) μm (n = 120, mean length = 7.52 ± 0.53, mean width = 4.68 ± 0.43, Q = (1.35-) 1.43-1.81 (-3.18), mean Q = 1.62 ± 0.19)，楕円形〜卵状楕円形，平坦，無色，非アミロイド，アルカリ溶液において無色，薄壁．担子器は (25.3-) 28.7-34.9 (-38.6) × (6.7-) 6.9-7.6 (-8.0) μm (本体)，(4.0-) 4.3-6.1 (-8.8) × (1.6-) 1.7-2.0 (-2.2) μm (ステリグマ)，こん棒形，4胞子性；偽担子器はこん棒形．縁シスチジアは (13.1-) 15.4-19.9 (-22.6) × (3.5-) 4.2-5.6 (-6.2) μm，群生し，類円柱形〜類こん棒形，不規則に分岐した多数の短指状分岐物を具え，無色，薄壁；短指状分岐物は (3.5-) 5.6-13.2 (-20.0) × (1.5-) 1.9-2.7 (-3.5) μm．側シスチジアを欠く．子実層托実質を構成する菌糸は幅 (1.9-) 2.6-3.4 (-4.1) μm，ほぼ平列し，円柱形，平滑，無色，壁は厚さ0.5-0.6 μm；子実下層は弱偽アミロイド．傘表皮組織は匍匐性の菌糸からなり，明瞭なラメアレス構造が発達する；菌糸は幅 (1.7-) 2.0-2.5 (-2.9) μm，円柱形，多数の瘤状〜短指状突起に被われ，平滑，無色，薄壁．傘実質の菌糸は幅 (1.8-) 2.5-3.4 (-4.0) μm，円柱形，平滑，無色，壁は厚さ 0.5-0.7 (-0.8) μm．偽柄実質を構成する菌糸は傘実質の菌糸に類似し，幅 (2.4-) 2.9-3.9 (-4.6) μm，壁は厚さ (0.5-) 0.6-0.8 (-1.0) μm．根元の菌糸体を構成する菌糸は幅 (1.7-) 1.8-2.2 (-2.4) μm，糸状，無分岐，平滑，無色，薄壁．全ての組織は多数のクランプを有し，非アミロイド（弱偽アミロイドの子実下層を除く），アルカリ溶液において無色．

生態および発生時期：広葉樹林内腐朽材上から群生または散生，5月〜6月．

分布：九州（大分）．

供試標本：TNS-F-52281（正基準標本），広葉樹林内腐朽材上，大分県湯布市狭間，2012年6月21日，波多野英治採集；TNS-F-61373，同上，2012年5月29日，波多野英治採集；TNS-F-61374，同上，2012年6月23日，波多野英治採集．

分子解析に用いた標本並びに GenBank 登録番号：TNS-F-52281, AB968236.

主な特徴：子実体はヒラタケ型で無柄または退化した側生の偽柄を持つ；傘は半球形〜腎臓形，淡褐色〜類白色，粉状〜微毛状；根元に白色綿毛状菌糸体が存在し，しばしば基質を取り巻くように発達する；ヒダは明瞭な側脈を有する；担子胞子は楕円形〜卵状楕円形，非アミロイド；縁シスチジアは類円柱形〜類こん棒形で，頂部に不規則に分岐する多数の樹枝状または短指状の突起を具える；傘の表皮組織は発達したラメアレス構造からなる；子実体および基質上の菌糸体に発光性がある；広葉樹林内腐朽材上から発生．

コメント：淡褐色〜類白色の傘を持つヒラタケ型の子実体，非アミロイドの全組織，傘表皮組織の発達したラメアレス構造，そしてクランプの存在は本種がツキヨタケ科 Omphalotaceae に分類されるシロホウライタケ属 *Marasmiellus* Murrill，ムラサキヤマンバ節 section *Marasmiellus*，ムラサキヤマンバ亜節 subsection *Inodermini* Singer（Singer 1973, 1986）に位置することを示唆している．

　本種の外観的特徴はスリランカから報告された *Marasmiellus purpureoalbus* (Petch) Singer（Petch 1947；Singer 1961；Pegler 1977, 1986）に極めて類似している．しかしながら *M.*

purpureoalbus は傘が淡紫色で平滑であること，ヒダが側脈を欠くこと，柄の根元に発達した菌糸体を持たないこと，担子胞子は長楕円形でより長いこと：長さ10-14 µm (Petch 1947)，そして縁シスチジアは平滑もしくは頂部に無分岐の短指状〜瘤状突起を具える特徴において明らかな有意差が認められる．

東アフリカから記載された *Marasmiellus goossensiae* (Beeli) Pegler (Beeli 1928；Petch 1947；Pegler 1977, 1986) は柄が著しく短く，小形のヒラタケ型子実体を形成する性質においてヒメヒカリタケと共通するが，前者は子実体が暗紫褐色であること，ヒダが側脈を欠くこと，根元に発達した菌糸体を持たないこと，担子胞子は類紡錘形で幅がより狭い：2.7-3.7 µm (Pegler 1986) こと，ラメアレス構造が不明瞭であること，そして頂部に分岐の少ない短指状〜瘤状突起を具えた縁シスチジアを持つ性質において本種とは異質である．

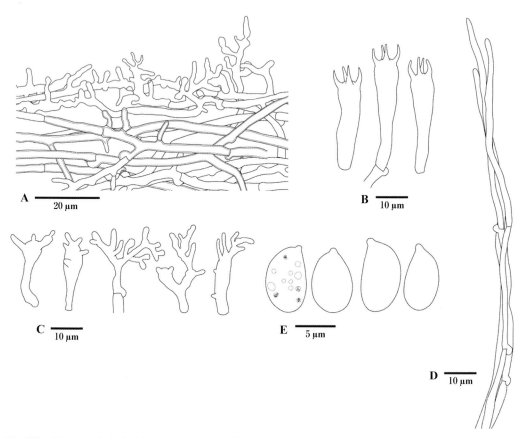

Fig. 130 – Micromorphological features of *Marasmiellus venosus* (holotype): **A**. Longitudinal cross section of the pileipellis showing the well-developed *Rameales*-structure. **B**. Basidia. **C**. Cheilocystidia. **D**. Elements of the basal mycelium. **E**. Basidiospores. Illustrations by Taneyama, Y.
ヒメヒカリタケの顕鏡図（正基準標本）：**A**. 傘表皮組織の縦断面，多数の樹枝状分岐物を持つ菌糸（ラメアレス構造）を示す．**B**. 担子器．**C**. 縁シスチジア．**D**. 根元の菌糸体を構成する菌糸．**E**. 担子胞子．図：種山裕一．

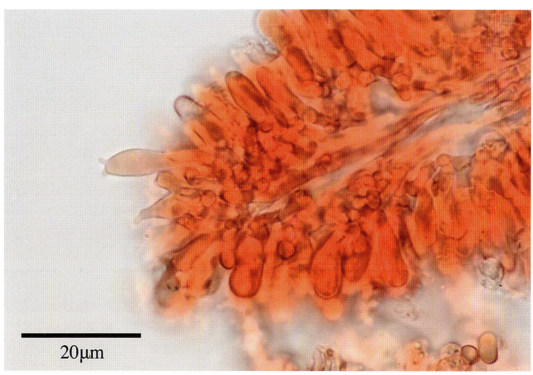

Fig. 131 – Lamella edge of *Marasmiellus venosus* (in 3%KOH and Congo red stain, holotype). Photo by Taneyama, Y.
ヒメヒカリタケのヒダ縁部（3%水酸化カリウム溶液で封入した後コンゴー赤染色，正基準標本）．写真：種山裕一．

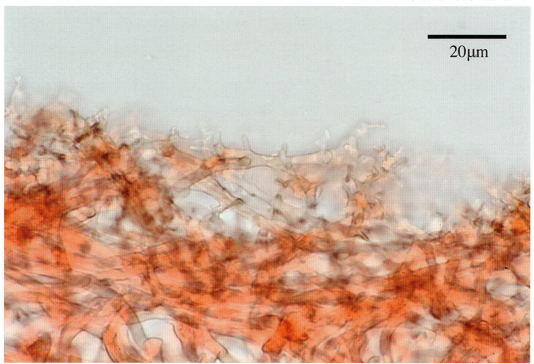

Fig. 132 – Pileipellis of *Marasmiellus venosus* (in 3%KOH and Congo red stain, holotype) showing the *Rameales*-structure. Photo by Taneyama, Y.
ヒメヒカリタケの傘表皮組織（3%水酸化カリウム溶液で封入した後コンゴー赤染色，正基準標本），ラメアレス構造を示す．写真：種山裕一．

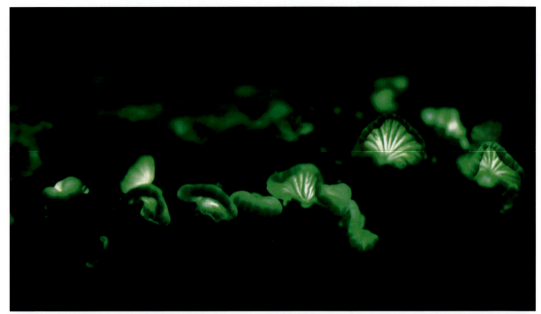

Fig. 133 – Basidiomata and mycelia of *Marasmiellus venosus* (holotype) emitting yellowish green light on a dead log, 21 Jun. 2012, Yufu-shi, Oita Pref. Photo by Takahashi, H.
腐朽材上において黄緑色に発光するヒメヒカリタケの子実体と菌糸体（正基準標本），2012年6月21日，大分県湯布市．写真：高橋春樹．

Fig. 134 – Basidiomata and mycelia of *Marasmiellus venosus* (holotype) emitting yellowish green light on a dead log, 21 Jun. 2012, Yufu-shi, Oita Pref. Photo by Takahashi, H.
腐朽材上において黄緑色に発光するヒメヒカリタケの子実体と菌糸体（正基準標本），2012年6月21日，大分県湯布市．写真：高橋春樹．

19. *Marasmiellus venosus* ヒメヒカリタケ —— 171

Fig. 135 – Basidiomata and mycelia of *Marasmiellus venosus* (holotype) on a dead log, 21 Jun. 2012, Yufu-shi, Oita Pref. Photo by Takahashi, H.
腐朽材上に発生したヒメヒカリタケの子実体と菌糸体（正基準標本），2012年6月21日，大分県湯布市．写真：高橋春樹．

Fig. 136 – Basidiomata of *Marasmiellus venosus* (holotype) on a dead log, 21 Jun. 2012, Yufu-shi, Oita Pref. Photo by Takahashi, H.
腐朽材上に発生したヒメヒカリタケの子実体（正基準標本），2012年6月21日，大分県湯布市．写真：高橋春樹．

Fig. 137 – Basidiomata and mycelia of *Marasmiellus venosus* (holotype) on a dead log, 21 Jun. 2012, Yufu-shi, Oita Pref. Photo by Takahashi, H.
腐朽材上に発生したヒメヒカリタケの子実体と菌糸体（正基準標本），2012年6月21日，大分県湯布市．写真：高橋春樹．

Fig. 138 – Basidiomata of *Marasmiellus venosus* (holotype) on a dead log, 21 Jun. 2012, Yufu-shi, Oita Pref. Photo by Takahashi, H.
腐朽材上に発生したヒメヒカリタケの子実体（正基準標本），2012年6月21日，大分県湯布市．写真：高橋春樹．

19. *Marasmiellus venosus* ヒメヒカリタケ ── 173

Fig. 139 – Basidiomata of *Marasmiellus venosus* (holotype) on a dead log, 21 Jun. 2012, Yufu-shi, Oita Pref. Photo by Takahashi, H.
腐朽材上に発生したヒメヒカリタケの子実体（正基準標本），2012年6月21日，大分県湯布市．写真：高橋春樹．

Fig. 140 – Basidiomata of *Marasmiellus venosus* (holotype) on a dead log, 21 Jun. 2012, Yufu-shi, Oita Pref. Photo by Takahashi, H.
腐朽材上に発生したヒメヒカリタケの子実体（正基準標本），2012年6月21日，大分県湯布市．写真：高橋春樹．

20. *Micropsalliota cornuta* Har. Takah. & Taneyama, **sp. nov.** ダイダイツノハラタケ

MycoBank no.: MB 809930.

Etymology: The specific epithet comes from the Latin word for "horned", referring to the erect to recurved, pointed scales which cover the pileus and stipe surfaces.

Distinctive features of this species are found in medium-sized, agaricoid basidiomata entirely shrouded by reddish orange, dry, floccose, erect to recurved, pointed scales; a fugacious, thin, floccose, reddish orange annulus; free, close, whitish then blackish lamellae; dark brown, dextrinoid and metachromatic basidiospores with thickened walls; broadly clavate to subcylindrical cheilocystidia; a dextrinoid stipe trama; and a terrestrial habitat.

Macromorphological characteristics (Figs. 144-148): Pileus 25-60 mm in diameter, at first hemispherical, then expanding to convex to broadly convex; margin involute when young, prominently floccose-appendiculate with velar remnants; surface covered with reddish orange (7A8), dry, floccose, erect to recurved, pointed scales that are more or less concentrically arranged when mature, pale orange in ground color. Flesh up to 3 mm thick in the center of pileus, whitish, soft, odor and taste indistinct. Stipe 40-60 ×2.5-5.5 mm, cylindrical, somewhat tapering toward the base, central, terete, fistulose; surface densely covered below the annulus with reddish orange (7A8), floccose, erect to recurved, pointed scales, pale orange at the apex, with a whitish tomentose basal mycelium; annulus fragile, fugacious thin, floccose, attached toward the stipe apex, reddish orange (7A8). Lamellae free, close, 70-90 reach the stipe, with 0-1 series of lamellulae, up to 1-3 mm broad, at first white, then becoming blackish blue (19F7-8) at maturity; edges even, concolorous. Spore print blackish blue (19F7-8).

Micromorphological characteristics (Figs. 141-143): Basidiospores (4.8-) 5.4-6.0 (-6.4)×(3.1-) 3.4-3.8 (-4.0) μm (n = 110, mean length = 5.70±0.34, mean width = 3.60±0.17, Q = (1.36-) 1.49-1.68 (-1.95), mean Q = 1.59±0.09), ovoid-ellipsoid to ellipsoid or obscurely phaseoliform, inequilateral with a shallow adaxial depression or applanation in profile, smooth, brown (7D6-7E6), distinctly dextrinoid, metachromatic in crecyl blue, somewhat thick-walled (0.5-0.6 (-0.8) μm), without a germ pore. Basidia (10.5-) 11.6-15.2 (-17.3)×(4.7-) 4.9-6.0 (-6.9) μm (n = 18, mean length = 13.43±1.80, mean width = 5.45±0.53) in main body, (1.2-) 1.3-1.8 (-2.1)×(0.8-) 0.8-1.0 (-1.1) μm (n = 12, mean length = 1.54±0.22, mean width = 0.88±0.12) in sterigmata, clavate, four-spored. Basidioles clavate to broadly clavate. Cheilocystidia (16.7-) 20.5-26.4 (-31.0)×(5.9-) 6.9-9.4 (-11.9) μm (n = 39, mean length = 23.47±2.97, mean width = 8.17±1.24), forming a compact sterile edge, projecting from the hymenium, broadly clavate to subcylindrical, smooth, hyaline, inamyloid, thin-walled. Pleurocystidia none. Hymenophoral trama composed of regularly arranged, cylindrical hyphae (3.3-) 3.8-9.9 (-15.6) μm wide (n = 31, mean width = 6.85±3.09), smooth, colorless, inamyloid, thin-walled; subhymenium consisting of subisodiametric elements (6.2-) 7.2-10.9 (-16.5)×(4.1-) 4.9-7.1 (-8.5) μm (n = 33, mean length = 9.06±1.88, mean width = 6.02±1.08). Pileipellis thickly coated with a velar layer made up of loosely interwoven, repent, cylindrical hyphae (3.0-) 3.9-5.5 (-7.9) μm wide (n = 68, mean width = 4.68±0.80), often branched, thinly encrusted with fine, brownish red (8C7-8) granules, which are decolorized under dry conditions, greenish yellow (1A6-7) in ammonium hydroxide, inamyloid, not gelatinized, thin-walled; terminal elements (17.1-) 24.3-41.2 (-46.0)×(4.3-) 4.7-6.1 (-6.8) μm (n = 28, mean length = 32.74±8.41, mean width = 5.42±0.73), cylindrical, undifferentiated. Pileitrama composed of subparallel to loosely intricate, cylindrical hyphae (3.7-) 5.6-9.4 (-11.8) μm wide (n = 51, mean width = 7.50±1.85), smooth, colorless, inamyloid, thin-walled. Stipitipellis similar to the pileipellis. Trama of stipe composed of longitudinally running, cylindrical hyphae (6.4-) 7.5-10.8 (-12.1) μm wide (n = 38, mean width = 9.18 ± 1.64), smooth, colorless, dextrinoid, thin-walled. Clamp connections absent in all tissues.

Habitat and phenology: Solitary or scattered on ground in *Castanopsis-Quercus* forests, late May to October, not common.

Known distribution: Okinawa (Ishigaki Island, Iriomote Island).

Holotype: TNS-F-52279, on ground in a *Castanopsis-Quercus* forest, Mt. Banna, Ishigaki-shi, Okinawa Pref., 16 Oct. 2011, coll. Takahashi, H.

Gene sequenced specimen and GenBank accession number: TNS-F-52279, AB968240 (ITS).

Japanese name: Daidai-tuno-haratake (named by H. Takahashi & Y. Taneyama).

Comments: On the basis of the metachromatic and dextrinoid, brown basidiospores without a germ pore and the pileipellis hyphae that are encrusted with brown pigment that turns greenish yellow in ammonium hydroxide, we recognize the new species as belonging to the genus *Micropsalliota* Höhn. (Höhnel 1914; Pegler and Rayner 1969; Heinemann 1977; Singer 1986; Zhao et al. 2010).

Apart from the dextrinoid basidiospores by which the genus *Micropsalliota* is characterized, *M. cornuta* is closely similar to *Agaricus crocopeplus* Berk. & Broome originally described from Sri Lanka (Berkeley and Broome 1871; Heinemann 1956; Pegler 1986) and *Agaricus trisulphuratus* Berk. from East Africa (Berkeley 1885; Singer 1947; Heinemann 1956; Pegler 1977). The latter two taxa mainly differ from *M. cornuta* in forming pyriform to inflated clavate, much broader cheilocystidia: (19-) 28-40 × (8-) 12-18 μm in *A. trisulphuratus* (Heinemann 1956); 16-30 × 10-15 μm in *A. crocopeplus* (Pegler 1986).

References 引用文献

Berkeley RE. 1885. Notices of fungi collected in Zanzibar. Ann Mag nat Hist 15: 384-387.

Berkeley MJ. Broome CE. 1871. The fungi of Ceylon. (Hymenomycetes, from *Agaricus* to *Cantharellus*). J Linn Soc, Bot 11: 494-567.

Heinemann P. 1956. *Agaricus* I. Flore Iconographique des Champignons du Congo 5: 99-119.

Heinemann P. 1977. The Genus *Micropsalliota*. Kew Bull 31 (3) 581-583.

Höhnel F. 1914. Fragmente zur Mykologie XVI (XVI. Mitteilung, Nr. 813 bis 875). Sber Akad Wiss Wien, Math-naturw Kl, Abt I 123: 49-155.

Pegler DN. 1977. A preliminary agaric flora of East Africa. Kew Bull, Addit Ser 6: 1-615.

Pegler DN. 1986. Agaric flora of Sri Lanka. Kew Bull, Addit Ser 12: 1-519.

Pegler DN, Rayner RW. 1969. A contribution to the agaric flora of Kenya. Kew Bull 23 (3): 347-412.

Zhao R, Desjardin DE, Soytong K, Perry BA, Hyde KD. 2010. A monograph of *Micropsalliota* in Northern Thailand based on morphological and molecular data. Fung Diver 45: 33-79.

Singer R. 1947. New genera of fungi. III. Mycologia 39 (1): 77-89.

Singer R. 1986. The Agaricales in modern taxonomy, 4th edn. Koeltz, Koenigstein.

20. ダイダイツノハラタケ（新種；高橋春樹 & 種山裕一新称） *Micropsalliota cornuta* Har. Takah. & Taneyama, **sp. nov.**

肉眼的特徴（Figs. 144-148）：傘は径25-60 mm，最初半球形，のち饅頭形，縁部は最初内側に巻き，顕著な綿毛状縁片膜が付着する；表面は乾性，淡橙色の地に帯赤橙色の直立した先の鋭い綿毛状鱗片に被われ，傘が開くと中央部は鱗片が密集し濃色を呈する．肉は傘の中央部において3 mm 以下，類白色，特別な味や匂いはない．柄は40-60×2.5-5.5 mm，円柱形，下方に向かってやや細くなり，中心生，中空；表面はツバより下において傘と同様の帯赤橙色のやや反り返った先の鋭い綿毛状鱗片に被われ，頂部は淡橙色；ツバは柄の上部に付着し，脆く，消失しやすく，薄い膜質，帯赤橙色．根本の菌糸体は白色．ヒダは離生，密，柄に到達するヒダは70-90，小ヒダは0-1，幅1-3 mm，最初白色，のち紫黒色；縁部は全縁，同色．胞子紋は紫黒色．

20. *Micropsalliota cornuta* ダイダイツノハラタケ

顕微鏡的特徴（Figs. 141-143）：担子胞子は (4.8-) 5.4-6.0 (-6.4) × (3.1-) 3.4-3.8 (-4.0) μm (n = 110, mean length = 5.70 ± 0.34, mean width = 3.60 ± 0.17, Q = (1.36-) 1.49-1.68 (-1.95), mean Q = 1.59 ± 0.09), 卵状楕円形〜楕円形またはややそら豆形, しばしば片側の下部になだらかな凹みを形成し, 平坦, 褐色, 偽アミロイド, クレシルブルーにより異性染色され, やや厚壁 (0.5-0.6 (-0.8) μm), 発芽孔を欠く. 担子器は (10.5-) 11.6-15.2 (-17.3) × (4.7-) 4.9-6.0 (-6.9) μm (本体), (1.2-) 1.3-1.8 (-2.1) × (0.8-) 0.8-1.0 (-1.1) μm (ステリグマ), こん棒形, 4胞子性. 偽担子器はこん棒形〜広こん棒形. 縁シスチジアは (16.7-) 20.5-26.4 (-31.0) × (5.9-) 6.9-9.4 (-11.9) μm, 子実層から突出して群生し, こん棒形〜亜円柱形, 平滑, 無色, 薄壁. 側シスチジアはない. 子実層托実質の菌糸は幅 (3.3-) 3.8-9.9 (-15.6) μm, 円柱形, 平列し, 平滑, 無色, 非アミロイド, 薄壁；子実下層を構成する菌糸細胞は短形で (6.2-) 7.2-10.9 (-16.5) × (4.1-) 4.9-7.1 (-8.5) μm. 傘表皮は緩く錯綜した匍匐性の菌糸からなる厚い被膜の層に被われる；菌糸は幅 (3.0-) 3.9-5.5 (-7.9) μm, 円柱形, しばしば分岐し, 平滑, 帯褐赤色の微細な結晶物 (乾燥時脱色する) が薄く凝着し, 色素はアルカリ (アンモニア) 溶液において溶解し帯緑黄色に染まり, 非アミロイド, 薄壁；末端細胞は (17.1-) 24.3-41.2 (-46.0) × (4.3-) 4.7-6.1 (-6.8) μm, 円柱形. 傘実質の菌糸は幅 (3.7-) 5.6-9.4 (-11.8) μm, 円柱形, 緩く錯綜〜やや並列し, 平滑, 無色, 非アミロイド, 薄壁. 柄表皮は傘表皮と同様. 柄実質の菌糸は幅 (6.4-) 7.5-10.8 (-12.1) μm, 円柱形, 平列し, 平滑, 無色, 偽アミロイド, 薄壁. 全ての組織において菌糸はクランプを欠く.

生態および発生時期：スダジイ, オキナワウラジロガシを主体とする常緑広葉樹林内地上に孤生または散生, 5月下旬〜10月.

分布：沖縄 (石垣島, 西表島).

供試標本：TNS-F-52279 (正基準標本), スダジイ, オキナワウラジロガシを主体とする常緑広葉樹林内地上, 沖縄県石垣市バンナ岳, 2011年10月16日, 高橋春樹採集.

分子解析に用いた標本並びに GenBank 登録番号：TNS-F-52279, AB968240 (ITS).

主な特徴：子実体はハラタケ型で, 全体に帯赤橙色の直立した先の鋭い綿毛状鱗片に被われ, 消失性綿毛状のツバを持つ；ヒダは離生, 密, 黒色；担子胞子は暗褐色で, 偽アミロイド, クレシルブルーにより異性染色され, 厚壁；縁シスチジアは広こん棒形〜亜円柱形；柄実質の菌糸は偽アミロイド；地上生.

コメント：発芽孔を欠き, 偽アミロイドに染まり且つクレシルブルーにより異性染色される褐色の担子胞子およびアンモニア溶液により帯緑黄色に変色する傘表皮組織の性質から判断して, 本種はネッタイハラタケ属 (高橋春樹 & 種山裕一新称) *Micropsalliota* Höhn. (Höhnel 1914；Pegler and Rayner 1969；Heinemann 1977；Singer 1986；Zhao et al. 2010) に位置すると考えられる.

ネッタイハラタケ属はジャワ産 *Micropsalliota pseudovolvulata* Höhn. を基準種として1914年Höhnelによって設立された. 肉眼的な外観はハラタケ属に酷似するが, 以下のような相違点を有する：子実体の類型は一般に小形で華奢なクヌギタケ型〜ウラベニガサ型またはキツネノカラカサ型 (ハラタケ属の子実体は一般に中〜大形で肉質)；胞子紋は暗褐色 (ハラタケ属の胞子紋は紫褐色)；担子胞子は偽アミロイド, 発芽孔を欠く (ハラタケ属の担子胞子は偽アミロイドに染まらず, しばしば発芽孔を持つ)；縁シスチジアは通常発達した頂部頭状形 (ハラタケ属の縁シスチジアは未分化もしくは広こん棒形〜球嚢状)；傘表皮組織の凝着型色素は通常アンモニア溶液で緑変する (ハラタケ属の傘表皮組織は通常アンモニア溶液で緑変しない). 地上生または腐植上に発生. 東南アジアおよびアフリカの熱帯〜亜熱帯地域を中心に分布し, 世界に68種 (Index Fungorum) 知られている.

ネッタイハラタケ属に特徴的な偽アミロイドに染まる担子胞子を別にすれば, スリランカ産 *Agaricus crocopeplus* Berk. & Broome (Berkeley and Broome 1871；Heinemann 1956；

Pegler 1986) 並びに東アフリカ産 *Agaricus trisulphuratus* Berk. (Berkeley 1885；Singer 1947；Heinemann 1956；Pegler 1977) はダイダイツノハラタケに酷似するが, 両種ともより広幅：(19-)28-40×(8-)12-18 µm in *A. trisulphuratus* (Heinemann 1956)；16-30×10-15 µm in *A. crocopeplus* (Pegler 1986) で膨大した洋梨形〜広こん棒形の縁シスチジアを有する点で有意差が認められる.

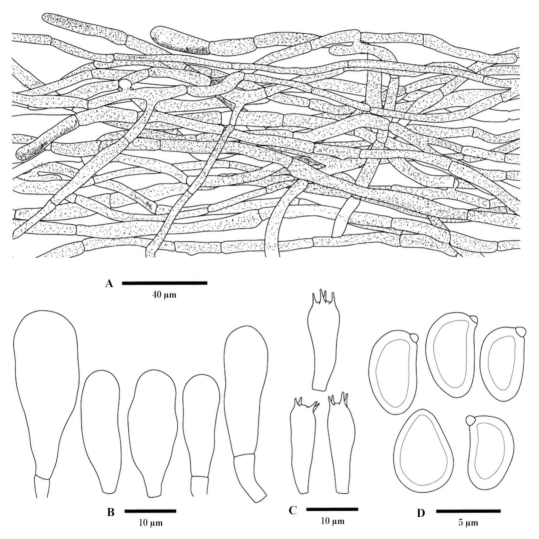

Fig. 141 – Micromorphological features of *Micropsalliota cornuta*. (holotype): **A.** Longitudinal cross section of the pileipellis. **B.** Cheilocystidia. **C.** Basidia. **D.** Basidiospores. Illustrations by Taneyama, Y.
ダイダイツノハラタケの顕鏡図（正基準標本）：**A**. 傘表皮組織の縦断面. **B**. 縁シスチジア. **C**. 担子器. **D**. 担子胞子. 図：種山裕一.

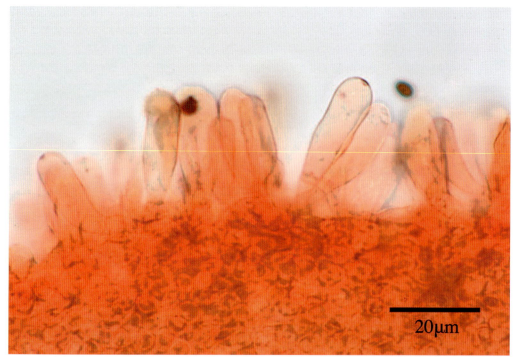

Fig. 142 – Cheilocystidia of *Micropsalliota cornuta* (in 3%KOH and Congo red stain, holotype). Photo by Taneyama, Y.
ダイダイツノハラタケの縁シスチジア（3％水酸化カリウム溶液で封入した後コンゴー赤染色，正基準標本）．写真：種山裕一．

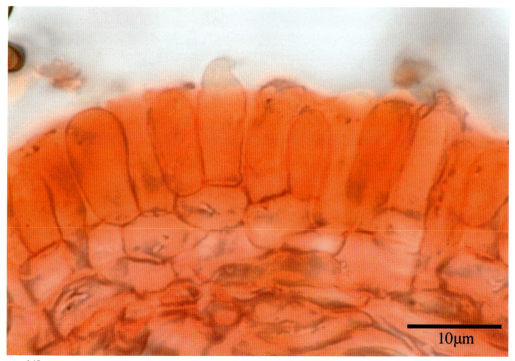

Fig. 143 – Hymenium of the lamella face in *Micropsalliota cornuta* (in 3%KOH and Congo red stain, holotype). Photo by Taneyama, Y.
ダイダイツノハラタケのヒダ側面の子実層（3％水酸化カリウム溶液で封入した後コンゴー赤染色，正基準標本）．写真：種山裕一．

20. *Micropsalliota cornuta* ダイダイツノハラタケ —— 179

Fig. 144 – Basidiomata of *Micropsalliota cornuta* (holotype) on ground in a *Castanopsis-Quercus* forest, 16 Oct. 2011, Mt. Banna, Ishigaki Island. Photo by Takahashi, H.
スダジイ-オキナワウラジロガシ林内地上から発生したダイダイツノハラタケの子実体（正基準標本），2011年10月16日，石垣島バンナ岳．写真：高橋春樹．

Fig. 145 – Immature basidioma of *Micropsalliota cornuta* (holotype), 16 Oct. 2011, Mt. Banna, Ishigaki Island. Photo by Takahashi, H.
ダイダイツノハラタケの幼菌（正基準標本）．2011年10月16日．石垣島バンナ岳．写真：高橋春樹．

20. *Micropsalliota cornuta* ダイダイツノハラタケ —— 181

Fig. 146 – Basidiomata of *Micropsalliota cornuta* (holotype), 16 Oct. 2011, Mt. Banna, Ishigaki Island. Photo by Takahashi, H.
ダイダイツノハラタケの子実体（正基準標本），2011年10月16日，石垣島バンナ岳．写真：高橋春樹．

Fig. 147 – Underside view of the mature basidioma of *Micropsalliota cornuta* (holotype), 16 Oct. 2011, Mt. Banna, Ishigaki Island. Photo by Takahashi, H.
ダイダイツノハラタケの成菌のヒダ（正基準標本），2011年10月16日，石垣島バンナ岳．写真：高橋春樹．

Fig. 148 – Pileus surface of *Micropsalliota cornuta* (holotype), 16 Oct. 2011, Mt. Banna, Ishigaki Island. Photo by Takahashi, H.
ダイダイツノハラタケの傘表面（正基準標本），2011年10月16日，石垣島バンナ岳．写真：高橋春樹．

21. *Mycena comata* Har. Takah. & Taneyama, sp. nov. キジムナハナガサ

MycoBank no.: MB 809931.

Etymology: The specific epithet means "shaggy" or "with hair tufts", referring to the hair-like flocci (dermatocysts) enveloping the entire basidioma.

Distinctive features of this species include pluteoid basidiomata with yellowish orange flocci on the pileus and stipe; delicate, plicate-striate pilei; colorless, weakly amyloid basidiospores; dimorphic cheilocystidia with or without long, hair-like, filiform appendages; thick-walled, dextrinoid dermatocysts tapering into long, hair-like, filiform appendages apically; and a lignicolous habit.

Macromorphological characteristics (Figs. 151-158): Primordium 0.5-1.5 mm in diameter, hemispherical, with a discoid stipe, rarely acutely umbonate, enveloped in orange (5A6-5A7) to yellowish orange (4A6-4A7) flocci. Pileus 10-20 (-30) mm in diameter, at first hemispherical, then subfusiform, expanding to campanulate or obtusely conical, finally plane but often narrowly subumbonate, radially plicate-striate almost to the disk, often split in places along the lines of the lamellae; surface dry, non-hygrophanous, pale yellow (4A3) under yellowish orange (4A6-4A7) flocci, yellowish orange (4A6-4A7) to orange (5A6-5A7) furfuraceous over the center, paler toward the orange (5A6-5A7), crenulate-pruinose margin. Flesh very thin (up to 1 mm in the center of the pileus), yellowish white (4A2) to whitish, soft; odor and taste indistinct. Stipe 15-35 × 1-2 mm, 1-1.5 mm above, almost equal or slightly enlarged at the base, central, slender, terete or somewhat compressed, brittle, hollow; surface dry, non-hygrophanous, covered with yellowish orange (4A6-4A7) flocci as on the pileus, yellowish orange (4A6-4A7) to orange (5A6-5A7) furfuraceous toward the dark orange (5A8) base, yellowish white (4A2) to white pruinose at the apex, without a basal mycelium and disk. Lamellae free, subcrowded (35-42 reach the stipe), with 0-1 series of lamellulae, up to 2.5 mm broad, thin, white; edges finely fimbriate under a lens, pale orange near the margin of the pileus. Spore print pure white.

Micromorphological characteristics (Figs. 149, 150): Basidiospores (4.6-) 5.2-6.1 (-7.1) × (2.9-) 3.1-3.4 (-3.7) μm (n = 111, mean length = 5.66 ± 0.42, mean width = 3.29 ± 0.15, Q = (1.44-) 1.59-1.86 (-2.18), mean Q = 1.72 ± 0.13), ovoid-ellipsoid to ellipsoid, smooth, colorless, bluish grey in Melzer's reagent, thin-walled, without germ pore. Basidia (12.0-) 12.5-16.2 (-20.8) × (5.3-) 6.7-8.0 (-9.3) μm (n = 31, mean length = 14.35 ± 1.84, mean width = 7.35 ± 0.61) in main body, (2.3-) 2.7-3.8 (-4.5) × (1.0-) 1.1-1.4 (-1.5) μm (n = 32, mean length = 3.23 ± 0.55, mean width = 1.26 ± 0.14) in sterigmata, clavate, four-spored. Cheilocystidia forming a compact sterile edge, smooth, unbranched, colorless or pale yellow, inamyloid, with slightly thickened walls 0.5-0.7 μm (n = 20, mean thickness = 0.58 ± 0.07), dimorphic: 1) pyriform or ventricose to lageniform cheilocystidia (11.6-) 13.5-17.9 (-21.1) × (4.2-) 5.2-7.4 (-8.8) μm (n = 55, mean length = 15.73 ± 2.21, mean width = 6.34 ± 1.11) in main cell bodies, with filiform, flexuous, simple appendages (13.3-) 29.5-52.2 (-73.2) × (1.3-) 1.8-2.9 (-4.4) μm (n = 50, mean length = 40.86 ± 11.38, mean width = 2.36 ± 0.52) tapering to an obtuse or subacute apex; 2) broadly clavate cheilocystidia (11.2-) 12.3-20.2 (-24.8) × (5.8-) 6.6-8.5 (-9.1) μm (n = 15, mean length = 16.25 ± 3.91, mean width = 7.53 ± 0.93). Pleurocystidia none. Trama of lamellae and pileus composed of long-celled inflated or subcylindrical hyphal cells (423.0-) 550.7-985.6 (-1223.2) × (16.3-) 25.5-51.1 (-62.7) μm (n = 19, mean length = 768.15 ± 217.42, mean width = 38.34 ± 12.80), with the ends attenuating into narrow septa, smooth, colorless, inamyloid, thin-walled, without uninflated and short-celled hyphae. Surface of pileus a poorly developed narrow cutis of not gelatinized, parallel, repent, cylindrical hyphal cells (18.6-) 22.7-58.6 (-93.6) × (4.7-) 6.0-12.1 (-18.1) μm (n = 56, mean length = 40.64 ± 17.93, mean width = 9.05 ± 3.10), often with indistinctly clamped septa, smooth, pale yellow, inamyloid or partially dextrinoid, thin-walled. Dermatocysts of pileus numerous, dispersed, decumbent, arising directly from the cutis hyphae, pale yellow, smooth, unbranched, dextrinoid or inamyloid, with thickened walls (0.7-) 1.0-1.7 (-2.1) μm (n = 36, mean thickness = 1.36 ± 0.33); main cell bodies (10.7-) 16.5-31.1 (-40.2) × (5.2-) 6.8-12.6 (-15.4) μm (n = 35, mean length = 23.80 ± 7.34, mean width = 9.71 ± 2.89) in simply clavate to pyriform type, (26.0-) 53.8-132.3 (-157.6) × (3.8-) 3.8-8.9 (-14.6) μm (n = 17, mean length = 93.03 ± 39.24, mean width = 6.36 ± 2.54) in irregularly cylindrical to strangulated type, occasionally with a secondary septum; appendages (61.7-) 87.0-159.5 (-199.5) × (1.9-) 2.7-4.0 (-4.8) μm (n = 41, mean length = 123.21 ± 36.24, mean width = 3.35 ± 0.65), long, flexuous, filiform, simple, tapering to an obtuse or subacute apex. Surface of stipe thinly coated with not gelatinized parallel, repent, cylindrical hyphae (3.5-) 4.7-9.8 (-13.4) μm wide (n = 40, mean width = 7.27 ± 2.54), smooth, pale yellow, inamyloid, thin-walled. Dermatocysts of stipe similar to those of the pileus but inamyloid. Trama of stipe composed of longitudinally running, long, inflated or subcylindrical hyphae (7.6-) 10.6-26.3 (-45.8) μm wide (n = 34, mean width = 18.44 ± 7.85), smooth, colorless, weakly dextrinoid, with thickened walls (1.5-) 1.6-2.2 (-2.6) μm (n = 14, mean thickness = 1.87 ± 0.30). Clamp connections absent except for the pileus surface.

Habitat and phenology: Solitary or scattered on dead fallen twigs of broad-leaved trees, from May to October, not common.

Known distribution: Okinawa (Ishigaki Island).

Holotype: TNS-F-61375, on dead fallen twigs of broad-leaved trees, Banna Park, Ishigaki-shi, Okinawa Pref., Japan, 20 Oct. 2011, coll. Takahashi, H.

Other specimens examined: TNS-F-61376, on dead fallen twigs of broad-leaved trees, Banna Park, Ishigaki-shi, Okinawa Pref., 25 May 2002, coll. Takahashi, H; KPM-NC0023869 (Isotype), same location, 20 Oct. 2011, coll. Takahashi, H.

Japanese name: Kijimuna-hanagasa (named by H. Takahashi & Y. Taneyama).

Comment: In addition to the ephemeral expanded basidiomata which are typified by the genus *Coprinus*, the pluteoid habit and the hair-like, dextrinoid, thick-walled dermatocysts reminiscent of the genus *Crinipellis* are quite alien to the genus *Mycena*. Considering the combination of features, such as the plicate-striate pilei, the insititious stipe lacking a basal disc, the white, free lamellae, the amyloid basidiospores, the non-diverticulate cystidia with a long, filiform appendage, and the lignicolous habit, we provisionally relegate the new species to the genus *Mycena*, section *Radiatae* in the sense of Singer (Singer 1986).

"*Trogia*" *crinipelliformis* Corner from Malaysia (Corner 1996) and *Mycena auricoma* Har. Takah. from central Honshu, Japan (Takahashi 1999), are considered to be closest relatives. Both taxa, however, differ markedly from *M. comata* in forming invariably inamyloid dermatocysts and appendages of cheilocystidia reaching as much as 150 μm long. The Malaysian species produces inamyloid basidiospores and cheilocystidia with occasionally branched appendages.

References 引用文献

Corner EJH. 1996. The agaric genera *Marasmius, Chaetocalathus, Crinipellis, Heimiomyces, Resupinatus, Xerula* and *Xerulina* in Malesia. Beih Nova Hedwigia 111: 1-175.
Singer R. 1986. The Agaricales in modern taxonomy, 4th edn. Koeltz, Koenigstein.
Takahashi H. 1999. *Mycena auricoma*, a new species of *Mycena* section *Radiatae* from Japan, and *Mycena spinosissima*, a new record in Japan. Mycoscience 40 (1): 73-80.

21. キジムナハナガサ（新種；高橋春樹 & 種山裕一新称）*Mycena comata* Har. Takah. & Taneyama, **sp. nov.**

和名の語源：「キジムナ」は沖縄県を代表する樹木の精霊キジムナー（キジムン）に由来する．

肉眼的特徴 (Figs. 151-158)：原基は径0.5-1.5 mm，半球形の傘と盤状の柄からなり，稀に鋭く尖った中丘を持ち，全体に橙黄色の毛被に被われる．傘は径10-20 (-30) mm，最初半球形，のち類紡錘形〜釣り鐘形になり，最後は平開し，しばしば中丘を具え，放射状に走る明瞭な扇状の溝線を表し，ヒダの線に沿ってしばしば裂け目を生じる；表面は粘性および吸水性を欠き，淡黄色の地に橙黄色の毛被に被われ，中央部は橙黄色〜橙色の細鱗片状をなし，周辺部に向かって淡色を呈する；縁部は橙色，細円鋸歯状〜微粉状．肉は非常に薄く（傘の中央部において1 mm以下），淡黄色〜類白色，特別な味や匂いはない．柄は 15-35×1-2 mm，頂部は1-1.5 mm，ほぼ上下同大または基部に向かってやや拡大し，中心生，やせ型，時にやや変圧され，脆く，中空；表面は乾性，非吸水性，傘と同様に橙黄色の毛被に被われ，下半部は橙黄色の細鱗片状をなし，頂部は淡黄色〜白色粉状，根本に発達した菌糸体は見られない．ヒダは離生，やや疎（柄に到達するヒダは35-42），幅2.5 mm 以下，白色；縁部は長縁毛状，傘の縁部周辺で淡橙色に縁取られる．胞子紋は白色．

顕微鏡的特徴 (Figs. 149, 150)：担子胞子は (4.6-) 5.2-6.1 (-7.1) × (2.9-) 3.1-3.4 (-3.7) µm (n = 111, mean length = 5.66 ± 0.42, mean width = 3.29 ± 0.15, Q = (1.44-) 1.59-1.86 (-2.18), mean Q = 1.72 ± 0.13), 卵状楕円形〜楕円形，平坦，無色，弱アミロイド（淡青灰色），薄壁，発芽孔を欠く．担子器は (12.0-) 12.5-16.2 (-20.8) × (5.3-) 6.7-8.0 (-9.3) µm（本体），(2.3-) 2.7-3.8 (-4.5) × (1.0-) 1.1-1.4 (-1.5) µm（ステリグマ），こん棒形，4胞子性．縁シスチジアはヒダ縁部において不稔帯を形成し，群生，平滑，無分岐，無色または淡黄色，非アミロイド，やや厚壁 (0.5-0.7 µm)，2形性：1) 洋梨形または片脹れ状〜フラスコ形，(11.6-) 13.5-17.9 (-21.1) × (4.2-) 5.2-7.4 (-8.8) µm，付属糸は (13.3-) 29.5-52.2 (-73.2) × (1.3-) 1.8-2.9 (-4.4) µm，糸状で曲がりくねり，頂部は鈍頭またはやや鋭端；2) 広こん棒形，(11.2-) 12.3-20.2 (-24.8) × (5.8-) 6.6-8.5 (-9.1) µm．側シスチジアはない．ヒダおよび傘実質の菌糸細胞は (423.0-) 550.7-985.6 (-1223.2) × (16.3-) 25.5-51.1 (-62.7) µm，膨大し，非常に長く，亜円柱形，隔壁の周囲で狭くなり，平滑，無色，非アミロイド，薄壁，短形の細い糸状菌糸は存在しない．傘の表皮は平行菌糸被をなし，非ゼラチン質；菌糸は (18.6-) 22.7-58.6 (-93.6) × (4.7-) 6.0-12.1 (-18.1) µm，円柱形，しばしば不明瞭なクランプが見られ，平滑，淡黄色，非アミロイドまたは部分的に偽アミロイド，薄壁．傘の皮嚢体は多生し，傾伏性，傘の表皮を形成する平行菌糸被に直接つながり，淡黄色，平滑，無分岐，偽アミロイドまたは非アミロイド，壁は厚さ (0.7-) 1.0-1.7 (-2.1) µm；皮嚢体の本体は (10.7-) 16.5-31.1 (-40.2) × (5.2-) 6.8-12.6 (-15.4) µm（こん棒形〜洋梨形），(26.0-) 53.8-132.3 (-157.6) × (3.8-) 3.8-8.9 (-14.6) µm（不規則な円柱形でしばしば不規則なくびれがある），時に二次隔壁を有する；付属糸は (61.7-) 87.0-159.5 (-199.5) × (1.9-) 2.7-4.0 (-4.8) µm，長く，曲がりくねり，糸状，頂部に向かって細くなり，鈍頭または鋭端．柄表皮組織の菌糸は幅 (3.5-) 4.7-9.8 (-13.4) µm，円柱形，匍匐性，非ゼラチン質，平滑，淡黄色，非アミロイド，薄壁．柄の皮嚢体は非アミロイドである以外傘と共通する．柄の実質は縦に沿って平列し，菌糸は幅 (7.6-) 10.6-26.3 (-45.8) µm，紡錘形，平滑，無色，弱偽アミロイド，壁は厚さ (1.5-) 1.6-2.2 (-2.6) µm．傘表皮組織を除きクランプを欠く．

生態および発生時期：広葉樹の落枝上に孤生または散生，5月〜10月，やや稀．

分布：沖縄（石垣島）

供試標本：TNS-F-61375（正基準標本），広葉樹の落枝上，沖縄県石垣市バンナ公園，2011年10月20日，高橋春樹採集；TNS-F-61376，同上，2002年5月25日，高橋春樹採集；KPM-NC0023869 (Isotype)，

21. *Mycena comata* キジムナハナガサ —— 187

同上，2011年10月20日，高橋春樹採集.

主な特徴：子実体はウラベニガサ型で橙黄色の毛被に被われ，華奢で脆い；傘は扇のヒダ状の明瞭な条線を表す；ヒダは白色，離生；担子胞子は無色，弱アミロイド；縁シスチジアは2形性，毛状の付属糸を持つかまたは欠く；表皮シスチジアは厚壁，偽アミロイド，毛状付属糸を持つ；広葉樹の枯れ枝上に発生.

コメント：傘が平開すると一晩で萎れるヒトヨタケ型の子実体に加え，ウラベニガサ型の子実体およびニセホウライタケ属を想起させる毛状，偽アミロイド，厚壁の皮嚢体を持つ性質はクヌギ

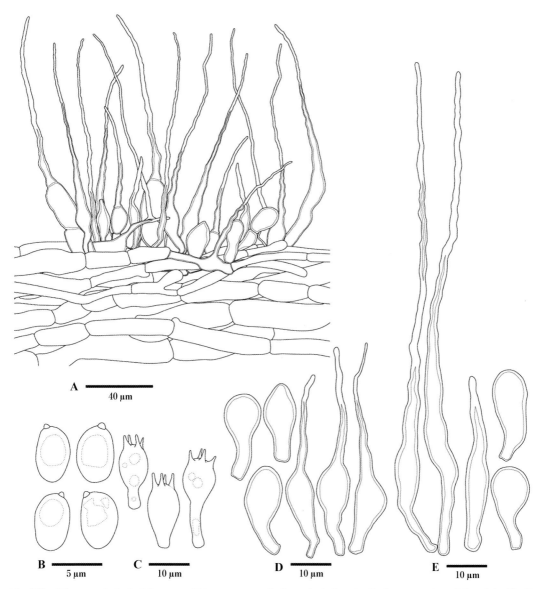

Fig. 149 – Micromorphological features of *Mycena comata* (holotype): **A**. Longitudinal cross section of the pileipellis. **B**. Basidiospores. **C**. Basidia. **D**. Cheilocystidia. **E**. Caulocystidia. Illustrations by Taneyama, Y.
キジムナハナガサの顕微鏡図（正基準標本）：**A**. 傘の表皮組織．**B**. 担子胞子．**C**. 担子器．**D**. 縁シスチジア．**E**. 柄シスチジア．図：種山裕一.

タケ属として異質であるが，クヌギタケ属 *Mycena*，コガネハナガサ節 section *Radiatae*（Singer 1986）と共通する特徴：扇のヒダ状の明瞭な条線を持つ傘；根元に発達した菌糸体並びに基盤を形成しない柄；白色，離生のヒダ；アミロイドの担子胞子；長い付属糸を持つ平滑なシスチジア；材上生の生態を考慮し，暫定的に本属に置いた．

マレーシア産 "*Trogia*" *crinipelliformis* Corner（Corner 1996）および本州産コガネハナガサ *Mycena auricoma* Har. Takah.（Takahashi 1999）はキジムナハナガサに最も近縁と考えられる．しかしながら両種とも非アミロイドの皮嚢体とより長い150 μm に達する縁シスチジアを有する点で有意差が認められる．さらにマレーシア産種は非アミロイドの担子胞子およびしばしば分岐した付属糸を持つ縁シスチジアを形成する．

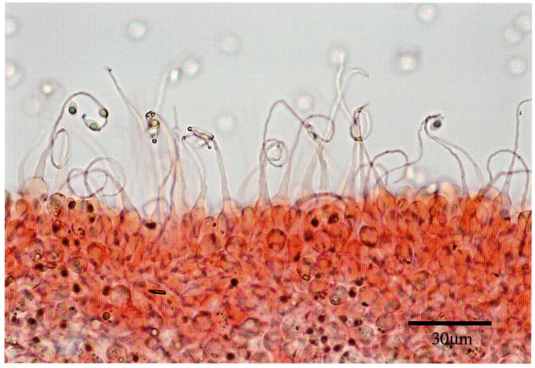

Fig. 150 - Lamella edge of *Mycena comata* (in 3%KOH and Congo red stain, holotype) showing the gregarious cheilocystidia. Photo by Taneyama, Y.
キジムナハナガサのヒダ縁部（3%水酸化カリウム溶液で封入した後コンゴー赤染色，正基準標本），縁シスチジアの群生状態を示す．写真：種山裕一．

Fig. 151 – Immature basidiomata of *Mycena comate*, 16 Oct. 2011, Banna Park, Ishigaki Island. Photo by Takahashi, H.
キジムナハナガサの幼菌，2011年10月16日．石垣島バンナ公園．写真：高橋春樹．

21. *Mycena comata* キジムナハナガサ

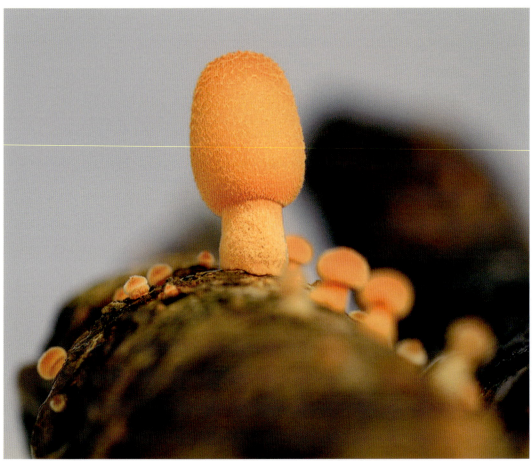

Fig. 152 – Immature basidiomata of *Mycena comata*, 16 Oct. 2011, Banna Park, Ishigaki Island. Photo by Takahashi, H.
キジムナハナガサの幼菌．2011年10月16日．石垣島バンナ公園．写真：高橋春樹．

21. *Mycena comata* キジムナハナガサ —— 191

Fig. 153 – Immature basidiomata of *Mycena comata*, 16 Oct. 2011, Banna Park, Ishigaki Island. Photo by Takahashi, H.
キジムナハナガサの幼菌，2011年10月16日，石垣島バンナ公園．写真：高橋春樹．

Fig. 154 – Mature basidiomata and primordia of *Mycena comata* (holotype), 20 Oct. 2011, Banna Park, Ishigaki Island. Photo by Takahashi, H.
キジムナハナガサの成菌と原基（正基準標本），2011年10月20日．石垣島バンナ公園．写真：高橋春樹．

Fig. 155 – Mature basidiomata and primordia of *Mycena comata* (holotype), 20 Oct. 2011, Banna Park, Ishigaki Island. Photo by Takahashi, H.
キジムナハナガサの成長した子実体と原基（正基準標本），2011年10月20日，石垣島バンナ公園．写真：高橋春樹．

Fig. 156 – Basidiomata and primordia of *Mycena comata* (holotype), 20 Oct. 2011, Banna Park, Ishigaki Island. Photo by Takahashi, H.
キジムナハナガサの成長した子実体と原基（正基準標本），2011年10月20日，石垣島バンナ公園．写真：高橋春樹．

Fig. 157 – Surface of the pileus of *Mycena comata* (holotype), 20 Oct. 2011, Banna Park, Ishigaki Island. Photo by Takahashi, H.
キジムナハナガサの傘表面（正基準標本），2011年10月20日，石垣島バンナ公園．写真：高橋春樹．

Fig. 158 – Underside view of the basidioma of *Mycena comata* (holotype), 20 Oct. 2011, Banna Park, Ishigaki Island. Photo by Takahashi, H.
キジムナハナガサのヒダ（正基準標本），2011年10月20日，石垣島バンナ公園．写真：高橋春樹．

22. *Mycena flammifera* Har. Takah. & Taneyama, **sp. nov.** モリノアヤシビ

MycoBank no.: MB 809932.

Etymology: The specific epithet comes from the Latin word for "flame-like", referring to the luminescence of basidiomata which emit a pale clear light.

Particularly notable features delimiting this species are mycenoid basidiomata with a typically pale reddish brown to whitish pileus and a slender, distinctly white pruinose stipe; a poroid hymenophore; broadly ellipsoid, amyloid basidiospores; basidia with relatively elongate sterigmata (5.0–) 6.2–9.8 (–11.8) μm long; narrowly fusiform to fusoid-ventricose cheilocystidia with an occasional nodulose apical outgrowths; ventricose-rostrate to subfusiform pleurocystidia without outgrowths; a well differentiated innermost layer of the pileipellis composed of highly inflated, short elements ("*Mycena*-structure"); narrowly subclavate to subcylindrical caulocystidia with several warty or short finger-like protuberances; greenish phosphorescence, especially in the lower part of the stipe; and a lignicolous habit.

Macromorphological characteristics (Figs. 162–168): Pileus 10–30 mm in diameter, at first conico-campanulate, then convex to broadly convex, occasionally subumbonate, smooth; surface subpruinose when young, glabrescent in age, hygrophanous, translucent when moist, dry, at first reddish brown (8D7–8 to 9D7–8), then paler from the margin, often almost whitish with age; margin not incurved, finely fimbriate. Flesh thin (up to 2.5 mm), white, odor and taste not distinctive. Stipe 20–50 × 3–5 mm, almost equal but somewhat thickened at the dilated base, central, slender, terete, fistulose, smooth; surface distinctly white pruinose overall, dry, ground color paler concolorous with the pileus toward the base, whitish at the apex; base covered with conspicuous white mycelioid bristles. Tubes adnate to subdecurrent, up to 6 mm deep, white; pores 1–2 per mm, angular, finely fimbriate under a lens, concolorous.

Luminescence: The entire basidioma, especially in the lower part of the stipe, emits a yellowish green light; cultured mycelia produce weak luminescence.

Micromorphological characteristics (Figs. 159–161): Basidiospores (5.3–) 6.4–7.8 (–8.6) × (3.5–) 4.8–5.8 (–6.4) μm (n = 152, mean length = 7.08 ± 0.70, mean width = 5.31 ± 0.53, Q = (1.18–) 1.25–1.42 (–1.64), mean Q = 1.34 ± 0.08), broadly ellipsoid, smooth, hyaline, bluish amyloid, colorless in KOH, thin-walled. Basidia (18.6–) 21.4–25.6 (–26.8) × (6.1–) 6.9–8.7 (–10.1) μm, (n = 29, mean length = 23.52 ± 2.10, mean width = 7.80 ± 0.90) in the main body, with relatively elongate sterigmata (5.0–) 6.2–9.8 (–11.8) × (1.6–) 1.9–2.4 (–2.6) μm (n = 31, mean length = 8.00 ± 1.82, mean width = 2.14 ± 0.27), clavate, 4-spored. Cheilocystidia (29.2–) 35.7–60.5 (–75.5) × (6.5–) 7.9–12.3 (–14.8) μm (n = 30, mean length = 48.11 ± 12.38, mean width = 10.11 ± 2.23), abundant, subclavate to subcylindrical, often with a few coralloid to nodulose apical outgrowths (3.3–) 3.9–6.3 (–8.5) × (1.9–) 2.2–3.8 (–5.1) μm (n = 23, mean length = 5.08 ± 1.20, mean width = 2.99 ± 0.82), inamyloid, with walls (0.6–) 0.7–0.9 (–1.1) μm (n = 20, mean thickness = 0.81 ± 0.12). Pleurocystidia (19.6–) 22.7–28.7 (–30.6) × (5.4–) 5.7–7.4 (–8.5) μm (n = 23, mean length = 25.68 ± 2.99, mean width = 6.53 ± 0.84), infrequent, ventricose-rostrate to subfusiform, with an obtuse or subacute apex, smooth, inamyloid, thin-walled. Hymenophoral trama subregular; element hyphae (9.0–) 11.7–18.6 (–25.1) μm wide (n = 44, mean width = 15.19 ± 3.45), subcylindrical, occasionally somewhat inflated, walls thin, smooth, inamyloid; hymenium and subhymenium dextrinoid or not. Uppermost layer of pileipellis a cutis of non-gelatinous, loosely interwoven, repent, cylindrical hyphae (3.5–) 5.2–8.8 (–12.5) μm wide (n = 51, mean width = 7.02 ± 1.83), occasionally branched, smooth or thinly encrusted, inamyloid, thin-walled; pilocystidia (71.0–) 76.4–98.1 (–103.5) × (6.3–) 6.4–8.2 (–9.1) μm (n = 8, mean length = 87.26 ± 10.84, mean width = 7.28 ± 0.91), infrequently present in young specimens, disrupted in age, subcylindrical, with several warty or short finger-like protuberances (3.5–) 4.0–6.9 (–8.5) × (3.6–) 3.7–5.7 (–7.3) µm (n = 18, mean length = 5.45 ± 1.45, mean width = 4.71 ± 1.01). Innermost layer of pileipellis well differentiated from the upper stratum and the pileitrama, made up of voluminous, short hyphal cells ("*Mycena*-structure") (36.8–) 41.6–62.9 (–74.5) × (23.1–) 26.9–39.0 (–43.2) µm (n = 18, mean length = 52.23 ± 10.66, mean width = 32.94 ± 6.04), subcylindrical to isodiametric, constricted at the septum, smooth, inamyloid, without clamps, thin-walled. Elements of pileitrama two types: 1) (14.2–) 17.2–25.5 (–33.0) µm wide (n = 43, mean width = 21.32 ± 4.15), loosely interwoven, cylindrical, smooth, inamyloid, without clamps, thin-walled; 2) (2.4–) 3.0–5.2 (–6.8) µm wide (n = 35, mean width = 4.09 ± 1.08), filiform, often branched, smooth, inamyloid, with clamps, thin-walled. Stipitipellis a cutis of parallel, repent hyphae (1.9–) 2.7–4.4 (–5.6) µm wide (n = 24, mean width = 3.59 ± 0.86), cylindrical, smooth, inamyloid, thin-walled; caulocystidia (56.9–) 63.0–84.2 (–104.4) × (7.9–) 9.3–14.4 (–18.7) µm (n = 26, mean length = 73.63 ± 10.58, mean width = 11.84 ± 2.52) at the apex of stipe, (51.9–) 64.8–106.1 (–130.5) × (8.3–) 9.6–14.7 (–16.3) µm (n = 17, mean length = 85.47 ± 20.67, mean width = 12.17 ± 2.53) at the base of stipe, abundant, narrowly subclavate to subcylindrical, in places with several warty or short finger-like protuberances (2.7–) 3.3–5.5 (–7.3) × (1.8–) 2.2–3.7 (–5.3) µm (n = 29, mean length = 4.40 ± 1.12, mean width = 2.92 ± 0.74) at the apex of stipe, (3.7–) 4.1–8.8 (–12.9) × (2.1–) 2.9–4.0 (–4.2) µm (n = 24, mean length = 6.42 ± 2.37, mean width = 3.44 ± 0.56) at the base of stipe, inamyloid, with slightly thickened walls (0.7–) 0.8–1.1 (–1.2) µm (n = 13, mean thickness = 0.98 ± 0.13). Stipe trama composed of longitudinally running, cylindrical hyphae (10.5–) 11.5–18.2 (–22.7) µm wide (n = 16, mean width = 14.81 ± 3.35), monomitic, unbranched, smooth, inamyloid, with thickened walls 1.0–1.7 (–1.9) µm (n = 9, mean thickness = 1.36 ± 0.32). Elements of basal mycelioid bristles (2.7–) 4.2–7.1 (–8.3) µm wide (n = 27, mean width = 5.65 ± 1.47), smooth, inamyloid, with thickened walls (0.8–) 1.0–1.4 (–1.5) µm (n = 21, mean thickness = 1.19 ± 0.16). All tissues colorless in water and KOH, with clamp connections.

Habitat and phenology: Gregarious or scattered, lignicolous on dead branches or logs (various kinds of unidentified substrata) in evergreen broad-leaved forests, almost year-round, common.

Known distribution: Okinawa (Ishigaki Island).

Holotype: TNS-F-52271, on dead logs in an evergreen broad-leaved forest, Mt. Banna, Ishigaki-shi, Okinawa Pref., 20 Oct. 2011, coll. Takahashi, H.

Other specimens examined: TNS-F-61364, on dead logs in an evergreen broad-leaved forest, Mt. Banna, Ishigaki-shi, Okinawa Pref., 6 Jun. 2012, coll. Takahashi, H.; TNS-F-61365, same location, 25 May 2013, coll. Takahashi, H; KPM-NC0023863 (Isotype), same location, 20 Oct. 2011, coll. Takahashi, H.

Japanese name: Morino-ayasibi (named by H. Takahashi & Y. Taneyama).

Comments: The broadly ellipsoid, amyloid basidiospores, the apically nodulose cheilocystidia, ventricose-rostrate to subfusiform pleurocystidia, and the non-gelatinous pileipellis suggest that the present species is a member of the genus *Mycena*, section *Fragilipedes* (Fr.) Quel. as defined by Maas (Maas 1988).

With its luminescent, lignicolous basidiomata, the hymenophore configuration, and the micromorphological characteristics, the new species has much in common with *Mycena manipularis* (Berk.) Sacc. (? *Poromycena hanedai* Kobayasi), originally described from Sri Lanka and widely distributed in the tropical and subtropical regions (Berkeley 1850; Singer 1945; Métrod 1949; Kobayasi 1951; Corner 1954; Miyagi 1960; Kobayasi et al. 1973; Maas 1982; Pegler 1986; Hongo 1987; Desjardin et al. 2008). Unlike *M. flammifera*, however, the latter has significantly shorter sterigmata: 6–8 µm long (Corner 1954) and lacks pleurocystidia.

References 引用文献

Berkeley MJ. 1850. Decades of fungi. Decades XXV. to XXX. Sikkim Himalaya fungi, collected by Dr. J.D. Hooker. Hooker's J Bot Kew Gard Misc 2: 76-88.

Corner EJH. 1954. Further descriptions of luminous agarics. Trans Br mycol Soc 37 (3): 256-271.

Desjardin DE, Oliveira AG, Stevani CV. 2008. "Fungi bioluminescence revisited". Photochemical & Photobiological Sciences 7 (2): 170-82.

Hongo T. 1987. Tricholomataceae. In: Imazeki R, Hongo T (eds), Colored illustrations of mushrooms of Japan II. Hoikusha, Osaka, pp 9-26 (in Japanese).

Kobayasi Y. 1951. Contribution to the luminous fungi from Japan. J Hattori bot Lab 5: 1-6.

Kobayasi Y, Otani Y, Hongo T. 1973. Some higher fungi found in New Guinea. Mycological reports from New Guinea and the Solomon Islands. 14. Rept Tottori Mycol Inst 10: 341-356.

Maas Geesteranus RA. 1982. Studies in Mycenas 72. Berkeley's fungi referred to Mycena - 2. Proc K Ned Akad Wet 85 (4): 527-539.

Maas Geesteranus RA. 1988. Conspectus of the Mycenas of the Northern Hemisphere - 9. Section *Fragilipedes,* species A-G. Proc K Ned Akad Wet 91 (1): 43-83.

Métrod G. 1949. Les Mycènes de Madagascar. Prodrome à une flore mycologique de Madagascar et dépendance 3: 1-146.

Miyagi G. 1960. Notes on luminous fungi, *Filoboletus manipularis*, on Okinawa (in Japanese). Bulletin of Arts and Science Division, University of the Ryukyus, Mathematics and sciences 4: 77-87.

Pegler DN. 1986. Agaric flora of Sri Lanka. Kew Bull, Addit Ser 12: 1-519.

Singer R. 1945. The *Laschia* complex (Basidiomycetes). Lloydia 8: 170-230.

22. モリノアヤシビ（森の怪火，新種；高橋春樹 & 種山裕一新称）*Mycena flammifera* Har. Takah. & Taneyama, **sp. nov.**

肉眼的特徴（Figs. 162-168）：傘は径10-30 mm，最初円錐状釣り鐘形，のち丸山形～中高偏平，時に中央部がやや突出し，平坦；表面は最初やや粉状のち平滑，吸水性，湿時半透明，粘性を欠き，幼時赤褐色，まもなく縁部から次第に褪色し，しばしば老成すると全体に類白色を呈する；縁部は長縁毛状で内側に巻かない．肉は厚さ2.5 mm以下，白色，特別な味や匂いはない．柄は20-50×3-5 mm，ほぼ上下同大，根元はやや拡大し，中心生，痩せ型，中空，平坦；表面は全体に白色粉状，乾生，下方に向かって淡褐色を帯び，頂部は類白色；根元は顕著な白色の毛状菌糸に被われる．管孔は直生～やや垂生，長さ6 mm以下，白色；孔口は0.5-1 mm，角形，長縁毛状，管孔と同色．

発光性：子実体全体，特に柄の下部が黄緑色に発光；培養菌糸体も微弱な発光性が確認されている．

顕微鏡的特徴（Figs. 159-161）：担子胞子は (5.3-) 6.4-7.8 (-8.6) × (3.5-) 4.8-5.8 (-6.4) μm (n = 152, mean length = 7.08±0.70, mean width = 5.31±0.53, Q = (1.18-) 1.25-1.42 (-1.64), mean Q = 1.34 ±0.08)，広楕円形，平坦，無色，アミロイド，アルカリ溶液において無色，薄壁．担子器は (18.6-) 21.4-25.6 (-26.8) × (6.1-) 6.9-8.7 (-10.1) μm（本体），相対的に長径のステリグマ (5.0-) 6.2-9.8 (-11.8) × (1.6-) 1.9-2.4 (-2.6) μmを有し，こん棒形，4胞子性．縁シスチジアは (29.2-) 35.7-60.5 (-75.5) × (6.5-) 7.9-12.3 (-14.8) μm，群生し，亜こん棒形～亜円柱形，しばしば上部に不規則な短指状～イボ状分岐物を具え，非アミロイド，壁は厚さ (0.6-) 0.7-0.9 (-1.1) μm；分岐物は (3.3-) 3.9-6.3 (-8.5) × (1.9-) 2.2-3.8 (-5.1) μm．側シスチジアは (19.6-) 22.7-28.7 (-30.6) × (5.4-) 5.7-7.4 (-8.5) μm，稀，上部がやや嘴状に伸びた片膨れ状～亜紡錘形，鈍頭，分岐物や突起を欠き，非アミロイド，薄壁．子実層托実質を構成する菌糸は幅 (9.0-) 11.7-18.6 (-25.1) μm，はば平列し，亜円柱形，時にやや膨大し，平滑，非アミロイド，薄壁；子実層並びに子実下層は偽アミロイドまたは非アミロイド．傘の表皮組織は非ゼラチン質で錯綜した匍匐性の菌糸からなる；菌糸は幅 (3.5-) 5.2-8.8 (-12.5) μm，円柱形，時折分岐し，平滑または結

晶物が薄く付着し，非アミロイド，薄壁；傘シスチジアは (71.0-) 76.4-98.1 (-103.5) × (6.3-) 6.4-8.2 (-9.1) μm，幼時稀に存在するが成熟時崩壊し，亜円柱形，不規則な短指状〜イボ状分岐物を具え，壁は厚さ (0.6-) 0.6-0.8 (-0.8) μm；分岐物は (3.5-) 4.0-6.9 (-8.5) × (3.6-) 3.7-5.7 (-7.3) μm．傘表皮組織の下層は最上層および実質から明瞭に分化し，著しく膨大した短形の菌糸細胞（ミセナ構造）からなる；菌糸細胞は (36.8-) 41.6-62.9 (-74.5) × (23.1-) 26.9-39.0 (-43.2) μm，亜円柱形〜ほぼ等径，隔壁の周囲において細くなり，平滑，非アミロイド，クランプを欠き，薄壁．傘実質の菌糸は2形性；1) 幅 (14.2-) 17.2-25.5 (-33.0) μm wide，緩く錯綜し，円柱形，平滑，非アミロイド，クランプを欠き，薄壁；2) 幅 (2.4-) 3.0-5.2 (-6.8) μm，しばしば分岐し，糸状，平滑，非アミロイド，クランプを持ち，薄壁．柄表皮を構成する菌糸は幅 (1.9-) 2.7-4.4 (-5.6) μm，並列し，円柱形，平滑，非アミロイド，薄壁；柄シスチジアは柄の頂部において (56.9-) 63.0-84.2 (-104.4) × (7.9-) 9.3-14.4 (-18.7) μm，柄の根元付近において (51.9-) 64.8-106.1 (-130.5) × (8.3-) 9.6-14.7 (-16.3) μm，多生し，痩せ型の亜こん棒形〜亜円柱形，ところどころに微疣または短指状分岐物を持ち，非アミロイド，壁は厚さ (0.7-) 0.8-1.1 (-1.2) μm；分岐物は柄の頂部において (2.7-) 3.3-5.5 (-7.3) × (1.8-) 2.2-3.7 (-5.3) μm，柄の下部において (3.7-) 4.1-8.8 (-12.9) × (2.1-) 2.9-4.0 (-4.2) μm．柄実質の菌糸は幅 (10.5-) 11.5-18.2 (-22.7) μm，縦に沿って平列し，一菌糸型，円柱形，無分岐，平滑，非アミロイド，厚壁 (1.0-1.7 (-1.9) μm)．根元の毛状菌糸は (2.7-) 4.2-7.1 (-8.3) μm，平滑，非アミロイド，壁は厚さ (0.8-) 1.0-1.4 (-1.5) μm．全ての組織において菌糸はクランプを有し，水封並びにアルカリ溶液で無色．

生態および発生時期：常緑広葉樹林内の腐木上から群生または散生し，ほぼ年間を通じて普通に見られる．

分布：沖縄（石垣島）．

供試標本：TNS-F-52271（正基準標本），広葉樹の腐木上，沖縄県石垣島バンナ岳，2011年10月20日，高橋春樹採集；TNS-F-61364，同上，2012年6月6日，高橋春樹採集；TNS-F-61365，同上，2013年5月25日，高橋春樹採集；KPM-NC0023863（複基準標本），同上，2011年10月20日，高橋春樹採集．

主な特徴：子実体はクヌギタケ型で，淡赤褐色〜類白色の傘，管孔状の子実層托および痩せ型，白粉状の柄を持つ；担子胞子は広楕円形，アミロイド；相対的に長いステリグマタ：長さ (5.0-) 6.2-9.8 (-11.8) μm を持つ担子器；縁シスチジアは狭紡錘形〜片膨れ状紡錘形で平滑または頂部にイボ状付属物が散在する；側シスチジアは上部が嘴状に伸びた片膨れ状〜亜紡錘形；傘表皮組織は下層が明瞭に分化し，著しく膨大した短形の菌糸細胞（ミセナ構造）からなる；柄シスチジアは痩せ型の亜こん棒形〜亜円柱形で，イボ状〜短指状突起が散在する；子実体全体，特に柄の下部が黄緑色に発光する；広葉樹の腐木上から発生．

コメント：広楕円形でアミロイドに染まる担子胞子，頂部に微疣状分岐物を持つ縁シスチジア，上部が嘴状に伸びた片膨れ状〜亜紡錘形の側シスチジア，そして非ゼラチン質の傘表皮組織は本種が Maas の分類概念 (Maas 1988) によるクヌギタケ属 *Mycena*，アクニオイタケ節 Sect. *Fragilipedes* (Fr.) Quel. に属することを示唆している．

　発光性がある材上生の子実体，子実層托の形状，そして顕微鏡的特徴において本種は，最初スリランカから報告され，熱帯〜亜熱帯に渡って広い分布域を持つとされているアミヒカリタケ *Mycena manipularis* (Berk.) Sacc. (Berkeley 1850；Singer 1945；Métrod 1949；Kobayasi 1951；Corner 1954；Kobayasi et al. 1973；Maas 1982；Pegler 1986；Hongo 1987；Desjardin et al. 2008) に最も近縁と思われるが，後者はより短いステリグマ：6-8 μm long (Corner 1954) を有し，側シスチジアを欠くとされている．

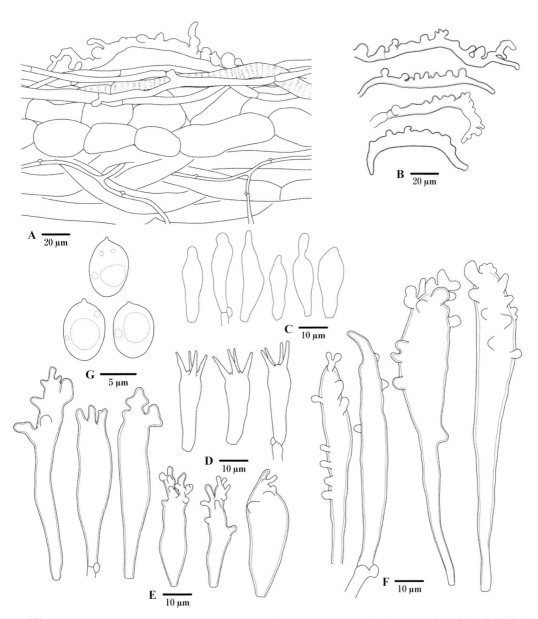

Fig. 159 – Micromorphological features of *Mycena flammifera* (holotype): **A**. Longitudinal cross section of the pileipellis. **B**. Terminal elements of the pileipellis. **C**. Pleurocystidia. **D**. Basidia. **E**. Cheilocystidea. **F**. Caulocystidia. **G**. Basidiospores. Illustrations by Taneyama, Y.

モリノアヤシビの顕鏡図（正基準標本）：**A**. 傘表皮組織の縦断面. **B**. 傘表皮組織の末端細胞. **C**. 側シスチジア. **D**. 担子器. **E**. 縁シスチジア. **F**. 柄シスチジア. **G**. 担子胞子. 図：種山裕一.

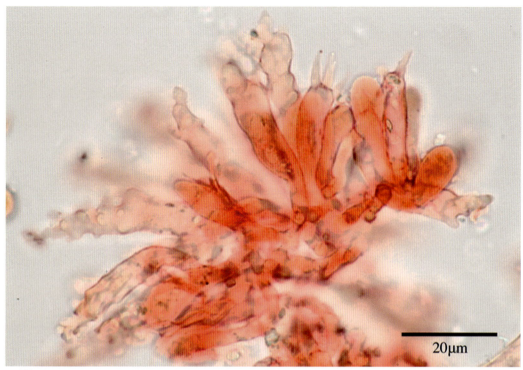

Fig. 160 – Cheilocystidea of *Mycena flammifera* (in 3%KOH and Congo red stain, holotype). Photo by Taneyama, Y.
モリノアヤシビの縁シスチジア（3％水酸化カリウム溶液で封入した後コンゴー赤染色，正基準標本）．写真：種山裕一．

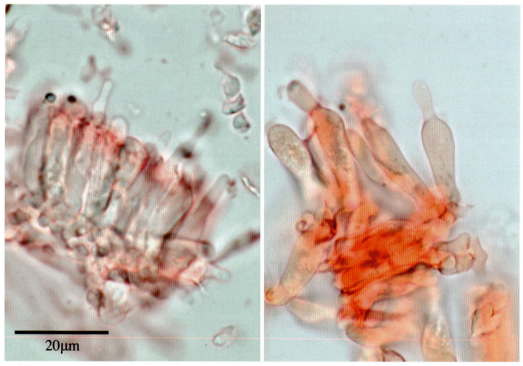

Fig. 161 – Pleurocystidia of *Mycena flammifera* (in 3%KOH and Congo red stain, holotype). Photo by Taneyama, Y.
モリノアヤシビの側シスチジア（3％水酸化カリウム溶液で封入した後コンゴー赤染色，正基準標本）．写真：種山裕一．

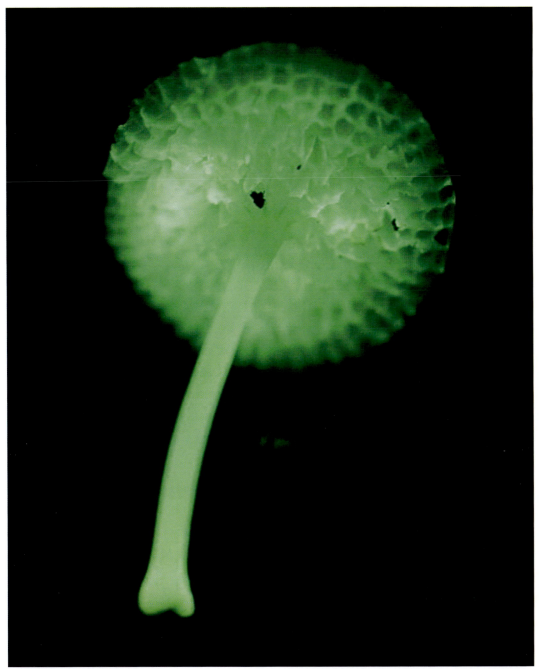

Fig. 162 – Basidioma of *Mycena flammifera* (holotype) emitting yellowish green light on a dead fallen log, 20 Oct. 2011, Mt. Banna, Ishigaki Island. Photo by Takahashi, H.
腐木上において黄緑色に発光するモリノアヤシビの子実体（正基準標本），2011年10月20日，石垣島バンナ岳．写真：高橋春樹．

204 —— 22. *Mycena flammifera* モリノアヤシビ

Fig. 163 – Basidiomata of *Mycena flammifera* (TNS-F-61365) emitting yellowish green light on a dead fallen log, 25 May 2013, Mt. Banna, Ishigaki Island. Photo by Takahashi, H.
腐木上において黄緑色に発光するモリノアヤシビの子実体（TNS-F-61365），2013年5月25日，石垣島バンナ岳．写真：高橋春樹．

Fig. 164 – Basidiomata of *Mycena flammifera* (TNS-F-61365) on a dead fallen log, 25 May 2013, Mt. Banna, Ishigaki Island. Photo by Takahashi, H.
腐木上に発生するモリノアヤシビの子実体（TNS-F-61365），2013年5月25日，石垣島バンナ岳．写真：高橋春樹．

Fig. 165 – Immature basidiomata of *Mycena flammifera* on a dead fallen log, 23 Dec. 2001, Mt. Banna, Ishigaki Island. Photo by Takahashi, H.
腐木上に発生するモリノアヤシビの未熟な子実体．2001年12月23日．石垣島バンナ岳．写真：高橋春樹．

Fig. 166 – Mature basidiomata of *Mycena flammifera*, 23 Dec. 2001, Mt. Banna, Ishigaki Island. Photo by Takahashi, H.
モリノアヤシビの成熟した子実体．2001年12月23日，石垣島バンナ岳．写真：高橋春樹．

22. *Mycena flammifera* モリノアヤシビ —— 207

Fig. 167 – Underside view of the basidiomata of *Mycena flammifera*, 23 Dec. 2001, Mt. Banna, Ishigaki Island. Photo by Takahashi, H.
モリノアヤシビの管孔．2001年12月23日，石垣島バンナ岳．写真：高橋春樹．

208 —— 22. *Mycena flammifera* モリノアヤシビ

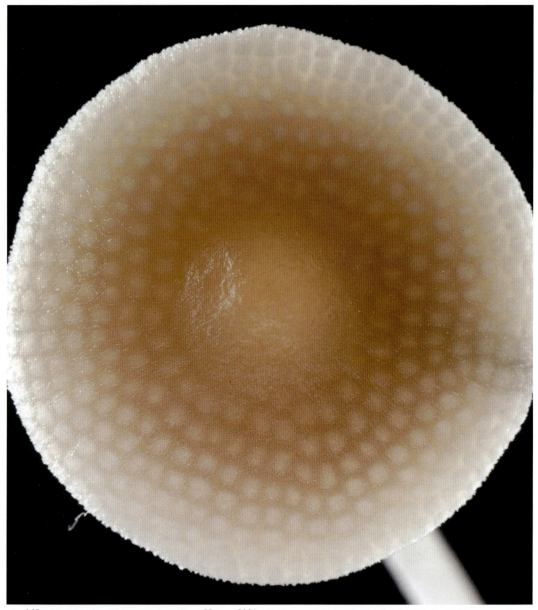

Fig. 168 – Pileus surface of *Mycena flammifera*, 23 Dec. 2001, Mt. Banna, Ishigaki Island. Photo by Takahashi, H.
モリノアヤシビの傘表面．2001年12月23日，石垣島バンナ岳．写真：高橋春樹．

23. *Mycena lazulina* Har. Takah., Taneyama, Terashima & Oba, **sp. nov.** コンルリキュウバンタケ

MycoBank no.: MB 809933.

Etymology: The specific epithet comes from the Latin word for "ultramarine" or "blue", referring to the vivid blue disk-like base.

Distinctive features of this species are found in bioluminescent, white basidiomata arising from a small, vivid blue, disk-like base that attaches to the substrate; a tiny plicate-striate pileus up to 2 mm in diameter; distinctly collariate, distant lamellae; amyloid basidiospores; apically echinulate, broadly clavate to subcylindrical cheilocystidia; a hymeniform pileipellis consisting of acanthocyst end cells; broadly clavate or narrowly ventricose-rostrate, apically echinulate caulocystidia (acanthocysts); dextrinoid stipe trama; luminescence of the basidiomata and mycelia; and a gregarious habit on a dead bamboo and palm.

Macromorphological characteristics (Figs. 174-178): Pileus 1-2 mm in diameter, at first subglobose, becoming convex, not plane; surface translucent, dry, pure white, pruinose overall, plicate-striate from midway to the white pruinose margin. Flesh very thin (up to 0.1 mm), membranous, white; odor and taste none. Stipe 3-6 × 0.1-0.5 mm, relatively short, subequal but somewhat bulbous at the disk-like base, central, terete; surface white, dry, opaque, entirely pruinose; base up to 0.3-0.7 mm in diameter, attaching directly to the substrate, small, disk-like, vivid blue (21A8) to blue (21B-C8) or at times deep blue (21D8), pruinose. Lamellae distinctly collariate, distant, L = 6-7, up to 0.3 mm broad, thin, white; edges even, pruinose, concolorous.

Luminescence: Basidiomata and mycelia emit bright yellowish green light.

Micromorphological characteristics (Figs. 170-173): Basidiospores (4.7-) 6.1-7.4 (-8.4) × (3.4-) 4.3-5.2 (-6.3) μm (n = 146, mean length = 6.77 ± 0.67, mean width = 4.73 ± 0.48, Q = (1.13-) 1.32-1.55 (-1.76), mean Q = 1.44 ± 0.11), ellipsoid to broadly ellipsoid, smooth, colorless in distilled water, violet grey in Melzer's reagent (amyloid), hyaline in KOH, thin-walled. Basidia (13.1-) 16.3-21.2 (-25.0) × (5.3-) 6.4-7.7 (-8.6) μm (n = 38, mean length = 18.76 ± 2.43, mean width = 7.05 ± 0.62), clavate, four-spored; sterigmata (3.2-) 3.7-5.6 (-7.9) × (1.0-) 1.4-2.0 (-2.9) μm (n = 44, mean length = 4.64 ± 0.99, mean width = 1.72 ± 0.27). Cheilocystidia (15.6-) 19.7-27.9 (-38.8) × (9.3-) 12.0-16.3 (-19.6) μm (n = 59, mean length = 23.82 ± 4.09, mean width = 14.18 ± 2.17), gregarious, broadly clavate to subcylindrical, echinulate above, thin-walled, inamyloid. Pleurocystidia none. Trama of lamellae composed of more or less regularly arranged, highly inflated and spindle-shaped hyphae (19.3-) 24.5-42.0 (-51.2) × (12.6-) 14.0-20.6 (-25.0) μm (n = 19, mean length = 33.28 ± 8.73, mean width = 17.28 ± 3.31), constricted at the septa, thin-walled, smooth, inamyloid, with clamp connections; hymenium vivid yellow in Melzer's reagent. Surface of pileus a hymeniform layer of acanthocyst end cells arising from repent, filiform hyphae; constituent hyphae (1.9-) 2.4-3.5 (-4.3) μm wide (n = 37, mean width = 2.97 ± 0.56), filiform, parallel with each other, thin-walled, smooth, inamyloid, not gelatinized; acanthocysts (15.9-) 20.1-33.6 (-45.6) × (9.9-) 14.1-25.3 (-34.0) μm (n = 63, mean length = 26.85 ± 6.70, mean width = 19.73 ± 5.59), subglobose to broadly clavate, echinulate in upper half, thin-walled, inamyloid. Trama of pileus made up of inflated and spindle-shaped hyphae (64.2-) 75.4-109.7 (-134.0) × (15.1-) 20.0-28.4 (-31.0) μm (n = 28, mean length = 92.55 ± 17.16, mean width = 24.20 ± 4.24), subparallel, thin-walled, smooth, inamyloid, not gelatinized. Elements of stipitipellis (2.4-) 3.1-4.9 (-6.9) μm wide (n = 35, mean width = 3.96 ± 0.89), parallel, filiform, smooth, inamyloid, not gelatinized, thin-walled; caulocystidia (acanthocysts) numerous, (16.3-) 22.0-47.0 (-66.2) × (5.1-) 9.1-19.6 (-22.5) μm (n = 33, mean length

= 34.50 ± 12.48, mean width = 14.36 ± 5.28), broadly clavate or narrowly ventricose-rostrate, echinulate above, inamyloid, thin-walled. Trama of stipe composed of longitudinally running, cylindrical to subfusiform hyphae (56.9–) 64.1–157.5 (–201.8) μm long (n = 16, mean length= 110.78 ± 46.73), (5.5–) 8.4–19.5 (–29.8) μm wide (n = 63, mean width = 13.97 ± 5.56), inflated or not, smooth, dextrinoid, with walls (0.7–) 0.8–1.2 (–1.5) μm, with clamped septa. Bulbous base consisting of perpendicularly arranged, cylindrical to subfusiform hyphae (26.3–) 32.2–47.5 (–59.0) × (4.2–) 5.0–7.6 (–8.5) μm (n = 19, mean length = 39.84 ± 7.68, mean width = 6.29 ± 1.32), smooth, unbranched, encrusted with blue pigment, with walls (0.4–) 0.6–0.9 (–1.0) μm (n = 18, mean thickness = 0.74 ± 0.16). All tissues hyaline in KOH and distilled water except the bulbous base.

Habitat and phenology: Gregarious on dead bamboo and palm, June to November.

Known distribution: Ishigaki Island, Iriomote Island and Yonaguni Island (?).

Holotype: TNS-F-52275, on a dead culm of bamboo, Banna Park, Ishigaki Island, Okinawa Pref., 7 Nov. 2012, coll. Nishino, Y.

Other specimens examined: TNS-F-61361, on a dead palm (*Arenga engleri* Becc.), Hunaura, Iriomote Island, Okinawa Pref., 15 Nov. 2013, coll. Wada, S.; KPM-NC0023861, same location, 15 Nov. 2013, coll. Wada, S.; TNS-F-52274 (*Mycena* aff. *lazulina*), on a dead palm (*Livistona chinensis* (Jacquin) R. Brown ex Martius), Yonaguni Island, Okinawa Pref., 16 Jun. 2012, coll. Terashima, Y.

Gene sequenced specimen and GenBank accession number: TNS-F-52275, AB971703.

Japanese name: Konruri-kyubantake (named by H. Takahashi, Y. Taneyama, Y. Terashima & Y. Oba).

Comments: The presence of acanthocysts, the non-gelatinized pileipellis hyphae, the amyloid basidiospores and the tiny white basidiomata with a deep blue disk-like base suggest that the present taxon belongs to the genus *Mycena*, section *Sacchariferae* Kühner ex Singer (Singer 1986; Desjardin 1995). Within the section, the new species closely resembles *Mycena arundinarialis* Pegler described from East Africa (Pegler 1977), which differs in having a purplish pigment in the hairs of the basal disk.

Mycena interrupta (Berk.) Sacc. (= *Mycena cyanocephala* Singer), widely distributed in Australasia and Chile (Hooker 1860; Singer 1969; Horak 1980; Grgurinovic 2003), is commonly known as a beautiful species with a flat discoid base and brilliant cyan blue pigments, which, unlike *M. lazulina*, produces much larger, non-luminescent basidiomata: up to 16 mm across in the viscid, brilliant cyan-blue pileus (Grgurinovic 2003), adnexed lamellae with blue margins, and much larger basidiospores: 8.4–11.6 × 5.7–8.8 μm (Grgurinovic 2003).

The identification of the materials from Yonaguni Island (TNS-F-52274) appears to be somewhat dubious because of the cylindrical, non-inflated hyphae of the stipe trama and the presence of brownish intracellular pigment in the basal disk.

References 引用文献

Desjardin DE. 1995. A preliminary accounting of the worldwide members of *Mycena* sect. *Sacchariferae*. Taxonomic monographs of Agaricales. Bibl Mycol 159: 1–89.

Grgurinovic CA. 2003. The genus *Mycena* in South-Eastern Australia. Fung Diver Research Series 9: 1–329.

Hooker JD. 1860. Botany of the Antarctic Voyage III. Fl Tasman 2: 1–422.

Horak E. 1980. Fungi, Basidiomycetes. Agaricales y Gasteromycetes secotioides. Fl criptog Tierra del Fuego (Buenos Aires) 11 (6): 1–524.

Pegler DN. 1977. A preliminary agaric flora of East Africa. Kew Bull, Addit Ser 6: 1–615.

Singer R. 1969. Mycoflora australis. Beih Nova Hedwigia 29: 1–405.

Singer R. 1986. The Agaricales in modern taxonomy, 4th edn. Koeltz, Koenigstein.

23. コンルリキュウバンタケ（紺瑠璃吸盤茸，新種；高橋春樹，種山裕一，寺嶋芳江 & 大場由美子新称）*Mycena lazulina* Har. Takah., Taneyama, Terashima & Oba, **sp. nov.**

肉眼的特徴（Figs. 174-178）：傘は径1-2 mm，最初半球形，のち饅頭形；表面は半透明，乾性，純白色，老成するとくすんだ淡黄色を帯び，放射状に走る明瞭な扇状の溝線を表し，周辺部は平滑またはやや白粉状．肉は極めて薄い膜質（0.1 mm以下），白色，特別な味や匂いはない．柄は3-6×0.1-0.5 mm，ほぼ上下同大，根元は球根状にやや膨らみ小型の台状基盤を形成し，中心生；表面は白色，乾性，粉状．根本の台状基盤は径0.3-0.7 mm，濃青色〜紺色，粉状．ヒダは明瞭な襟帯を形成し，疎，L = 6-7，幅0.3 mm以下，白色，縁部は粉状，同色．

発光性：菌糸体および子実体が黄緑色に発光．

顕微鏡的特徴（Figs. 170-173）：担子胞子は (4.7-) 6.1-7.4 (-8.4) × (3.4-) 4.3-5.2 (-6.3) μm（n = 146, mean length = 6.77 ± 0.67, mean width = 4.73 ± 0.48, Q = (1.13-) 1.32-1.55 (-1.76), mean Q = 1.44 ± 0.11)，楕円形〜広楕円形，無色，平坦，アミロイド，アルカリ溶液において無色，薄壁．担子器は (13.1-) 16.3-21.2 (-25.0) × (5.3-) 6.4-7.7 (-8.6) μm，こん棒形，4胞子性；ステリグマは (3.2-) 3.7-5.6 (-7.9) × (1.0-) 1.4-2.0 (-2.9) μm．縁シスチジアは (15.6-) 19.7-27.9 (-38.8) × (9.3-) 12.0-16.3 (-19.6) μm，群生し，広こん棒形〜円柱形，上部が微疣に被われ，薄壁，非アミロイド．側シスチジアはない．子実層托実質は並列形；菌糸細胞は (19.3-) 24.5-42.0 (-51.2) × (12.6-) 14.0-20.6 (-25.0) μm，膨大し，紡錘形，隔壁の周囲において急激に収縮し，無色，非アミロイド，薄壁，隔壁にクランプを持つ；子実層はMelzer溶液において鮮黄色を呈する．傘表皮は匍匐性の糸状菌糸に起源を持つ末端細胞（アカントシスト様細胞）が子実層状被を成す；菌糸は幅 (1.9-) 2.4-3.5 (-4.3) μm，ほぼ平列し，平滑，非アミロイド，薄壁；アカントシスト様細胞は (15.9-) 20.1-33.6 (-45.6) × (9.9-) 14.1-25.3 (-34.0) μm，類球形〜広こん棒形，上部を中心に微疣に被われ，非アミロイド，薄壁．傘実質の菌糸細胞は紡錘形に膨大し，(64.2-) 75.4-109.7 (-134.0) × (15.1-) 20.0-28.4 (-31.0) μm，ほぼ平列し，平滑，薄壁，非アミロイド．柄表皮を構成する菌糸は幅 (2.4-) 3.1-4.9 (-6.9) μm，並列し，糸状，平滑，非アミロイド，薄壁；柄シスチジア（アカントシスト様細胞）は多生し，(16.3-) 22.0-47.0 (-66.2) × (5.1-) 9.1-19.6 (-22.5) μm，広こん棒形または上部が嘴状に伸びた片膨れ状，上部は微疣に被われ，非アミロイド，薄壁．柄実質の菌糸は長さ (56.9-) 64.1-157.5 (-201.8) μm，幅 (5.5-) 8.4-19.5 (-29.8) μm，平列し，円柱形または紡錘形，しばしば膨大し，平滑，偽アミロイド，壁は厚さ (0.7-) 0.8-1.2 (-1.5) μm，隔壁にクランプを持つ．台状の基盤を構成する菌糸は (26.3-) 32.2-47.5 (-59.0) × (4.2-) 5.0-7.6 (-8.5) μm，垂直に配列し，円柱形または亜紡錘形，無分岐，平滑，青色色素が凝着し，壁は厚さ (0.4-) 0.6-0.9 (-1.0) μm．根元の菌糸を除く全ての組織はアルカリ溶液および水封において無色．

生態および発生時期：朽ちた竹およびコミノクロツグなどのヤシ科植物に群生，6月〜11月．

分布：沖縄（八重山諸島：石垣島，西表島，？与那国島）．

供試標本：TNS-F-52275（正基準標本），朽ちた竹の稈上，沖縄県石垣島バンナ公園，2012年11月7日，西野嘉憲採集；TNS-F-61361，朽ちたコミノクロツグから発生，沖縄県八重山郡竹富町西表島船浦，2013年11月15日，和田匠平採集；KPM-NC0023861，同上，2013年11月15日，和田匠平採集；TNS-F-52274 (*Mycena* aff. *lazulina*)，朽ちたビロウから発生，沖縄県八重山郡竹富町与那国島，2012年6月16日，寺嶋芳江採集．

分子解析に用いた標本並びにGenBank登録番号：TNS-F-52275, AB971703.

主な特徴：子実体は微小（傘の径2 mm以下）で全体に白色を呈し，濃青色の台状基盤から発生する；傘は扇のヒダ状の溝線を表す；ヒダは明瞭な襟帯を形成し，疎で互いの間隔がきわめて広い；担子胞子はアミロイド；縁シスチジアは広こん棒形〜類円柱形で，上部が微突起に被われ

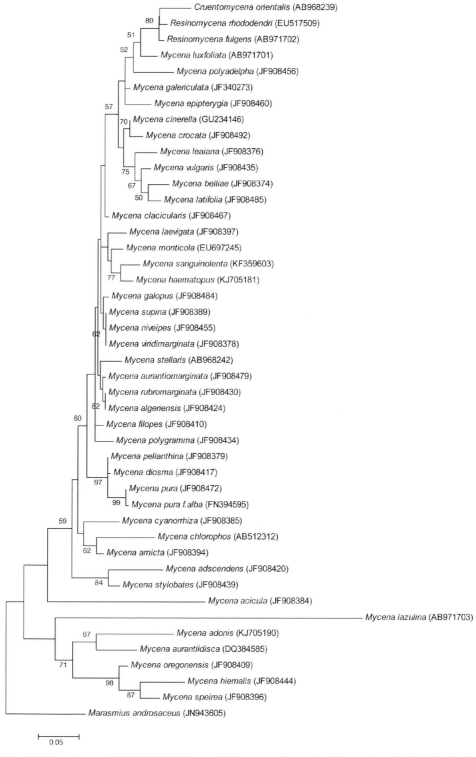

Fig. 169 – A phylogenetic tree of ITS1-5.8S-ITS2 dataset for *Mycena* species, analyzed by maximum likelihood method based on the Tamura-Nei model (Tamura and Nei 1993) in MEGA5 (Tamura et al. 2011). Numbers on the branches represent bootstrap values obtained from 2000 replications (only values greater than 50% are shown).
核リボソーム ITS1-5.8S-ITS2を用いた最尤法によるクヌギタケ属系統樹.

23. *Mycena lazulina* コンルリキュウバンタケ —— 213

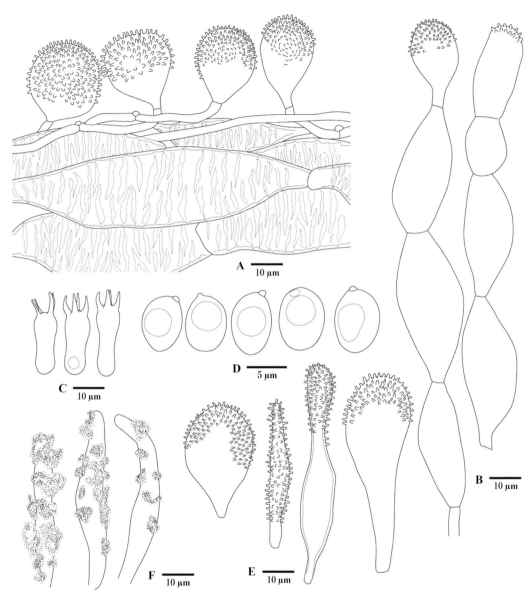

Fig. 170 – Micromorphological features of *Mycena lazulina* (holotype): **A**. Longitudinal cross section of the pileipellis with pilocystidia (acanthocysts). **B**. Cheilocystidia arising directly from the inflated hyphae of the hymenophoral trama. **C**. Basidia. **D**. Basidiospores. **E**. Terminal cells (acanthocysts) of the stipitipellis. **F**. Terminal cells of the deep blue disk-like base. Illustrations by Taneyama, Y.

コンルリキュウバンタケの顕鏡図（正基準標本）：**A**. 傘シスチジア（アカントシスト様細胞）を持つ傘表皮組織の縦断面．**B**. 膨大したヒダ実質菌糸につながる縁シスチジア．**C**. 担子器．**D**. 担子胞子．**E**. 柄表皮の末端細胞（アカントシスト様細胞）．**F**. 濃青色の台状基盤を構成する菌糸の末端細胞．図：種山裕一．

る；傘表皮組織はアカントシスト様末端細胞が子実層状被を成す；柄シスチジア（アカントシスト様細胞）は広こん棒形または上部が嘴状に伸びた片膨れ状で，頂部は微疣に被われる；柄実質は偽アミロイド；菌糸体並びに子実体全体が黄緑色に発光する；朽ちた竹およびコミノクロツグ，ビロウなどのヤシ科植物に群生する．

コメント：アカントシスト型の表皮シスチジアの存在，非ゼラチン質の傘の表皮組織，アミロイドの担子胞子，そして台状基盤を有する微小で白色の子実体は，本種がクヌギタケ属 *Mycena*，シロコナカブリ節 section *Sacchariferae* Kühner ex Singer（Singer 1986；Desjardin 1995）に属することを示唆している．節内において東アフリカ産 *Mycena arundinarialis* Pegler（Pegler 1977）は本種に最も近縁と思われるが，紫色の毛状基盤を持つとされている．

オーストラレーシアおよび南米チリに分布する *Mycena interrupta* (Berk.) Sacc.（= *Mycena cyanocephala* Singer）（Hooker 1860；Singer 1969；Horak 1980；Grgurinovic 2003）は吸盤形の根元と濃青色の色素を持つ美しい種類として知られているが，発光性を欠き，傘の径16 mm（Grgurinovic 2003）に達するより大型の子実体を形成し，湿時粘性を表す濃青色の傘，青色の縁取りがある上生のヒダ，そしてより大型の担子胞子：8.4-11.6×5.7-8.8 μm（Grgurinovic 2003）を有する．

与那国島から採取された標本（TNS-F-52274）については，柄実質の菌糸が膨大しないこと，また台状の基盤を構成する組織に褐色の細胞内色素が存在する点で同定にやや疑問がある．

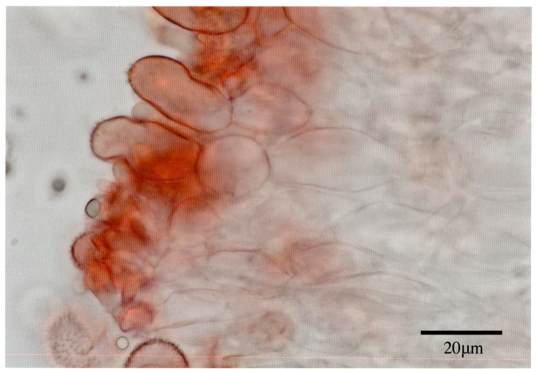

Fig. 171 – Longitudinal cross section of the edge of lamellae of *Mycena lazulina* (in 3%KOH and Congo red stain, holotype). Photo by Taneyama, Y.
コンルリキュウバンタケのヒダ縁部の縦断面（3％水酸化カリウム溶液で封入した後コンゴー赤染色，正基準標本）．写真：種山裕一．

216 —— 23. *Mycena lazulina* コンルリキュウバンタケ

Fig. 174 – Primordia and developed basidiomata of *Mycena lazulina* on a dead culm of bamboo, 2 Nov. 2013, Banna Park, Ishigaki Island. Photo by Takahashi, H.
朽ちた竹の稈上に群生するコンルリキュウバンタケの原基と成長した子実体．2013年11月2日，石垣島バンナ公園．写真：高橋春樹．

Fig. 175 – Primordia and developed basidiomata of *Mycena lazulina* on a dead culm of bamboo, 2 Nov. 2013, Banna Park, Ishigaki Island. Photo by Takahashi, H.
朽ちた竹の稈上に群生するコンルリキュウバンタケの原基と成長した子実体．2013年11月2日，石垣島バンナ公園．写真：高橋春樹．

23. *Mycena lazulina* コンルリキュウバンタケ —— 215

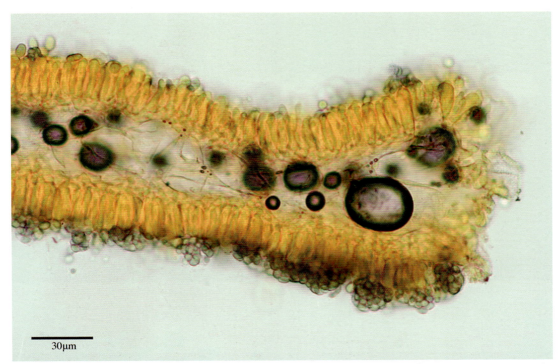

Fig. 172 – Longitudinal cross section of the edge of lamellae of *Mycena lazulina* (in Melzer's reagent, holotype). Photo by Taneyama, Y.
コンルリキュウバンタケのヒダ縁部の縦断面（メルツァー溶液で封入，正基準標本）．写真：種山裕一．

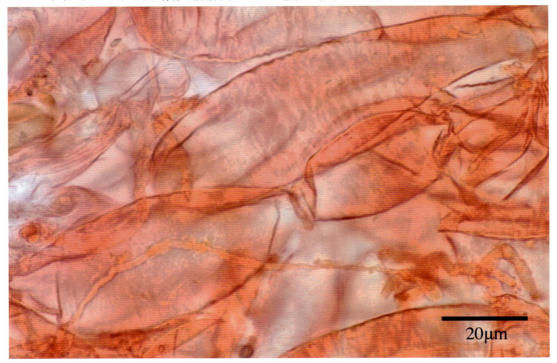

Fig. 173 – Hyphal cells in the pileitrama of *Mycena lazulina* (in 3%KOH and Congo red stain, holotype). Photo by Taneyama, Y.
コンルリキュウバンタケの傘実質の菌糸細胞（3％水酸化カリウム溶液で封入した後コンゴー赤染色，正基準標本）．写真：種山裕一．

23. *Mycena lazulina* コンルリキュウバンタケ —— 217

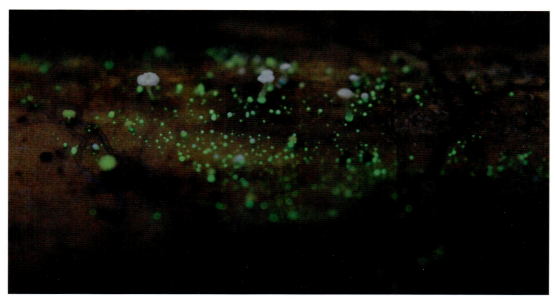

Fig. 176 – Primordia and developed basidiomata of *Mycena lazulina* emitting yellowish green light on a dead culm of bamboo, 18 Nov. 2013, Banna Park, Ishigaki Island. Photo by Takahashi, H.
朽ちた竹の稈上において黄緑色に発光するコンルリキュウバンタケの原基および成長した子実体．2013年11月18日，石垣島バンナ公園．写真：高橋春樹．

Fig. 177 – Primordia and developed basidiomata of *Mycena lazulina* emitting yellowish green light on a dead culm of bamboo. 2 Nov. 2013, Banna Park, Ishigaki Island. Photo by Takahashi, H.
朽ちた竹の稈上において黄緑色に発光するコンルリキュウバンタケの原基と成長した子実体．2013年11月2日，石垣島バンナ公園．写真：高橋春樹．

Fig. 178 – Basidiomata of *Mycena* aff. *lazulina* (TNS-F-52274) on a dead palm (*Livistona chinensis*), 16 Jun. 2012, Yonaguni Island. Photo by Terashima, Y.
朽ちたビロウに群生するコンルリキュウバンタケ近縁種の子実体（TNS-F-52274），2012年6月16日，与那国島．写真：寺嶋芳江．

24. *Mycena luxfoliata* Har. Takah., Taneyama & Terashima, sp. nov. カレハヤコウタケ

MycoBank no.: MB 809934.

Etymology: The specific epithet means "*lux*- (light) + *foliata* (leaved)", referring to the luminescent mycelia in dead leaves.

Distinctive features of this species are found in minute, pure white, mycenoid basidiomata with a puberulous pileus and pruinose stipe; weakly amyloid, ellipsoid to oblong-ellipsoid basidiospores; two-typed cheilocystidia; thick-walled, long, hair-like pileipellis elements; dimorphic caulocystidia; mycelium luminescence in dead leaves; and an invariably foliicolous habit.

Macromorphological characteristics (Figs. 182-185): Pileus 1-2.5 mm in diameter, at first convex then plano-convex, smooth or slightly striatulate at a minutely crenulate to eroded margin; surface non-hygrophanous, dry, finely puberulous overall under a lens, pure white. Flesh thin (up to 0.1 mm thick), pure white, soft; odor not distinctive, taste undetermined. Stipe 3-5(-9) × 0.1-0.3 mm, almost equal, often slightly enlarged at the base, central, terete, smooth; surface dry, pure white, entirely pruinose under a lens; basal mycelium none. Lamellae adnate to subdecurrent, arched, distant (7-11 reach the stipe), not intervenose, up to 0.2 mm broad, thin, pure white; edges finely fimbriate under a lens, concolorous.

Luminescence: Basidiomata non-luminescent; mycelia luminescent in the substrata (dead leaves).

Micromorphological characteristics (Figs. 180, 181): Basidiospores (5.6-) 6.6-7.9 (-8.6) × (3.1-) 3.6-4.3 (-4.7) μm (n = 82, mean length = 7.26 ± 0.64, mean width = 3.96 ± 0.34, Q = (1.50-) 1.70-1.98 (-2.08), mean Q = 1.84 ± 0.14), ellipsoid to oblong-ellipsoid, smooth, hyaline, weakly amyloid, colorless in KOH, thin-walled. Basidia (12.1-) 14.4-17.8 (-18.5) × (7.8-) 8.0-8.7 (-8.8) μm (n = 14, mean length = 16.07 ± 1.69, mean width = 8.35 ± 0.33) in main body, (3.3-) 3.5-4.7 (-4.9) × (1.7-) 1.8-2.0 (-2.2) μm (n = 15, mean length = 4.07 ± 0.61, mean width = 1.91 ± 0.13) in sterigmata, clavate, 4-spored; basidioles clavate. Cheilocystidia forming a compact sterile edge, dimorphic: 1) (8.9-) 11.8-17.2 (-22.6) × (5.9-) 6.6-8.5 (-10.3) μm (n = 40, mean length = 14.47 ± 2.69, mean width = 7.54 ± 0.92), subclavate, with coralloid to several finger-like apical appendages (2.6-) 3.1-5.1 (-6.3) × (1.3-) 1.7-2.4 (-3.0) μm, hyaline, inamyloid, with walls 0.8-1.2 (-1.4) μm; 2) (23.1-) 25.6-34.3 (-40.5) × (7.1-) 8.1-9.9 (-10.6) μm (n = 30, mean length = 29.94 ± 4.31, mean width = 9.01 ± 0.89), fusoid-ventricose, at times subcapitulate, smooth, inamyloid, with walls (0.7-) 0.8-1.2 (-1.4) μm. Hymenophoral trama regular, similar to the pileitrama. Pileipellis a cutis of loosely entangled hyphae (2.0-) 2.2-3.0 (-3.5) μm wide (n = 33, mean width = 2.60 ± 0.39), cylindrical, smooth, occasionally branched, inamyloid, with walls (0.8-) 0.9-1.3 (-1.4) μm; pilocystidia none. Elements of pileitrama (10.5-) 18.1-31.8 (-42.5) × (9.0-) 9.9-12.6 (-16.4) μm (n = 26, mean length = 24.98 ± 6.87, mean width = 11.25 ± 1.38), parallel, cylindrical, smooth, hyaline, weakly dextrinoid, thin-walled. Stipitipellis a cutis of parallel, repent hyphae (2.2-) 2.5-3.6 (-4.0) μm wide (n = 27, mean width = 3.05 ± 0.51), filiform, smooth, inamyloid, with walls (0.5-) 0.7-1.2 (-1.4) μm. Caulocystidia dimorphic: 1) fusoid-ventricose caulocystidia with a median constriction at the upper portion of stipe, (16.2-) 17.8-24.1 (-32.5) × (3.7-) 5.8-8.8 (-11.3) μm (n = 34, mean length = 20.93 ± 3.12, mean width = 7.32 ± 1.48), scattered but often aggregating and occurring in bunches, smooth, inamyloid, with walls (0.8-) 0.9-1.4 (-1.7) μm; 2) irregularly cylindrical to coralloid caulocystidia at the lower portion of stipe, (25.6-) 30.4-54.5 (-75.5) × (2.3-) 3.0-4.3 (-5.0) μm (n = 27, mean length = 42.46 ± 12.09, mean width = 3.67 ± 0.65), scattered, often irregularly branched, smooth, inamyloid, with walls (0.6-) 0.8-1.1 (-1.2) μm. Basal elements of stipe surface made up of clusters of tightly adherent hyphae (1.5-) 2.3-3.2 (-3.7) μm wide (n = 32, mean width = 2.76 ± 0.46), cylindrical, smooth, inamyloid, with walls (0.5-) 0.6-0.9 (-1.0) μm (n = 18, mean thickness = 0.73 ± 0.13). Stipe trama composed of longitudinally running, cylindrical hyphae (42.1-) 50.1-98.1 (-144.4) µm long (n = 43, mean width = 74.08 ± 23.98), (6.1-) 8.0-11.3 (-13.0) µm wide (n = 27, mean width = 9.65 ± 1.65), smooth, unbranched, septate, dextrinoid, with walls (1.1-) 1.3-2.0 (-2.2) µm, intermixed with much narrower, occasionally branched, thin-walled hyphae (1.4-) 1.8-2.8 (-3.2) µm wide (n = 17, mean width = 2.34 ± 0.49). Clamp connections present in the base of basidia, the stipitipellis and the stipe trama. All tissues colorless in water and KOH.

Habitat and phenology: Scattered on dead fallen leaves of *Castanopsis sieboldii* (Makino) Hatus. ex T. Yamaz. et Mashiba, April to October.

Known distribution: Okinawa (Yaeyama Islands: Ishigaki Island and Iriomote Island).

Holotype: TNS-F-52268, on dead fallen leaves of *C. sieboldii,* Banna Park, Ishigaki Island, Okinawa Pref., 4 Apr. 2013, coll. Takahashi, H.

Other specimen examined: TNS-F-52269, on dead fallen leaves of *C. sieboldii,* Oomijya, Iriomote Island, Taketomi-cho, Yaeyama-gun, Okinawa Pref., 14 Oct. 2011, coll. Terashima, Y.

GenBank accession number: TNS-F-52268, AB971701.

Japanese name: Kareha-yakoutake (named by H. Takahashi, Y. Taneyama & Y. Terashima).

Comments: The mycenoid basidiomata, the dextrinoid hymenophoral and stipe trama, and the weakly amyloid basidiospores place the new species in the genus *Mycena* (Pers.) Roussel (Singer 1986; Maas Geesteranus 1992a, 1992b). With its very small basidiomata with thick-walled elements in the pileipellis, the new species appears similar to *Mycena brevisetosa* Corner from Malaysia (Corner 1994), which belongs to the genus *Mycena*, section *Longisetae* A.H.Sm. ex Maas Geest. (Maas Geesteranus 1983). The Malaysian species, however, is significantly different from *M. luxfoliata* in having a greyish pileus, collariate lamellae, inamyloid basidiospores, echinulate pileipellis elements, ventricose-filiform, thin-walled caulocystidia at the base of stipe, and a lignicolous habitat.

Its long, thick-walled, hair-like elements over the pileus surface parallel those seen in the genus *Crinipellis* Pat. (Patouillard 1889; Singer 1943, 1986). Unlike *Crinipellis*, the present fungus produces weakly amyloid basidiospores and lacks the dextrinoid hyphae on the pileipellis and stipitipellis.

References 引用文献

Corner EJH. 1994. Agarics in Malesia. II. Mycenoid. Beih Nova Hedwigia 109: 165-271.

Maas Geesteranus RA. 1983. Conspectus of the Mycenas of the northern hemisphere - 1. Sections *Sacchariferae, Basipedes, Bulbosae, Clavulares, Exiguae* and *Longisetae*. Proc K Ned Akad Wet 86 (3): 401-421.

Maas Geesteranus RA. 1992a. Mycenas of the Northern Hemisphere. I. Studies in Mycenas and other papers. Royal Netherlands Academy of Arts and Sciences, Amsterdam.

Maas Geesteranus RA. 1992b. Mycenas of the Northern Hemisphere. II. Conspectus of the Mycenas of the Northern Hemisphere. Royal Netherlands Academy of Arts and Sciences, Amsterdam.

Patouillard NT. 1889. Fragments mycologiques. Notes sur quelques champignons de la Martinique. J Bot, Paris 3: 335-343.

Singer R. 1942. A monographic study of the genera *Crinipellis* and *Chaetocalathus*. Lilloa 8: 441-534 (published in 1943).

Singer R. 1986. The Agaricales in modern taxonomy, 4th edn. Koeltz, Koenigstein.

24. カレハヤコウタケ（新種：高橋春樹，種山裕一＆寺嶋芳江新称）*Mycena luxfoliata* Har. Takah., Taneyama & Terashima, **sp. nov.**

肉眼的特徴（Figs. 182-185）：傘は径1-2.5 mm，最初半球形，のち饅頭形〜中高偏平（平凸），周縁部

において浅い条線を表す；縁部は細鋸歯状；表面は全体に柔毛に被われ，非吸水性，粘性を欠き，純白色；縁部は細鋸歯状で内側に巻かない．肉は厚さ0.1 mm以下，純白色，特別な匂いはなく，味は不明．柄は3-5 (-9) ×0.1-0.3 mm，ほぼ上下同大，しばしば根元がやや拡大し，中心生，痩せ型；表面は全体に微粉状，純白色，乾生，根元に発達した菌糸体は見られない．ヒダは直生〜やや垂生，弓状，疎 (柄に到達するヒダは7-11)，連絡脈を欠き，幅0.2 mm以下，純白色；縁部は同色，長縁毛状．

発光性：子実体は発光しない；基質（枯れ葉）上の菌糸体が弱く発光する．

顕微鏡的特徴 (Figs. 180, 181)：担子胞子は (5.6-) 6.6-7.9 (-8.6) × (3.1-) 3.6-4.3 (-4.7) µm (n = 82, mean length = 7.26±0.64, mean width = 3.96±0.34, Q = (1.50-) 1.70-1.98 (-2.08), mean Q = 1.84 ±0.14)，楕円形〜長楕円形，平坦，無色，弱アミロイド，アルカリ溶液において無色，薄壁．担子器は (12.1-) 14.4-17.8 (-18.5) × (7.8-) 8.0-8.7 (-8.8) µm (本体)，(3.3-) 3.5-4.7 (-4.9) × (1.7-) 1.8-2.0 (-2.2) µm (ステリグマ)，こん棒形，4胞子性．縁シスチジアは不実帯をなさし，2形性：1) (8.9-) 11.8-17.2 (-22.6) × (5.9-) 6.6-8.5 (-10.3) µm，亜こん棒形，複数の不規則な珊瑚の枝状〜短指状の頂生付属糸 (2.6-) 3.1-5.1 (-6.3) × (1.3-) 1.7-2.4 (-3.0) µm を持ち，無色，非アミロイド，壁は厚さ 0.8-1.2 (-1.4) µm；2) (23.1-) 25.6-34.3 (-40.5) × (7.1-) 8.1-9.9 (-10.6) µm，片膨れ状紡錘形，時に頂部がやや頭状形になり，平滑，無色，非アミロイド，壁は厚さ (0.7-) 0.8-1.2 (-1.4) µm．子実層托実質は平列し，傘実質に類似する．傘の表皮組織は緩くもつれて錯綜した匍匐性の菌糸からなる；菌糸細胞は幅 (2.0-) 2.2-3.0 (-3.5) µm，円柱形，しばしば分岐し，平滑，非アミロイド，壁は厚さ0.9-1.3 µm；傘シスチジアはない．傘実質の菌糸は (10.5-) 18.1-31.8 (-42.5) × (9.0-) 9.9-12.6 (-16.4) µm，並列し，円柱形，平滑，無色，弱アミロイド，薄壁．柄表皮を構成する菌糸は幅 (2.2-) 2.5-3.6 (-4.0) µm，並列し，円柱形，平滑，無色，非アミロイド，壁は厚さ 0.6-0.9 µm．柄シスチジアは2形性：1) 柄上部に存在する途中にくびれがある片膨れ状紡錘形の柄シスチジアは (16.2-) 17.8-24.1 (-32.5) × (3.7-) 5.8-8.8 (-11.3) µm，散在し，しばしば束状に集り，平滑，無色，非アミロイド，壁の厚さは0.9-1.4 (-1.7) µm；2) 柄下部に存在する不規則な円柱形〜珊瑚の枝状柄シスチジアは (25.6-) 30.4-54.5 (-75.5) × (2.3-) 3.0-4.3 (-5.0) µm，散在し，不規則に枝分かれし，平滑，無色，非アミロイド，(0.6-) 0.8-1.1 (-1.2) µm．柄実質の菌糸は (42.1-) 50.1-98.1 (-144.4) × (6.1-) 8.0-11.3 (-13.0) µm，平列し，円柱形，無分岐，隔壁を有し，平滑，無色，偽アミロイド，厚壁 (1.1-) 1.3-2.0 (-2.2) µm，時折分岐する薄壁な菌糸 (1.4-) 1.8-2.8 (-3.2) µm が混在する．担子器と縁シスチジアの根元にクランプを有する．全ての組織の菌糸は水およびアルカリ溶液において無色．

生態および発生時期：スダジイの枯れ葉上に散生，4月〜10月．

分布：沖縄（八重山諸島：石垣島，西表島）．

供試標本：TNS-F-52268（正基準標本），スダジイの枯れ葉上，沖縄県石垣市バンナ公園，2013年4月4日，高橋春樹採集；TNS-F-52269，スダジイの枯れ葉上，沖縄県八重山郡竹富町西表島大見謝，2011年10月14日，寺嶋芳江採集．

GenBank登録番号：TNS-F-52268，AB971701．

主な特徴：子実体は柔毛に被われた傘と粉状の柄を持つクヌギタケ型で，微小，全体に純白色を呈する；胞子は弱アミロイド，楕円形〜長楕円形；縁シスチジアは2形性；傘の表皮組織は厚壁な毛状菌糸からなる；柄シスチジアは2形性；基質（枯れ葉）上の菌糸体発光性；スダジイの枯れ葉から発生．

コメント：クヌギタケ型の子実体，偽アミロイドに染まる実質，そして弱アミロイドの担子胞子は本種がクヌギタケ属 *Mycena* (Singer 1986；Maas Geesteranus 1992a, 1992b) に近縁であることを示唆している．微小な子実体と厚壁な菌糸からなる傘表皮の特徴はクヌギタケ属 *Mycena*，ロン

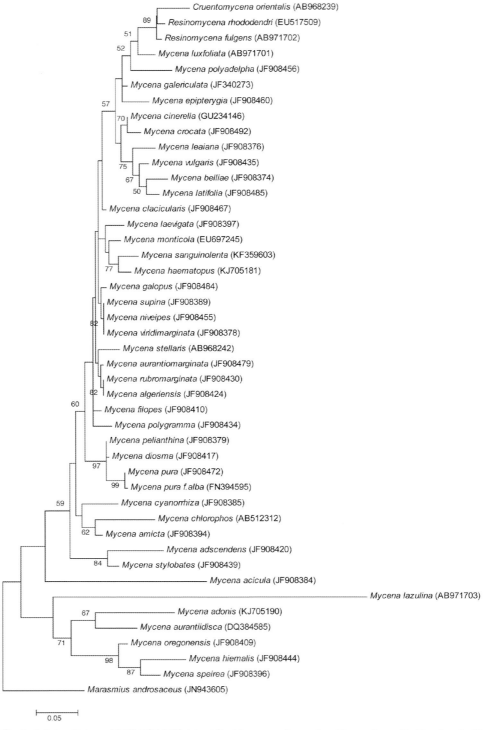

Fig. 179 – A phylogenetic tree of ITS1-5.8S-ITS2 dataset for *Mycena* species, analyzed by maximum likelihood method based on the Tamura-Nei model (Tamura and Nei 1993) in MEGA5 (Tamura et al. 2011). Numbers on the branches represent bootstrap values obtained from 2000 replications (only values greater than 50% are shown).
核リボソーム ITS1-5.8S-ITS2を用いた最尤法によるクヌギタケ属系統樹.

ギセタエ節 section *Longisetae* A.H.Sm. ex Maas Geest.（Maas Geesteranus 1983）に属するマレーシア産 *Mycena brevisetosa* Corner（Corner 1994）と共通する．しかしながらマレーシア産種は傘が灰色を呈すること，ヒダが襟帯を形成すること，担子胞子が非アミロイドであること，傘表皮組織が微突起に被われた細胞からなること，柄シスチジアが薄壁であること，そして材上生の生態の特徴において相違が見られる．

傘表皮組織の厚壁な毛状菌糸はニセホウライタケ属 *Crinipellis* Pat.（Patouillard 1889；Singer 1943, 1986）を想起させるが，カレハヤコウタケは弱アミロイドの担子胞子を有し，傘および柄の表皮組織の菌糸が偽アミロイドに染まらない点でニセホウライタケ属とは異質である．

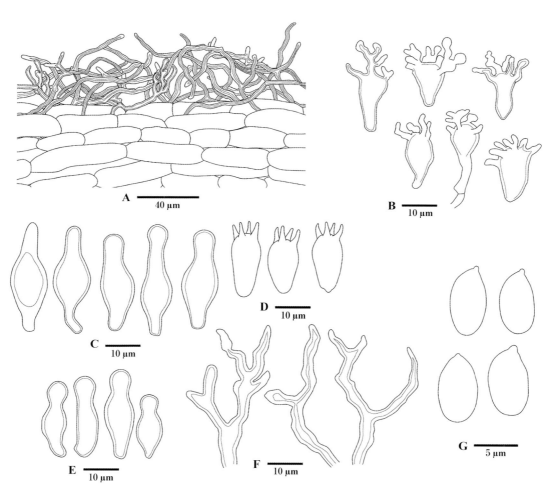

Fig. 180 – Micromorphological features of *Mycena luxfoliata* (holotype): **A**. Longitudinal cross section of the pileipellis and pileitrama. **B**. Subclavate cheilocystidia with coralloid to several finger-like apical appendages. **C**. Fusoid-ventricose or subcapitulate cheilocystidia. **D**. Basidia. **E**. Fusoid-ventricose caulocystidia with a median constriction at the upper portion of stipe. **F**. Irregularly cylindrical to coralloid caulocystidia at the lower portion of stipe. **G**. Basidiospores. Illustrations by Taneyama, Y.

カレハヤコウタケの顕微鏡図（正基準標本）：**A**. 傘表皮組織および傘実質の縦断面．**B**. 複数の不規則な珊瑚の枝状〜短指状の頂生付属糸を持つ縁シスチジア．**C**. 片膨れ状紡錘形または頂部がやや頭状形の縁シスチジア．**D**. 担子器．**E**. 柄の上部に存在する途中にくびれがある片膨れ状紡錘形の柄シスチジア．**F**. 柄の下部に存在する円柱形〜珊瑚の枝状柄シスチジア．**G**. 担子胞子．図：種山裕一．

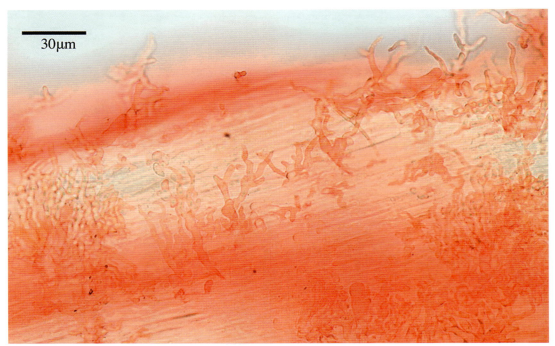

Fig. 181 – Irregularly cylindrical to coralloid caulocystidia of *Mycena luxfoliata* (holotype) at the lower portion of stipe (in 3% KOH and Congo red stain). Photo by Taneyama, Y.
柄の下部におけるカレハヤコウタケの珊瑚の枝状〜不規則な円柱形の柄シスチジア（3%水酸化カリウム溶液で封入した後コンゴー赤染色，正基準標本）．写真：種山裕一．

Fig. 182 – Basidiomata of *Mycena luxfoliata* (holotype) on a dead fallen leaf of *C. sieboldii*, 4 Apr. 2013, Banna Park, Ishigaki Island. Photo by Takahashi, H.
スダジイの枯れ葉上に発生したカレハヤコウタケの子実体（正基準標本），2013年4月4日，石垣島バンナ公園．写真：高橋春樹．

Fig. 183 – Basidiomata of *Mycena luxfoliata* (holotype) on a dead fallen leaf of *C. sieboldii*, 4 Apr. 2013, Banna Park, Ishigaki Island. Photo by Takahashi, H.
スダジイの枯れ葉上に発生したカレハヤコウタケの子実体（正基準標本），2013年4月4日，石垣島バンナ公園．写真：高橋春樹．

Fig. 184 – Basidiomata of *Mycena luxfoliata* (holotype) on dead fallen leaves of *C. sieboldii*, 4 Apr. 2013, Banna Park, Ishigaki Island. Photo by Takahashi, H.
スダジイの枯れ葉上に発生したカレハヤコウタケの子実体（正基準標本），2013年4月4日，石垣島バンナ公園．写真：高橋春樹．

24. *Mycena luxfoliata* カレハヤコウタケ

Fig. 185 – Mycelia of *Mycena luxfoliata* (TNS-F-52269) emitting yellowish green light in dead fallen leaves of *C. sieboldii*, 14 Oct. 2011, Oomijya, Iriomote Island. Photo by Takahashi, H.
黄緑色に発光するスダジイの枯れ葉上の菌糸体（TNS-F-52269），2011年10月14日，西表島大見謝．写真：高橋春樹．

25. *Mycena stellaris* Har. Takah., Taneyama & A. Hadano, **sp. nov.** ホシノヒカリタケ
MycoBank no.: MB 809935.

Etymology: The specific epithet comes from the Latin word for "star-like", referring to the luminescence of basidiomata.

The important combination of features delimiting this species is small, white omphalinoid basidiomata with somewhat concave mature pilei and arcuate-decurrent lamellae; subglobose, weakly amyloid basidiospores (mean 7.59 × 6.05 μm); fusoid-ventricose cheilocystidia with one or two short mucrones or digitate appendages up to 5 μm long; a pileipellis and stipitipellis of repent, cylindrical, smooth hyphae; thick-walled generative hyphae in the stipe trama; luminescence in the pileus, lamellae and stipe; and a lignicolous habit.

Macromorphological characteristics (Figs. 189–193): Pileus 2–5 mm in diameter, at first hemispherical, then convex with depressed center, at first smooth but soon radially pellucid-striate; surface glabrous, hygrophanous, dry, pure white but sometimes pale greyish brown toward the center in older specimens; margin entire. Flesh very thin (up to 0.2 mm), pure white, odor not distinctive, taste undetermined. Stipe 6–18 × 1–1.5 mm, almost equal but slightly thickened toward the apex, central, slender, terete, smooth; surface glabrous, dry, pure white; base covered with conspicuous white mycelioid bristles. Lamellae arcuate-decurrent, subdistant (12–14 reach the stipe), with 1–2 series of lamellulae, not intervenose, narrow (up to 0.3 mm broad), thin, pure white; edges finely fimbriate under a lens, concolorous.

Luminescence: Pileus, lamellae and stipe strongly emit yellowish green light; cultured mycelia produce weak luminescence.

Micromorphological characteristics (Figs. 187, 188): Basidiospores (6.1–) 7.0–8.2 (–9.0) × (4.3–) 5.6–6.5 (–7.7) μm (n = 116, mean length = 7.59 ± 0.62, mean width = 6.05 ± 0.49, Q = (1.06–) 1.17–1.35 (–1.62), mean Q = 1.26 ± 0.09), subglobose, smooth, hyaline, weekly amyloid, colorless in KOH, thin-walled. Basidia (23.5–) 25.3–31.9 (–35.9) × (6.4–) 7.3–9.2 (–10.1) μm (n = 28, mean length = 28.58 ± 3.30, mean width = 8.22 ± 0.95) in main body, (4.0–) 4.5–6.2 (–7.1) × (1.6–) 1.9–2.3 (–2.5) μm (n = 17, mean length = 5.33 ± 0.83, mean width = 2.11 ± 0.21) in sterigmata, clavate, four-spored, weekly dextrinoid. Cheilocystidia (23.4–) 25.2–31.3 (–37.6) × (8.0–) 8.6–12.9 (–16.2) μm, (n = 33, mean length = 28.28 ± 3.03, mean width = 10.74 ± 2.17), abundant, fusoid-ventricose, occasionally with one or two short mucrones or digitate apical appendages (3.5–) 4.3–8.3 (–11.9) μm length (n = 32, mean width = 6.30 ± 2.02), hyaline, inamyloid, thin-walled. Pleurocystidia absent. Hymenophoral trama subregular; element hyphae (5.6–) 7.7–12.4 (–14.5) μm wide (n = 23, mean width = 10.05 ± 2.35), cylindrical, occasionally somewhat inflated, walls thin, smooth, hyaline, dextrinoid; subhymenium dextrinoid. Pileipellis a cutis of interwoven, repent hyphae (3.8–) 5.2–14.7 (–22.3) μm wide (n = 29, mean width = 9.97 ± 4.76), cylindrical or somewhat inflated, smooth or thinly encrusted with crystalline granules, hyaline, inamyloid, thin-walled, intermixed with occasionally branched, non-encrusted, filiform, sometimes sinuate hyphae (2.0–) 2.2–3.3 (–4.3) μm wide (n = 32, mean width = 2.75 ± 0.52); pilocystidia none. Elements of pileitrama (6.2–) 8.8–14.1 (–17.3) μm wide (n = 32, mean width = 11.42 ± 2.65), loosely intricate, cylindrical, smooth, hyaline, dextrinoid, thin-walled. Stipitipellis a cutis of parallel, repent hyphae (2.1–) 2.7–4.6 (–6.6) μm wide (n = 36, mean width = 3.68 ± 0.94), cylindrical, with hyaline, thin, granular incrustations, inamyloid, with walls 0.4–0.5 (–0.6) μm (n = 13, mean thickness = 0.49 ± 0.05); caulocystidia absent. Stipe trama composed of longitudinally running, cylindrical hyphae (5.9–) 7.1–11.6 (–13.6) μm wide (n = 27, mean width = 9.36 ± 2.22), monomitic, unbranched, smooth, hyaline, inamyloid, with walls (0.8–) 1.1–1.6 (–1.8) μm (n = 16, mean thickness = 1.39 ± 0.25). Elements of basal mycelioid bristles (2.4–) 2.6–3.4 (–3.8) μm wide (n = 26, mean width = 2.99 ± 0.39), cylindrical, hyaline, inamyloid, with walls (0.5–) 0.6–0.9 (–0.9) μm (n = 16, mean thickness = 0.76 ± 0.13). Clamps present and colorless in KOH in all tissues.

Habitat and phenology: Gregarious or scattered on dead fallen logs in *Castanopsis-Quercus* forests, May to June.

Known distribution: Okinawa (Ishigaki Island, Okinawa Island).

Holotype: TNS-F-52282, on dead fallen logs (unidentified substrata) in an evergreen broad-leaved forest dominated by *Q. miyagii* Koidz. and *C. sieboldii* (Makino) Hatus. ex T. Yamaz. et Mashiba, Banna Park, Ishigaki-shi, Okinawa Pref., 21 May 2005, coll. Hadano, E.

Other specimen examined: TNS-F-52283, on dead fallen logs in an evergreen broad-leaved forest, Okinawa Island, Okinawa Pref., 2 Jun. 2012, coll. Hadano, A.

Gene sequenced specimen and GenBank accession number: TNS-F-52283, AB968242.

Japanese name: Hosino-hikaritake (named by A. Hadano).

Comments: The weekly amyloid basidiospores, the smooth cheilocystidia, and the non-spinulose, non-gelatinous pileipellis and stipitipellis place this species in the genus *Mycena*, section *Fragilipedes* (Fr.) Quel. (Maas Geesteranus 1992a, 1992b), where it appears to be closely related to *Mycena silvaelucens* B.A. Perry & Desjardin from Malaysia (Desjardin et al. 2010) largely due to having small, lignicolous, luminescent basidiomata and fusoid-ventricose, mucronate cheilocystidia. The latter species, however, is significantly different from *M. stellaris* in having mycenoid basidiomata with convex to applanate (not concave), greyish brown to pale grey pilei and subdecurrent, intervenose lamellae, strongly amyloid, broadly ellipsoid, slightly longer basidiospores: mean 8.3×5.8 μm in *M. silvaelucens* (Desjardin et al. 2010) compared to mean 7.59×6.05 μm in *M. stellaris*, erect, cylindrical caulocystidia, and thin-walled, much broader medullary elements of the stipe: up to 20 μm diam. (Desjardin et al. 2010).

The small, white omphalinoid basidiomata of *M. stellaris* bear a superficial resemblance to four Malaysian taxa described by Corner (Corner 1994), viz. *Mycena obscuritatis* Corner, *Mycena parsimonia* Corner, *Mycena putroris* Corner, and *Mycena pocilliformis* Corner, the bioluminescence characteristics of which are unknown. The latter four taxa differ primarily from *M. stellaris* in yielding inamyloid basidiospores and lacking clamp connections.

References 引用文献

Corner EJH. 1994. Agarics in Malesia. II Mycenoid. Beih Nova Hedwigia 109: 165–271.
Desjardin DE, Perry BA, Lodge DJ, Stevani CV, Nagasawa E. 2010. Luminescent *Mycena*: new and noteworthy species. Mycologia 102 (2): 459–477.
Maas Geesteranus RA. 1992a. Mycenas of the Northern Hemisphere. I. Studies in Mycenas and other papers. Royal Netherlands Academy of Arts and Sciences, Amsterdam.
Maas Geesteranus RA. 1992b. Mycenas of the Northern Hemisphere. II. Conspectus of the Mycenas of the Northern Hemisphere. Royal Netherlands Academy of Arts and Sciences, Amsterdam.

25. ホシノヒカリタケ（新種；波多野敦子新称）*Mycena stellaris* Har. Takah., Taneyama & A. Hadano, **sp. nov.**

肉眼的特徴（Figs. 189-193）：傘は径2-5 mm，最初半球形，のち中央部がやや へこんだ饅頭形，最初平坦，まもなく半透明の条線を放射状に表す；表面は平滑，吸水性，粘性を欠き，純白色，時に老成すると中心部に向かって淡灰褐色を帯びる；縁部は全縁で内側に巻かない．肉は厚さ0.2

mm 以下，白色，特別な味や匂いはない．柄は6-18×1-1.5 mm，円柱形で頂部に向かってやや太くなり，中心生，痩せ型；表面は平滑，純白色，乾生，根元は白色の毛状菌糸に被われる．ヒダは垂生，やや疎（柄に到達するヒダ12-14），1-2の小ヒダを伴い，連絡脈を欠き，幅0.3 mm以下，純白色；縁部は同色，長縁毛状．

発光性：子実体全体が黄緑色に強く発光；培養菌糸体も微弱な発光性が確認されている．

顕微鏡的特徴（Figs. 187, 188）：担子胞子は (6.1-) 7.0-8.2 (-9.0) × (4.3-) 5.6-6.5 (-7.7) μm（n = 116, mean length = 7.59±0.62, mean width = 6.05±0.49, Q = (1.06-) 1.17-1.35 (-1.62), mean Q = 1.26±0.09），亜球形，平坦，無色，弱アミロイド，アルカリ溶液において無色，薄壁．担子器は (23.5-) 25.3-31.9 (-35.9) × (6.4-) 7.3-9.2 (-10.1) μm（本体），(4.0-) 4.5-6.2 (-7.1) × (1.6-) 1.9-2.3 (-2.5) μm（ステリグマ），こん棒形，4胞子性，弱偽アミロイド．縁シスチジアは (23.4-) 25.2-31.3 (-37.6) × (8.0-) 8.6-12.9 (-16.2) μm，群生し，片膨れ状紡錘形，しばしば頂部に1-2個の短指状突起（長さ (3.5-) 4.3-8.3 (-11.9) μm）を具え，無色，非アミロイド，薄壁．側シスチジアを欠く．子実層托実質を構成する菌糸は幅 (5.6-) 7.7-12.4 (-14.5) μm，ほぼ平列し，円柱形またはやや膨大し，平滑，無色，偽アミロイド，薄壁；子実下層は偽アミロイド．傘の表皮組織は錯綜した匍匐性の菌糸からなる；菌糸は幅 (3.8-) 5.2-14.7 (-22.3) μm，円柱形またはやや膨大し，平滑または透明な結晶物が薄く付着し，無色，非アミロイド，薄壁，しばしば分岐する細い糸状の平滑な菌糸（幅 (2.0-) 2.2-3.3 (-4.3) μm）が混在する；傘シスチジアはない．傘実質の菌糸は幅 (6.2-) 8.8-14.1 (-17.3) μm，緩く錯綜し，円柱形，平滑，無色，偽アミロイド，薄壁．柄表皮組織は並列型；菌糸は幅 (2.1-) 2.7-4.6 (-6.6) μm，円柱形，平滑または透明な結晶物に薄く被われ，無色，非アミロイド，壁は厚さ 0.4-0.5 μm；柄シスチジアはない．柄実質の菌糸は幅 (5.9-) 7.1-11.6 (-13.6) μm，平列し，円柱形，平滑，無色，非アミロイド，隔壁を有し，厚壁 (1.1-1.6 μm)．根元の毛状菌糸を校正する菌糸は幅 (2.4-) 2.6-3.4 (-3.8) μm，円柱形，無色，非アミロイド，壁は厚さ (0.5-) 0.6-0.9 (-0.9) μm．全ての組織はクランプを有し，アルカリ溶液において無色．

生態および発生時期：常緑広葉樹林内の腐朽した材上に群生または散生，5月〜6月．

分布：沖縄（石垣島，沖縄本島）．

供試標本：TNS-F-52282（正基準標本），スダジイとオキナワウラジロガシを主体とする常緑広葉樹林内の腐朽した材上，沖縄県石垣市バンナ公園，2005年5月21日，波多野英治採集；TNS-F-52283，常緑広葉樹林内の腐朽した材上，沖縄県沖縄島，2012年6月2日，波多野敦子採集．

分子解析に用いた標本並びにGenBank登録番号：TNS-F-52283, AB968242.

主な特徴：子実体は小型（傘の径 3-8 mm）のヒダサカズキタケ型で，全体に白色を呈する；傘は白色，老成すると部分的に淡褐色を帯びる；担子胞子は類球形で弱アミロイド；縁シスチジアは片膨れ状紡錘形で頂部に微突起を持つ；傘と柄の表皮組織の菌糸は平滑；柄の実質は隔壁を有する厚壁（厚さ1.1-1.6 μm）な菌糸細胞からなる；子実体全体が黄緑色に発光する；広葉樹の朽木から発生．

コメント：アミロイドに染まる担子胞子，平滑な縁シスチジア，そして平滑な菌糸からなる傘と柄の表皮組織の特徴は本種が Maas の分類概念（Maas Geesteranus 1992a, 1992b）によるクヌギタケ属 *Mycena*，アクニオイタケ節 section *Fragilipedes* (Fr.) Quel. に属することを示唆している．節内に所属する分類群の中で，最近マレーシアから記載された発光菌 *Mycena silvaelucens* B.A. Perry & Desjardin（Desjardin et al. 2010）は形態学的特徴において本種に最も近縁と思われるが，子実体がクヌギタケ型であること，傘と柄が灰褐色を帯びること，ヒダに連絡脈を有すること，強アミロイドに染まりやや大形：平均 8.3×5.8 μm（Desjardin et al. 2010）（ホシノヒカリタケは平均7.59×6.05 μm）で広楕円形の担子胞子を持つこと，柄の表皮組織に直立した円柱形の柄シ

25. *Mycena stellaris* ホシノヒカリタケ

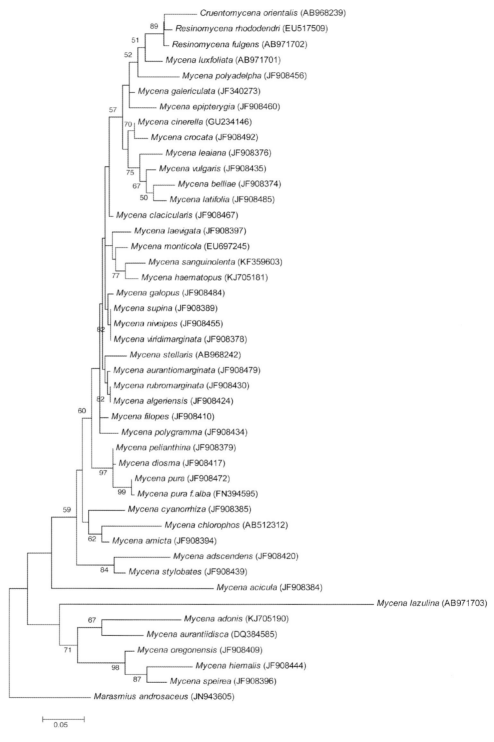

Fig. 186 – A phylogenetic tree of ITS1-5.8S-ITS2 dataset for *Mycena* species, analyzed by maximum likelihood method based on the Tamura-Nei model (Tamura and Nei 1993) in MEGA5 (Tamura et al. 2011). Numbers on the branches represent bootstrap values obtained from 2000 replications (only values greater than 50% are shown).
核リボソーム ITS1-5.8S-ITS2を用いた最尤法によるクヌギタケ属系統樹.

スチジアが存在すること，そして柄実質の菌糸がより太く薄壁な点で明らかな形態学的有意差が認められる．

　Corner（1994）によってマレーシアから報告された *Mycena obscuritatis*, *Mycena parsimonia*, *Mycena putroris*, *Mycena pocilliformis* の4種（発光性は不明）は，ヒダサカズキタケ型で白色，微小な子実体を形成する点においてホシノヒカリタケに類似するが，菌糸にクランプを欠き，非アミロイドの担子胞子を持つとされている．

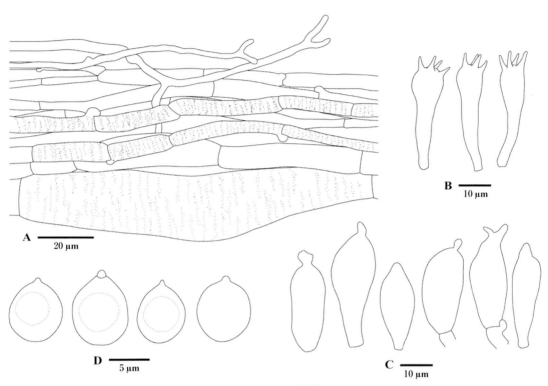

Fig. 187 − Micromorphological features of *Mycena stellaris* (TNS-F-52283): **A**. Longitudinal cross section of the pileipellis. **B**. Basidia. **C**. Cheilocystidia. **D**. Basidiospores. Illustrations by Taneyama, Y.
ホシノヒカリタケの顕微鏡図（TNS-F-52283）：**A**. 傘表皮組織の縦断面．**B**. 担子器．**C**. 縁シスチジア．**D**. 担子胞子．図：種山裕一．

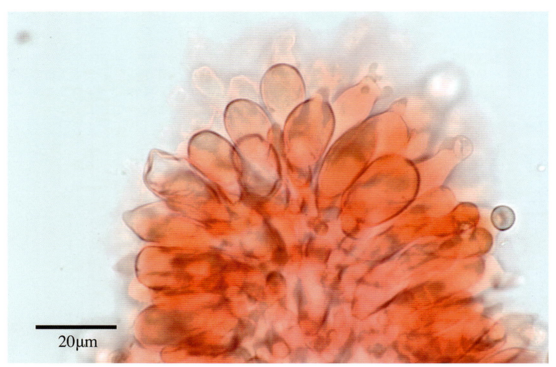

Fig. 188 – Lamella edge of *Mycena stellaris* (in 3%KOH and Congo red stain, TNS-F-52283) showing gregarious cheilocystidia. Photo by Taneyama, Y.
ホシノヒカリタケのヒダ縁部（3%水酸化カリウム溶液で封入した後コンゴー赤染色，TNS-F-52283）．群生した縁シスチジアを示す．写真：種山裕一．

Fig. 189 – Basidiomata of *Mycena stellaris* (holotype) on a dead fallen log in an evergreen broad-leaved forest, 21 May 2005, Banna Park, Ishigaki Island. Photo by Takahashi, H.
広葉樹林内の腐木上に発生したホシノヒカリタケの子実体(正基準標本), 2005年5月21日, 石垣島バンナ公園. 写真: 高橋春樹.

Fig. 190 – Basidiomata of *Mycena stellaris* (holotype) on a dead fallen log in an evergreen broad-leaved forest, 21 May 2005, Banna Park, Ishigaki Island. Photo by Takahashi, H.
広葉樹林内の腐木上に発生したホシノヒカリタケの子実体（正基準標本），2005年5月21日，石垣島バンナ公園．写真：高橋春樹．

Fig. 191 – Basidiomata of *Mycena stellaris* (holotype) on a dead fallen log in an evergreen broad-leaved forest, 21 May 2005, Banna Park, Ishigaki Island. Photo by Takahashi, H.
広葉樹林内の腐木上に発生したホシノヒカリタケの子実体（正基準標本），2005年5月21日，石垣島バンナ公園．写真：高橋春樹．

25. *Mycena stellaris* ホシノヒカリタケ

Fig. 192 – Basidiomata of *Mycena stellaris* (holotype) on a dead fallen log in an evergreen broad-leaved forest, 21 May 2005, Banna Park, Ishigaki Island. Photo by Takahashi, H.
広葉樹林内の腐木上に発生したホシノヒカリタケの子実体（正基準標本），2005年5月21日，石垣島バンナ公園．写真：高橋春樹．

25. *Mycena stellaris* ホシノヒカリタケ —— 237

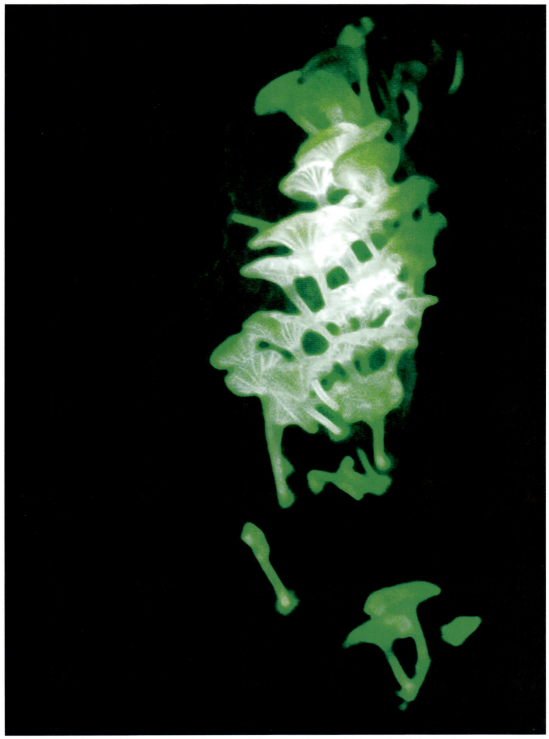

Fig. 193 – Basidiomata of *Mycena stellaris* (holotype) emitting yellowish green light on a dead fallen log in an evergreen broad-leaved forest, 21 May 2005, Banna Park, Ishigaki Island. Photo by Takahashi, H.
広葉樹林内の腐木上において黄緑色に発光するホシノヒカリタケの子実体（正基準標本），2005年5月21日，石垣島バンナ公園．写真：高橋春樹．

26. *Pleurotus nitidus* Har. Takah. & Taneyama, **sp. nov.** シロヒカリタケ

MycoBank no.: MB 809936.

Etymology: The specific epithet comes from the Latin meaning "shining", referring to the luminescence of the basidiomata.

Distinctive features of this species are found in white, pleurotoid basidiomata with a short, lateral stipe and deeply decurrent lamellae often stained with dark purplish brown spots; ellipsoid basidiospores; non-inflated, dextrinoid tramal hyphae with thickened walls (up to 1.2 μm) in the pileus and hymenophore; occasionally nodulose terminal elements of the pileipellis mostly consisting of cylindrical, 3-9 short cells with clampless septa; gregarious, subcylindrical, often strangulated caulocystidia protruding from the stipitipellis; strong luminescence of the basidiomata; and basidiome formation on dead corticated logs.

Macromorphological characteristics (Figs. 197-204): Basidiomata pleurotoid. Pileus 20-60 (-90) mm in diameter, semiorbicular to reniform, at first convex, soon applanate and depressed at the center, sometimes subumbonate, shallowly sulcate-striate toward the crenulate margin; surface dry, subhygrophanous, fibrillose toward the center, pure white when young, then tinged with light yellow (1A4) to greyish yellow (1B4-5) near the center, occasionally greyish yellow (1B4-5) overall with age, often with dark Magenta (13F7-8) stains or spots in places when mature. Flesh up to 2 mm thick, whitish, soft; odor and taste not distinctive. Stipe 10-20×3-8 mm, very short, cylindrical, eccentric to almost lateral, terete, solid; surface dry, fibrillose, pure white; base covered with white, strigose mycelial hairs. Lamellae deeply decurrent, subcrowded (17-22 reach the stipe), with 1-2 series of lamellulae, up to 5 mm broad, at times slightly intervenose, white, sometimes with dark Magenta (13F7-8) spots here and there when mature; edges entire, concolorous. Spore print pure white.

Luminescence: Pileus, lamellae and stipe strongly emit yellowish green light; mycelium luminescence unknown.

Micromorphological characteristics (Figs. 194-196): Basidiospores (4.0-) 4.9-6.1 (-7.1)×(2.4-) 2.9-3.4 (-3.7) μm (n = 34, mean length = 5.52±0.61, mean width = 3.13±0.24, Q = (1.50-) 1.63-1.90 (-1.99), mean Q = 1.77±0.14), ellipsoid, smooth, hyaline, colorless in KOH, inamyloid, thin-walled. Basidia (16.7-) 16.8-23.6 (-25.2)×(3.5-) 3.9-5.3 (-5.5) μm (n = 10, mean length = 20.23±3.41, mean width = 4.59±0.69) in main body, (3.8-) 4.2-5.8 (-6.4)×(1.3-) 1.4-1.8 (-1.8) μm (n = 12, mean length = 4.97 ±0.81, mean width = 1.55±0.20) in sterigmata, clavate, 4-spored; basidioles clavate. Cheilocystidia and pleurocystidia absent. Hymenophoral trama subregular; element hyphae (2.6-) 3.4-5.7 (-6.3) μm wide (n = 24, mean width = 4.55±1.14), cylindrical, smooth, hyaline, pale yellow in KOH, dextrinoid, with thickened walls (1.1-) 1.3-2.1 (-2.5) μm; hymenium pale yellow in KOH, inamyloid or weakly dextrinoid. Pileipellis a cutis of repent, cylindrical hyphae (2.1-) 2.5-3.5 (-4.3) μm wide (n = 37, mean width = 3.01±0.53), occasionally with nodulose outgrowths, not encrusting, hyaline, pale yellow in KOH, inamyloid, thin-walled; terminal elements mostly made up of 3-9 short, cylindrical cells (3.4-) 3.7-8.2 (-15.8)×(2.4-) 2.9-3.8 (-4.4) μm (n = 30, mean length = 5.94±2.25, mean width = 3.38± 0.44), with secondary septa. Pileitrama monomitic; constituent hyphae (2.6-) 4.0-5.3 (-6.1) μm wide (n = 32, mean width = 4.62±0.67), parallel, cylindrical, not inflated, smooth, hyaline, pale yellow in KOH, dextrinoid, with walls 0.8-1.2 μm. Outermost layer of stipitipellis consisting of crowded, vertically arranged terminal elements (caulocystidia) projecting from the surface; caulocystidia (33.4-) 38.8-61.7 (-75.5)×(3.9-) 4.4-5.7 (-6.3) μm (n = 22, mean length = 50.21±11.46, mean width = 5.05±0.64), gregarious, subcylindrical, often strangulated, smooth, hyaline, colorless in KOH, inamyloid, with walls 0.9-1.1 μm, arising directly from the underlying layer made up of densely interwoven hyphae (3.3-)

3.8–4.8 (–5.3) μm wide (n = 46, mean width = 4.28 ± 0.53) with thickened walls (0.9–) 1.1–1.7 (–2.1) μm. Stipe trama composed of longitudinally running, cylindrical hyphae (3.1–) 3.8–5.1 (–6.1) μm wide (n = 43, mean width = 4.47 ± 0.64), monomitic, smooth, hyaline, colorless in KOH, inamyloid, with thickened walls 1.3–2.1 (–2.4) μm. Elements of basal mycelium (2.6–) 3.2–4.3 (–5.2) μm wide (n = 36, mean width = 3.78 ± 0.55), cylindrical, often sinuate, occasionally with nodulose outgrowths, hyaline, colorless in KOH, inamyloid, with thickened walls (0.7–) 0.9–1.5 (–1.8) μm. All tissues with clamp connections.

Habitat and phenology: Scattered to gregarious, on dead corticated logs (various kinds of unidentified substrata) in evergreen broad-leaved forests, April to October.

Known distribution: Okinawa (Yaeyama Islands: Ishigaki Island and Iriomote Island).

Holotype: TNS-F-52277, on dead corticated logs in an evergreen broad-leaved forest, Banna Park, Ishigaki-shi, Okinawa, 15 Jul. 2011, coll. Takahashi, H.

Other specimens examined: TNS-F-61366, on dead corticated logs in an evergreen broad-leaved forest, Banna Park, Ishigaki-shi, Okinawa, 29 May 2012, coll. Takahashi, H.; TNS-F-61367, same location, 6 Jun. 2012, coll. Takahashi, H; KPM-NC0023865, same location, 16 Oct. 2011, coll. Takahashi, H; KPM-NC0023864, on dead corticated logs in an evergreen broad-leaved forest, Funaura, Iriomote Island, Okinawa Pref., 30 May 2002, coll. Takahashi, H.

Japanese name: Siro-hikaritake (named by Dr. G. Miyagi).

Comments: Although the ellipsoid basidiospores, the nodulose elements of the pileipellis, and the dextrinoid tramal hyphae appear to be out of place, we provisionally relegate the present fungus to the genus *Pleurotus* sensu lato based on a combination of characteristics, such as the gymnocarpic, pleurotoid basidiomata with a short, lateral stipe and deeply decurrent lamellae, the hyaline, inamyloid basidiospores, the poorly developed pileipellis, the thick-walled tramal hyphae with clamped septa, and a lignicolous habit.

The new species is closely related to the following four taxa: *Pleurotus eugrammus* (Mont.) Dennis originally described from Cuba (Montagne 1837; Singer 1944; Dennis 1953; Corner 1981; Pegler 1983); *P. eugrammus* var. *radicicola* Corner from Malaysia (Corner 1981); *P. eugrammus* var. *brevisporus* Corner from Malaysia (Corner 1981); and *Pleurotus hygrophanus* (Earle) Sacc. & Traverso from Cuba (Earle 1906; Saccardo and Traverso 1911; Pegler 1983). The tramal hyphae in *P. eugrammus* var. *eugrammus*, *P. eugrammus* var. *brevisporus* and *P. hygrophanus* are inflated (usually up to 13 μm wide: Corner 1981) and constricted at the septa. Moreover, *P. eugrammus* var. *eugrammus* has much longer, ellipsoid to subcylindrical basidiospores: 6–9 μm (Corner 1981) and inamyloid hyphae in the hymenophoral trama (Singer 1944). Even though *P. eugrammus* var. *radicicola* shares non-inflated, narrow tramal hyphae: 2–7 μm wide (Corner 1981) with *P. nitidus*, it is distinct in possessing a pale brownish tan or pale pinkish tan pileus when young, pileipellis elements encrusted with thin, pale brown, granular pigments, and is invariably found on the roots (living and dead) of trees and palms in the open.

Using topotypic materials that included Malaysian specimens, Petersen and Krisai-Greilhuber (1999) carefully examined the differences between *P. eugrammus* and *Neonothopanus nambi* (Speg.) R.H. Petersen & Krisai from Brazil (Spegazzin 1883; Kuntze 1898; Petersen and Krisai-Greilhuber 1999; Capelari et al. 2011); they found that both taxa are nearly identical.

The earliest report of the present species in Japan was given by Miyagi (1964) from Yaeyama Islands. He identified the fungus as *Pleurotus lunaillustris* Kawamura (nom. nud.) from Malaysia, but the micromorphological features have not yet been fully described.

References 引用文献

Capelari M, Desjardin DE, Perry BA, Asai T, Stevani CV. 2011. *Neonothopanus gardneri*: a new combination for a bioluminescent agaric from Brazil. Mycologia 103 (6): 1433-1440.
Corner EJH. 1981. The agaric genera *Lentinus, Panus* and *Pleurotus*. Beih Nova Hedwigia 69: 1-169.
Dennis RWG. 1953. Some pleurotoid fungi from the West Indies. Kew Bull 8 (1): 31-45.
Earle FS. 1906. Algunos hongos cubanos. Inf An Estac Cent Agr Cuba 1: 225-242.
Kuntze, O. 1898. Revisio generum plantarum 3: 1-576.
Miyagi G. 1964. On a luminous fungus, *Pleurotus lunaillustris* from the Yaeyama Islands (in Japanese). Bulletin of Arts and Science Division, University of the Ryukyus, Mathematics and sciences. 7: 54-56.
Montagne JPFC. 1837. Centurie de plantes exotiques nouvelles. Annls Sci Nat, Bot, sér 2 8: 345-370.
Pegler DN. 1983. Agaric flora of the Lesser Antilles. Kew Bull, Addit Ser 9: 1-668.
Petersen RH, Krisai-Greilhuber I. 1999. Type specimen studies in *Pleurotus*. Persoonia 17 (2): 201-219.
Saccardo PA, Traverso JB. 1911. Index Iconum Fungorum, Vol. II. Syll fung (Abellini) 20: 1-1310.
Singer R. 1944. New genera of fungi. I. Mycologia 36: 358-368.
Spegazzini C. 1883. Fungi Guaranitici. Pugillus 1. Anal Soc Cient Argent 16 (5): 242-248.

26. シロヒカリタケ（新種）*Pleurotus nitidus* Har. Takah. & Taneyama, **sp. nov.**

= ? *Pleurotus lunaillustris* Kawamura nom. nud.（sensu Miyagi 1964）

肉眼的特徴（Figs. 197-204）：子実体はヒラタケ型．傘は径20-60 (-90) mm，半球形〜腎臓形，最初丸山形，まもなく平らに開いて中央部がへこみ，時に中丘を形成し，周縁部に向かって放射状の浅い溝線を表す；縁部は細円鋸歯状；表面は粘性を欠き，やや吸水性，繊維状，最初純白色，のち中央部が淡黄色〜灰黄色を帯び，時に老成すると全体に灰黄色を呈し，成熟するとしばしば暗紫褐色の染みまたは斑点を所々に生じる．肉は厚さ2 mm以下，類白色，軟質，特別な味や匂いはない．柄は10-20×3-8 mm，短形，円柱形，偏在生または側生，中実；表面は粘性を欠き，繊維状，純白色；根本は剛毛状の白色菌糸体に被われる．ヒダは深く垂生し，やや密（柄に到達するヒダは17-22），1-2の小ヒダを伴い，幅5 mm以下，時にヒダの間にかすかな連絡脈を表し，類白色，成熟するとしばしば暗紫褐色の染みまたは斑点を所々に生じる；縁部は同色，全縁．胞子紋は純白色．

発光性：子実体全体が黄緑色に強く発光；菌糸体の発光性は不明．

顕微鏡的特徴（Figs. 194-196）：担子胞子は (4.0-) 4.9-6.1 (-7.1) × (2.4-) 2.9-3.4 (-3.7) μm (n = 34, mean length = 5.52±0.61, mean width = 3.13±0.24, Q = (1.50-) 1.63-1.90 (-1.99), mean Q = 1.77±0.14)．楕円形，平坦，無色，非アミロイド，アルカリ溶液において無色，薄壁．担子器は (16.7-) 16.8-23.6 (-25.2) × (3.5-) 3.9-5.3 (-5.5) μm（本体），(3.8-) 4.2-5.8 (-6.4) × (1.3-) 1.4-1.8 (-1.8) μm（ステリグマ），こん棒形，4胞子性；偽担子器はこん棒形．縁シスチジアおよび側シスチジアを欠く．子実層托実質の菌糸は幅 (2.6-) 3.4-5.7 (-6.3) μm，ほぼ平列し，円柱形，平滑，無色，アルカリ溶液において淡黄色，偽アミロイド，厚壁 (1.3-2.1 μm)；子実層はアルカリ溶液において淡黄色，非アミロイドまたは弱偽アミロイド．傘表皮組織の菌糸は幅 (2.1-) 2.5-3.5 (-4.3) μm，匍匐性，円柱形，ところどころに瘤状分岐物を有し，凝着物を欠き，平滑，無色，アルカリ溶液において淡黄色，非アミロイド，薄壁；末端細胞は (3.4-) 3.7-8.2 (-15.8) × (2.4-) 2.9-3.8 (-4.4) μm，円柱形，一般に短い間隔で3-9のクランプを持たない二次隔壁で仕切られる．傘実質は一菌糸型；菌糸は幅 (2.6-) 4.0-5.3 (-6.1) μm，並列し，円柱形，平滑，無色，アルカリ溶液において淡黄色，偽アミロイド，壁は厚さ0.8-1.2 μm．柄表皮組織の上層は垂直に立ち上がり，表皮から突出した末端細胞（柄シスチジア）が密集する；柄シスチジアは (33.4-) 38.8-61.7 (-75.5) × (3.9-) 4.4-5.7 (-6.3) μm，群生し，亜円柱形，しばしば不規則なく

びれがあり，平滑，無色，アルカリ溶液において無色，非アミロイド，壁は厚さ0.9-1.1 μm；表皮下層の菌糸は錯綜し，幅 (3.3-) 3.8-4.8 (-5.3) μm，厚壁 (1.1-1.7 (-2.1) μm)．柄実質は一菌糸型；菌糸は縦に沿って並列し，幅 (3.1-) 3.8-5.1 (-6.1) μm，平滑，無色，アルカリ溶液において無色，非アミロイド，厚壁 (1.3-2.1 (-2.4) μm)．根元の菌糸体を構成する菌糸は幅 (2.6-) 3.2-4.3 (-5.2) μm，円柱形，しばしばやや屈曲し，ところどころに瘤状分岐物を有し，無色，アルカリ溶液において無色，非アミロイド，厚壁 (0.9-1.5 (-1.8) μm)．全ての組織において菌糸はクランプを持つ．

生態および発生時期：常緑広葉樹林内の腐朽した材上から群生または散生，4月〜10月．

分布：沖縄（八重山諸島：石垣島，西表島）．

供試標本：TNS-F-52277（正基準標本），常緑広葉樹林内の腐木上，沖縄県石垣市バンナ公園，2011年7月15日，高橋春樹採集；TNS-F-61366，同上，2012年5月29日，高橋春樹採集；TNS-F-61367，同上，2012年6月6日，高橋春樹採集；KPM-NC0023865，同上，2011年10月16日，高橋春樹採集；KPM-NC0023864，常緑広葉樹林内の腐木上，沖縄県西表島船浦，2002年5月30日，高橋春樹採集．

主な特徴：子実体は全体に白色で，短い側生の柄と深く垂生するヒダからなるヒラタケ型；ヒダはしばしば暗紫褐色の染みまたは斑点を所々に生じる；担子胞子は楕円形；傘および子実層托の実質は円柱形で偽アミロイドの厚壁（1.2 μm 以下）菌糸からなる；傘表皮組織の末端細胞はところどころに瘤状分岐物を有し，一般に短い間隔で3-9のクランプを持たない隔壁で仕切られる；垂直に柄表皮組織から突出したしばしば括れがある亜円柱形の柄シスチジアが密生する；子実体に強い発光性がある；材上生．

コメント：楕円形の担子胞子，瘤状分岐物を持つ傘表皮組織の菌糸，そして偽アミロイドの実質の菌糸は異質であるが，被膜を持たず，短い側生の柄と深く垂生するヒダからなるヒラタケ型の子実体，無色，非アミロイドの担子胞子，発達の悪い傘表皮組織，厚壁でクランプを持つ実質の菌糸，そして材上生の生態に基づき暫定的に広義のヒラタケ属 Pleurotus に本種を置いた．

シロヒカリタケは以下の4種に極めて類似する：最初キューバから報告された *Pleurotus eugrammus* (Mont.) Dennis (Montagne 1837；Singer 1944；Dennis 1953；Corner 1981；Pegler 1983)；マレーシア産 *P. eugrammus* var. *radicicola* Corner (Corner 1981)；マレーシア産 *P. eugrammus* var. *brevisporus* Corner (Corner 1981)；キューバ産 *Pleurotus hygrophanus* (Earle) Sacc. & Traverso (Earle 1906；Saccardo and Traverso 1911；Pegler 1983)．*P. eugrammus* var. *eugrammus*，*P. eugrammus* var. *brevisporus*，*P. hygrophanus* の実質は隔壁周辺で急激にくびれ，膨大した菌糸（通常径 13 μm に達する：Corner 1981) からなる．また *P. eugrammus* var. *eugrammus* はより長径：6-9 μm (Corner 1981) の担子胞子を有し，子実層托実質は非アミロイドの菌糸からなる (Singer 1944)．*Pleurotus eugrammus* var. *radicicola* はシロヒカリタケと同様実質において幅の狭い円柱形の菌糸：幅2-7 μm (Corner 1981) を有するが，幼菌の傘は淡黄褐色または淡紅色を帯び，傘表皮組織の菌糸は淡褐色の粒状色素に被覆され，通常ヤシまたは樹木の根から発生する．

Petersen and Krisai-Greilhuber (1999) は *P. eugrammus* とブラジル産 *Neonothopanus nambi* (Speg.) R.H. Petersen & Krisai (Spegazzin 1883；Kuntze 1898；Petersen and Krisai-Greilhuber 1999；Capelari et al. 2011) の類似性に着目し，マレーシア産の標本を含む現地模式標本（トポタイプ）を用いた検討結果から，両種がほぼ同一であることを明らかにした．

宮城 (1964) は八重山諸島の標本に基づきシロヒカリタケの日本における分布を初めて確認し，マレーシア産 *P. lunaillustris* Kawamura (nom. nud.) と同一種として報告したが，顕微鏡的特徴に関する詳細なデータは記載されなかった．

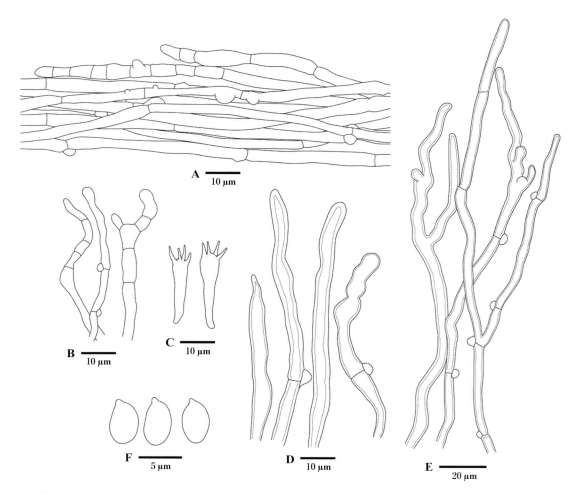

Fig. 194 – Micromorphological features of *Pleurotus nitidus* (holotype): **A**. Longitudinal cross section of the pileipellis. **B**. Terminal elements of the pileipellis. **C**. Basidia. **D**. Caulocystidia. **E**. Elements of the basal mycelium. **F**. Basidiospores. Illustrations by Taneyama, Y.

シロヒカリタケの顕鏡図（正基準標本）：**A**. 傘表皮組織の縦断面．**B**. 傘表皮組織の末端細胞．**C**. 担子器．**D**. 柄シスチジア．**E**. 根元の菌糸体を構成する菌糸．**F**. 担子胞子．図：種山裕一．

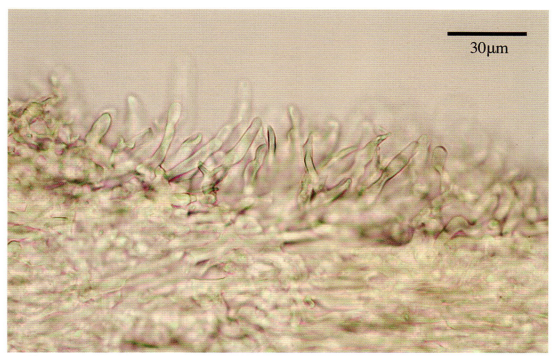

Fig. 195 − Longitudinal cross section of the stipitipellis of *Pleurotus nitidus* (in 3%KOH, holotype) showing gregarious caulocystidia protruding from the stipitipellis. Photo by Taneyama, Y.
シロヒカリタケの柄表皮組織の縦断面（3％水酸化カリウム溶液で封入，正基準標本）．柄表皮組織から突出した群生する柄シスチジアを示す．写真：種山裕一．

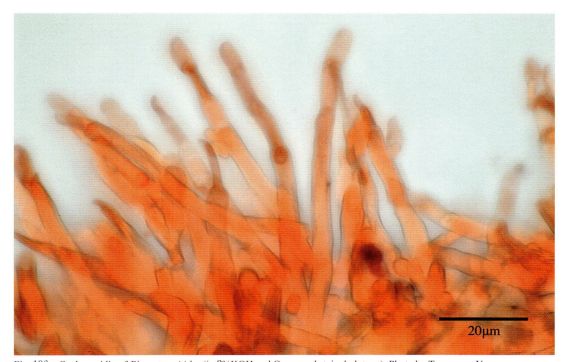

Fig. 196 − Caulocystidia of *Pleurotus nitidus* (in 3%KOH and Congo red stain, holotype). Photo by Taneyama, Y.
シロヒカリタケの柄シスチジア（3％水酸化カリウム溶液で封入した後コンゴー赤染色，正基準標本）．写真：種山裕一．

244 —— 26. *Pleurotus nitidus* シロヒカリタケ

Fig. 197 − Basidiomata of *Pleurotus nitidus* emitting yellowish green light on a dead fallen log, 24 Aug. 2013, Banna Park, Ishigaki Island. Photo by Takahashi, H.
腐木上において黄緑色に発光するシロヒカリタケの子実体．2013年8月24日．石垣市バンナ公園．写真：高橋春樹．

Fig. 198 − Basidiomata of *Pleurotus nitidus* on a dead fallen log, 5 Jun. 2010, Banna Park, Ishigaki Island. Photo by Takahashi, H.
腐木上から発生するシロヒカリタケの子実体．2010年6月5日．石垣市バンナ公園．写真：高橋春樹．

26. *Pleurotus nitidus* シロヒカリタケ —— 245

Fig. 199 – Basidiomata of *Pleurotus nitidus* on a dead fallen log, 5 Jun. 2010, Banna Park, Ishigaki Island. Photo by Takahashi, H. 腐木上から発生するシロヒカリタケの子実体，2010年6月5日，石垣市バンナ公園．写真：高橋春樹．

Fig. 200 – Basidiomata of *Pleurotus nitidus* (KPM-NC0023865) on a dead fallen log, 16 Oct. 2011, Banna Park, Ishigaki Island. Photo by Takahashi, H.
腐木上から発生するシロヒカリタケの子実体（KPM-NC0023865），2011年10月16日，石垣市バンナ公園．写真：高橋春樹．

26. *Pleurotus nitidus* シロヒカリタケ —— 247

Fig. 201 – Basidiomata of *Pleurotus nitidus* (KPM-NC0023865) on a dead fallen log, 16 Oct. 2011, Banna Park, Ishigaki Island. Photo by Takahashi, H.
腐木上から発生するシロヒカリタケの子実体（KPM-NC0023865），2011年10月16日，石垣市バンナ公園．写真：高橋春樹．

Fig. 202 – Basidiomata of *Pleurotus nitidus* (KPM-NC0023865) on a dead fallen log, 16 Oct. 2011, Banna Park, Ishigaki Island. Photo by Takahashi, H.
腐木上から発生するシロヒカリタケの子実体（KPM-NC0023865），2011年10月16日，石垣市バンナ公園．写真：高橋春樹．

26. *Pleurotus nitidus* シロヒカリタケ —— 249

Fig. 203 – Basidiomata of *Pleurotus nitidus* (TNS-F-48309) on a dead fallen log, 10 Jun. 2012, Arakawa, Ishigaki Island. Photo by Taneyama, Y.
腐木上から発生するシロヒカリタケの子実体（TNS-F-48309），2012年6月10日．石垣島荒川．写真：種山裕一．

Fig. 204 – Basidiomata of *Pleurotus nitidus* on a dead fallen log (TNS-F-48309), 10 Jun. 2012, Arakawa, Ishigaki Island. Photo by Taneyama, Y.
腐木上から発生するシロヒカリタケの子実体（TNS-F-48309），2012年6月10日．石垣島荒川．写真：種山裕一．

27. *Psilocybe capitulata* Har. Takah. ナンヨウシビレタケ

Mycoscience 52 (6): 397 (2011) [MB#519033].

Etymology: The specific epithet means "capitulate", referring to the capitulate pilocystidia.

Macromorphological characteristics (Figs. 206–213): Primordium 1–2 mm in diameter, ovoid to oblong-ellipsoid, enveloped in white, floccose universal veil. Pileus 42–93 mm in diameter, at first hemispherical to conic-campanulate, expanding to broadly convex to almost plane, sometimes broadly umbonate, with straight margin, not appendiculate; surface at first spotted with fugacious, white remnants of the veil, soon glabrescent, eventually covered overall with whitish to brownish, furfuraceous squamules in age, subviscid when wet, hygrophanous, brownish orange (7C6–7) to brown (7D6–7) when moist, mottled with darker areas (near brownish red: 8C6–7) at the center, drying to paler (6C4–5) or whitish from the margin, gradually changing to blue where handled. Flesh up to 9 mm, whitish, gradually changing to blue when cut; odor and taste not distinctive. Stipe 48–63 × 5–20 mm, subequal or at times somewhat thickened toward the base, central, terete, hollow; smooth; surface fibrillose above, squamulose below, sometimes striate above the annulus, whitish to pale brownish, gradually changing to blue where handled; annulus 8–15 mm wide, thin, membranous, persistent, fragile, white, striate; base covered with white strigose mycelial hairs. Lamellae adnate to adnexed, close (42–51 reach the stipe) with 1–3 series of lamellulae, up to 12 mm broad, dark-brown (8F6–8) to reddish-brown (9F7–8); edges minutely floccose under a lens, greyish.

Micromorphological characteristics (Fig. 205): Basidiospores (13–) 13.5–15 (–19) × 8–9 (–9.5) µm (n = 36, Q = 1.66–1.68), subellipsoid in side view, subhexagonal in face view, smooth, greyish red (10D5) to brownish red (10D6) in H2O, orange-yellow (4A7–4B7) in KOH, thick-walled (0.5–1 µm), truncated at the apex, with a distinct germ pore. Basidia 25–35 × 10–12 µm, clavate, 2- to 4-spored. Cheilocystidia 24–35 × 3–9 µm, abundant, forming a compact sterile edge, fusoid-ventricose with a subcapitate apex, smooth, hyaline in KOH, thin-walled. Pleurocystidia none or infrequent, 28–39 × 10–15 µm, basidiomorphous, broadly clavate, with a rounded apex, sometimes with a slight median constriction, smooth, hyaline in KOH, thin-walled. Hyphae of hymenophoral trama 5–22 µm wide, subcylindrical to fusoid, often inflated, parallel, smooth, hyaline in KOH, thin-walled. Pileipellis a loose cutis of entangled, gelatinous, subcylindrical hyphae 2–10 µm wide, encrusted with orange (6A7–6B7) pigment in KOH; pilocystidia 40–60 × 2–13 µm, clavatecylindrical, often with a capitulate apex, smooth, hyaline in KOH, thin-walled. Squamules of the pileus similar to the pileipellis. Hyphae of pileitrama 5–10 µm wide, parallel, subcylindrical, not inflated, smooth, hyaline in KOH, thin-walled. Stipitipellis a loose cutis of entangled, non-gelatinous, cylindrical hyphae 2–5 µm wide, smooth, hyaline in KOH, thin-walled, lacking differentiated terminal cells. Stipe trama composed of longitudinally running, cylindrical cells 5–12 µm wide, smooth, hyaline in KOH, thin-walled. Elements of annulus 2–3 µm wide, loosely interwoven, subcylindrical, not inflated, smooth, hyaline in KOH, thin-walled. Clamp connections present in all tissues.

Habitat and phenology: Solitary to scattered, coprophilous on cow dung, almost year-round, common.

Known distribution: Okinawa (Ishigaki Island).

Specimens examined: Specimens examined: KPM-NC0017300 (holotype), Ishigaki-shi, Okinawa Pref., on cow dung, 9 Feb. 2002, coll. Takahashi, H., Uehara, S., & Sakamoto, H.; KPM-NC0017301, same place, 2 Mar. 2002, coll. Takahashi, H.; KPM-NC0017302, same location, 26 Feb. 2008, coll. Takahashi, H. & Sakamoto, H.; KPM-NC0017303, same location, 28 Feb. 2008, coll. Takahashi, H.; KPM-NC0017304, same location, 3 Mar. 2008, coll. Takahashi, H.; KPM-NC0017305, same location, 2 Jun. 2008, coll.

Takahashi, H.; KPM-NC0017306, same location, 14 Jun. 2008, coll. Takahashi, H.; KPM-NC0017307, same location, 2 Feb. 2009, coll. Takahashi, H.; KPM-NC0017308, same location, 9 Mar. 2010, coll. Takahashi, H.

Japanese name: Nanyou-sibiretake.

Comments: *Psilocybe capitulata* is well characterized by having the brownish orange to brown pileus covered overall with whitish to brownish, furfuraceous squamules in age; the cyanescent flesh; the persistent, white, membranous annulus; the subhexagonal, thick-walled basidiospores; the capitulate pilocystidia; and the coprophilous habit on the cow dung.

The cyanescent basidiomata on the cow dung, the subhexagonal, thick-walled basidiospores, and the well-developed, persistent annulus indicate alignment of the present fungus with the genus *Psilocybe*, section *Cubensae* Guzmán (Guzmán 1983). Because of its typically furfuraceous-squamulose pileus in age, the capitulate pilocystidia, and the infrequent, less-developed pleurocystidia, *P. capitulata* can be distinguished from the other previously known taxa in the section *Cubensae* such as pantropical *Psilocybe cubensis* (Earle) Singer (Earle 1906; Guzmán 1978, 1983; Pegler 1983; Stamets 1996; Thomas et al. 2002; Cortez and Coelho 2004) and *Psilocybe subcubensis* Guzmán from Mexico (Guzmán 1978, 1983).

Apart from the section *Cubensae*, *Psilocybe subaeruginascens* Höhn. from Indonesia (Höhnel 1914; Guzmán 1983, Horak and Desjardin 2006) and Japan (Nagasawa 1987) bears a superficial resemblance to *P. capitulata,* though it is distinct in possessing rhomboid basidiospores, producing well-developed, broadly fusoid-ventricose pleurocystidia, and lacking pilocystidia. *Psilocybe subannulata* E. Horak & Guzmán from Puerto Rico (Guzmán et al. 2009) is also similar to *P. capitulata* in appearance, but the former is easily separated from *P. capitulata* in having rhombic basidiospores and lacking pilocystidia. The present species may possibly be related to *Psilocybe magnispora* E. Horak, Guzmán & Desjardin from Thailand (Horak et al. 2009) in having a submembranous annulus, cyanescent flesh, and a coprophilous habit. *Psilocybe magnispora*, however, is differentiated from *P. capitulata* in possessing rhomboid basidiospores, forming well developed pleurocystidia with refringent incrustations, and lacking pilocystidia.

References 引用文献

Cortez VG, Coelho G. 2004. The Stropharioideae (Strophariaceae, Agaricales) from Santa Maria, Rio Grande do Sul, Brazil. Mycotaxon 89: 355-378.
Earle FS. 1906. Algunos hongos cubanos. Inf An Estac Centr Agron Cuba 1: 225-242.
Guzmán G. 1978. The species of *Psilocybe* known from Central and South America. Mycotaxon 7: 225-255.
Guzmán G. 1983. The Genus *Psilocybe*, a systematic revision of the known species including the history, distribution and chemistry of the hallucinogenic species. Beih Nova Hedwigia 74: 1-439.
Guzmán G, Horak E, Halling R, Ramírez-Guillén F. 2009. Further studies on *Psilocybe* from the Caribbean, Central America and South America, with descriptions of new species and remarks to new records. Sydowia 61 (2): 215-242.
Höhnel FXR. 1914. Fragmente zur Mykologie, XVI. Mitteilung, Nr. 813-875. Sitzungsber Kais Akad Wiss Wien Math Naturwiss Kl 123: 49-155.
Horak E, Desjardin DE. 2006. Agaricales of Indonesia. 6. *Psilocybe* (Strophariaceae) from Indonesia (Java, Bali, Lombok). Sydowia 58 (1): 15-37.
Horak E, Guzmán G, Desjardin DE. 2009. Four new species of *Psilocybe* from Malaysia and Thailand, with a key to the species of sect. *Neocaledonicae* and discussion on the distribution of the tropical and temperate species. Sydowia 61 (1): 25-37.
Nagasawa E. 1987. Strophariaceae. In: Imazeki R, Hongo T (eds) Colored Illustrations of Mushrooms of Japan I (in

Japanese). Hoikusha, Osaka, pp 190-211.
Pegler DN. 1983. Agaric flora of the Lesser Antilles. Kew Bull, Addit Ser 9: 1-668.
Stamets P. 1996. Psilocybin Mushrooms of the World. Ten Speed Press, Berkeley.
Takahashi H. 2011. Two new species of Agaricales and a new Japanese record for *Chaetocalathus fragilis* from Ishigaki Island, a southwestern island of Japan. Mycoscience 52 (6): 392-400.
Thomas KA, Manimohan P, Guzmán G, Tapia F, Ramírez-Guillén F. 2002. The genus *Psilocybe* in Kerala State, India. Mycotaxon 83: 195-207.

27. ナンヨウシビレタケ *Psilocybe capitulata* Har. Takah.

Mycoscience 52 (6): 397 (2011) [MB#519033].

肉眼的特徴（Figs. 206-213）：子実体原基は径1-2 mm，卵形～楕円形で，全体に白色の綿毛状被膜に被われる．傘は径42-93 mm，最初半球形～円錐状釣り鐘形，のち山型～ほぼ平に開き，時に中丘を具える；表面は最初白色の消失性被膜の名残が点在するが間もなくほぼ平滑になり，老成すると全体が類白色～帯褐色小鱗片に被われ，湿時やや粘性，吸水性，湿時橙褐色～褐色を帯び部分的に暗色を呈し，乾くと周辺部から淡色～類白色になり，触れた部分は徐々に青変する．肉は厚さ9 mm以下，類白色，空気に触れると徐々に青変し，特別な味や臭いはない．柄は48-63×5-20 mm，ほぼ上下同大，時に基部に向かってやや拡大し，中心生，中空；表面は上部が繊維状，下部は小鱗片，しばしばツバより上部に条線を表し，類白色～淡褐色，触れた部分は徐々に青変する；ツバは幅8-15 mm，薄い膜質，永存性，脆く，白色，条線を表す；根元は白色の剛毛状菌糸体に被われる．ヒダは直生～上生，密（柄に到達するヒダは42-51），1-3の小ヒダを交え，幅12 mm以下，暗紫褐色；縁部は微細な綿毛状，帯灰色．

顕微鏡的特徴（Fig. 205）：担子胞子は (13-) 13.5-15 (-19) ×8-9 (-9.5) µm (n = 36, Q = 1.66-1.68)，側面観において亜楕円形，正面観においてやや角張った六角形，平滑，帯灰赤色～帯褐赤色，水酸化カリウム溶液において橙黄色，やや厚壁 (0.5-1 µm)，頂部において切形，明瞭な発芽孔を持つ．担子器は25-35×10-12 µm，こん棒形，2-4胞子性．縁シスチジアは24-35×3-9 µm，群生し，片脹れ状紡錘形で頂部はやや頭状になり，平滑，水酸化カリウムによる反応は陰性，薄壁．側シスチジアはないかまたは稀に存在し，28-39×10-15 µm，偽担子器形，広こん棒形，鈍頭，時に中央部が僅かに収縮し，平滑，水酸化カリウムによる反応は陰性，薄壁．子実層托実質の菌糸は幅5-22 µm，亜円柱形～紡錘形，しばしば膨大し，並列型，平滑，水酸化カリウムによる反応は陰性，薄壁．傘表皮組織は緩く錯綜した平行菌糸被を成す；菌糸は幅2-10 µm，亜円柱形，水酸化カリウム溶液において橙色に染まる粒状色素が凝着する；傘シスチジアは40-60×2-13 µm，こん棒形～円柱形，しばしば頂部が頭状形になり，平滑，水酸化カリウムによる反応は陰性，薄壁．小鱗片を形成する菌糸は傘表皮組織に類似する．傘実質の菌糸は幅5-10 µm，平列し，亜円柱形，膨大せず，平滑，水酸化カリウムによる反応は陰性，薄壁．柄表皮組織は緩く錯綜した平行菌糸被；菌糸は幅2-5 µm，円柱形，平滑，水酸化カリウムによる反応は陰性，薄壁；分化した末端細胞は存在しない．柄実質の菌糸は幅5-12 µm，縦に沿って配列し，円柱形，平滑，無色，非アミロイド，水酸化カリウムによる反応は陰性．ツバを構成する菌糸は幅2-3 µm，緩く錯綜し，亜円柱形，膨大せず，平滑，水酸化カリウムによる反応は陰性，薄壁．全ての組織において菌糸にクランプが見られる．

生態および発生時期：牛糞上に孤生～散生し，年間を通じて普通に見られる．

分布：沖縄（石垣島）．

供試標本：KPM-NC0017300（正基準標本），牛糞上，沖縄県石垣市，2002年2月9日，採集者：高橋春樹，上原貞美，坂本晴雄；KPM-NC0017301，同上，2002年3月2日，coll. Takahashi, H.；KPM-

NC0017302，同上，26 Feb. 2008，coll. Takahashi, H. & Sakamoto, H.；KPM-NC0017303，同上，2008年2月28日，coll. Takahashi, H.；KPM-NC0017304，同上，2008年3月3日，coll. Takahashi, H.；KPM-NC0017305，同上，2008年1月2日，coll. Takahashi, H.；KPM-NC0017306，同上，2008年1月14日，coll. Takahashi, H.；KPM-NC0017307，同上，2009年2月2日，coll. Takahashi, H.；KPM-NC0017308，同上，2010年3月9日，coll. Takahashi, H.

主な特徴：1) 傘の表面は老成時において帯褐色の小鱗片に被われる．2) 肉は青変性を有する．3) 白色で膜質の発達したツバを持つ．4) 担子胞子は厚壁で正面観においてやや角張った六角形をなす．5) 頂部頭状形の傘シスチジアが存在する．6) 牛糞上に発生する．

コメント：糞生の生態，青変性を有する子実体，発達した永存性のツバ，やや角張った六角形の担子胞子はシビレタケ属 *Psilocybe*，ナンヨウシビレタケ節（高橋春樹新称）section *Cubensae* Guzmán（Guzmán 1983）の特徴と考えられる．節内において本種は熱帯〜亜熱帯に広く分布するシビレタケモドキ（宮城 1967；別名：ミナミシビレタケ，ニライタケ）*Psilocybe cubensis*（Earle）Singer（Earle 1906；Guzmán 1978, 1983；Pegler 1983；Stamets 1996；Thomas et al. 2002；Cortez and Coelho 2004）およびメキシコ産 *Psilocybe subcubensis* Guzmán（Guzmán 1978, 1983）に最も近縁と考えられる．これら2種は発達した片脹れ状〜洋梨形の側シスチジアを有し（ナンヨウシビレタケは側シスチジアを欠くかもしくは発達の悪い偽担子器形の側シスチジアを形成する），頂部頭状形の傘シスチジアを欠き，老成時において通常傘が平滑になる点でナンヨウシビレタケと異なる．*Cubensae* 節以外では，インドネシア（ジャワ島）と日本に分布するオオシビレタケ *Psilocybe subaeruginascens* Höhn.（Hohnel 1914, Guzmán 1983, Nagasawa 1987, Horak and Desjardin 2006）が近縁種と思われる．オオシビレタケはレンズ形の担子胞子と発達した片脹れ状紡錘形の側シスチジアを有し，傘シスチジアを欠く点でナンヨウシビレタケと容易に区別できる．またプエルトリコ産 *Psilocybe subannulata* E. Horak & Guzmán（Guzmán et al. 2009）もナンヨウシビレタケにやや類似するが，傘シスチジアを欠き，レンズ形の担子胞子を持つとされている．タイ産 *Psilocybe magnispora* E. Horak, Guzmán & Desjardin（Horak et al. 2009）はやや膜質のツバ，青変性を有する肉，糞生の生態においてナンヨウシビレタケと共通する．しかしタイ産種は凝着物に被われた側シスチジアを有し，傘シスチジアを持たない点で有意差が認められる．

Psilocybe cubensis については八重山諸島および沖縄本島産の標本に基づき元琉球大学教授宮城元助博士によりシビレタケモドキの仮称が与えられた（宮城元助1967．沖縄島産マツタケ目について（II）．琉球大文理学部紀要（理学編）10：38-45）．

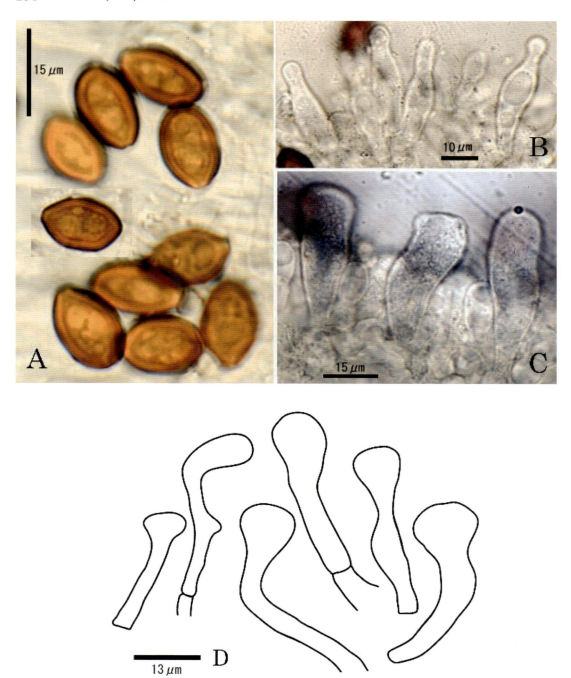

Fig. 205 – Micromorphological features of *Psilocybe capitulata* (KPM-NC0017300): **A**. Basidiospores (in 5%KOH). **B**. Cheilocystidia (in water). **C**. Pleurocystidia (in water). **D**. Pilocystidia. Photos and illustrations by Takahashi, H.
ナンヨウシビレタケの顕鏡図（KPM-NC0013139）：**A**. 担子胞子（5％水酸化カリウムで封入）. **B**. 縁シスチジア（水封）. **C**. 側シスチジア（水封）. **D**. 傘シスチジア. 顕鏡写真および線画：高橋春樹.

27. *Psilocybe capitulata* ナンヨウシビレタケ —— 255

Fig. 206 – Basidiomata of *Psilocybe capitulata* (KPM-NC0017302) on cow dung, 26 Feb. 2008, Ishigaki Island. Photo by Takahashi, H.
牛糞上に発生したナンヨウシビレタケの子実体（KPM-NC0017302），2008年2月26日，石垣島．写真：高橋春樹．

Fig. 207 – Basidiomata of *Psilocybe capitulata* (KPM-NC0017302) on cow dung, 26 Feb. 2008, Ishigaki Island. Photo by Takahashi, H.
牛糞上に発生したナンヨウシビレタケの子実体 (KPM-NC0017302),2008年2月26日,石垣島.写真:高橋春樹.

27. *Psilocybe capitulata* ナンヨウシビレタケ —— 257

Fig. 208 – Basidiomata of *Psilocybe capitulata* (KPM-NC0017302) on cow dung, 26 Feb. 2008, Ishigaki Island. Photo by Takahashi, H.
牛糞上に発生したナンヨウシビレタケの子実体（KPM-NC0017302），2008年2月26日，石垣島．写真：高橋春樹．

Fig. 209 – Immature basidiomata of *Psilocybe capitulata* (KPM-NC0017302) on cow dung, 26 Feb. 2008, Ishigaki Island. Photo by Takahashi, H.
牛糞上に発生したナンヨウシビレタケの幼菌（KPM-NC0017302），2008年2月26日，石垣島．写真：高橋春樹．

27. *Psilocybe capitulata* ナンヨウシビレタケ —— 259

Fig. 210 – Basidiomata of *Psilocybe capitulata* (KPM-NC0017304) on cow dung, 3 Mar. 2008, Ishigaki Island. Photo by Takahashi, H.
牛糞上に発生したナンヨウシビレタケの子実体（KPM-NC0017304），2008年3月3日，石垣島．写真：高橋春樹．

Fig. 211- Pileus surface of *Psilocybe capitulata* (KPM-NC0017304), 3 Mar. 2008, Ishigaki Island. Photo by Takahashi, H. ナンヨウシビレタケの傘表面（KPM-NC0017304），2008年3月3日，石垣島．写真：高橋春樹．

27. *Psilocybe capitulata* ナンヨウシビレタケ —— 261

Fig. 212– Immature basidioma of *Psilocybe capitulata* (KPM-NC0017302) on cow dung, 26 Feb. 2008, Ishigaki Island. Photo by Takahashi, H.
牛糞上に発生したナンヨウシビレタケの幼菌（KPM-NC0017302），2008年2月26日，石垣島．写真：高橋春樹．

262 —— 27. *Psilocybe capitulata* ナンヨウシビレタケ

Fig. 213 – Primordia and immature basidiomata of *Psilocybe capitulata* (KPM-NC0017302) on cow dung, 26 Feb. 2008, Ishigaki Island. Photo by Takahashi, H.
牛糞上に発生したナンヨウシビレタケの原基と幼菌（KPM-NC0017302），2008年2月26日，石垣島．写真：高橋春樹．

28. *Psilocybe definita* Har. Takah. & Taneyama, **sp. nov.** ハマシビレタケ

MycoBank no.: MB 809937.

Etymology: The specific epithet comes from the Latin word for "definite".

Distinctive features of this species are found in small, anellarioid to galerinoid basidiomata with a brown pileus and a slender stipe attached with fibrillose to submembranous annulus; unchanging flesh; subhexagonal, greyish red to brownish red basidiospores with a distinct germ pore; fusoid-ventricose cheilo- and pleurocystidia; and a fimicolous habit.

Macromorphological characteristics (Figs. 219-225): Pileus 10-15 (-20) mm in diameter, at first hemispherical, expanding to convex to broadly convex, sometimes broadly umbonate or shallowly depressed at the center, when young with fugacious, fibrillose veil remnants hanging from the slightly appendiculate margin; surface dry, hygrophanous, when wet radially translucently striate toward the margin, brown (6D7-8 to 7D7-8) when moist, often mottled with darker areas, becoming paler from the center when drying. Flesh up to 1 mm, pale brownish, unchanging when cut; odor and taste not distinctive. Stipe 10-25 × 1-4 mm, cylindrical or at times somewhat thickened toward the base, slender, central, terete, fistulose; surface at first floccose then woolly-fibrillose below the fibrillose to submembranous annuliform zone, minutely pruinose to fibrillose-striate at apex, brownish orange (5C5-6) or paler concolorous with the pileus; base covered with whitish mycelial tomentum. Lamellae adnate to subdecurrent, subdistant (17-24 reach the stipe) with 1-3 series of lamellulae, up to 3 mm broad, brownish orange (5C5-6) when young, dark-brown (8F6-8) to reddish-brown (9F7-8) with age; edges finely fimbriate, concolorous.

Micromorphological characteristics (Figs. 214-218): Basidiospores (11.4-) 13.0-14.1 (-14.8)×(7.3-) 8.0-8.9 (-9.4) μm (n = 121, mean length = 13.52±0.57, mean width = 8.48±0.46, Q = (1.44-) 1.52-1.67 (-1.85), mean Q = 1.60±0.08), ellipsoid to oblong-ellipsoid in side view, subhexagonal in face view, smooth, greyish brown (8D3-4) in water, yellowish brown (5D8-5E8) in KOH, dextrinoid or inamyloid, truncated at the apex, with a distinct germ pore, with thickened walls (1.1-) 1.3-1.8 (-2.1) μm. Basidia (24.1-) 27.7-32.9 (-37.7)×(10.3-) 11.0-12.1 (-12.6) μm (n = 39, mean length = 30.29± 2.61, mean width = 11.55±0.58) in main body, (5.0-) 5.5-6.8 (-7.5)×(2.3-) 2.6-3.1 (-3.5) μm (n = 40, mean length = 6.16±0.62, mean width = 2.84±0.25) in sterigmata, clavate, 2-4-spored. Cheilocystidia (19.5-) 24.2-31.5 (-34.6)×(5.4-) 6.2-8.5 (-11.1) μm (n = 39, mean length = 27.86±3.69, mean width = 7.31±1.15), abundant, forming a compact sterile edge, fusoid-ventricose, smooth, inamyloid, hyaline in water and KOH, thin-walled. Pleurocystidia (28.1-) 33.0-41.5 (-46.4)×(7.9-) 9.3-11.2 (-13.0) μm (n = 47, mean length = 37.22±4.26, mean width = 10.24±0.95), scattered, similar to the cheilocystidia. Hyphae of hymenophoral trama (3.8-) 6.1-13.3 (-17.8) μm wide (n = 53, mean width = 9.70±3.63), subcylindrical, constricted at septa, subparallel, loosely interwoven, smooth, inamyloid, pale brownish in water and KOH, thin-walled. Pileipellis a loose cutis of prostrate, subcylindrical hyphae (2.9-) 3.9-6.6 (-9.3) μm wide (n = 74, mean width = 5.26±1.34), often spirally encrusted with brownish pigments which are soluble in KOH, colorless in KOH, inamyloid, thin-walled; pilocystidia undifferentiated. Elements of pileitrama (34.1-) 37.3-62.9 (-72.3)×(13.8-) 15.6-21.8 (-26.4) μm (n = 19, mean length = 50.10±12.82, mean width = 18.69±3.08), subcylindrical, constricted at septa, loosely intricate, smooth, inamyloid, hyaline in water and KOH, thin-walled. Stipitipellis a loose cutis of prostrate, cylindrical hyphae (2.4-) 2.9-4.3 (-5.5) μm wide (n = 58, mean width = 3.60±0.66), often spirally encrusted with brownish pigments which are soluble in KOH, pale yellow in water, inamyloid, thin-walled; caulocystidia scattered above the annulus, dimorphic: 1) clavate to capitulate, (14.7-) 17.7-26.6 (-31.4)

×(7.4−) 8.6−11.7 (−14.3) μm (n = 28, mean length = 22.14 ± 4.44, mean width = 10.15 ± 1.54), with slightly thickened walls 0.7−0.9 (−1.1) μm; 2) subcylindrical, often strangulated, sometimes sinuous, (27.1−) 33.5−50.6 (−61.6)×(4.4−) 5.0−7.7 (−9.7) μm (n = 28, mean length = 42.06 ± 8.58, mean width = 6.34 ± 1.31), with somewhat thickened walls (0.6−) 0.7−1.0 (−1.1) μm. Stipe trama composed of longitudinally running, cylindrical hyphae; constituent hyphal cells (16.6−) 27.6−62.3 (−90.5)×(9.1−) 12.3−16.8 (−19.8) μm (n = 46, mean length = 44.95 ± 17.36, mean width = 14.54 ± 2.22), smooth, pale yellow in water, inamyloid, with slightly thickened walls (0.6−) 0.7−1.1 (−1.3) μm. Elements of annulus (3.2−) 4.1−5.8 (−6.8) μm wide, loosely interwoven, subcylindrical, not inflated, often spirally encrusted with brownish pigments which are soluble in KOH, thin-walled. Clamp connections present in all tissues.

Habitat and phenology: Solitary to scattered, coprophilous on cow dung in coastal dunes, almost year-round, common.

Known distribution: Okinawa (Ishigaki Island).

Holotype: TNS-F-39706, on cow dung, Ishigaki-shi, Okinawa Pref., Japan, 21 Jan. 2011, coll. Takahashi, H.

Other specimens examined: TNS-F-39707, on cow dung, Ishigaki-shi, Okinawa Pref., Japan, 24 Jan. 2011, coll. Takahashi, H.; TNS-F-39708, same location, 29 Jun. 2011, coll. Takahashi, H.; TNS-F-52280, same location, 21 Jun. 2011, coll. Takahashi, H; KPM-NC0023875 (Isotype), same location, 21 Jan. 2011, coll. Takahashi, H.

Gene sequenced specimen and GenBank accession number: TNS-F-39706, AB968235 (ITS).

Japanese name: Hama-sibiretake (named by H. Takahashi & Y. Taneyama).

Comments: The unchanging flesh, the slender stipe with a woolly-fibrillose to submembranous annulus, the greyish brown, angular basidiospores, and the fimicolous habit suggest placement of the new species in the genus *Psilocybe*, section *Merdariae* (Fr.) Singer (Singer 1986).

Within the section, *Psilocybe merdaria* (Fr.) Ricken, a widespread taxon in temperate zones (Fries 1821; Guzmán 1983; Watling 1987; Noordeloos 1999), appears to be aligned with *P. definita*, despite having somewhat shorter basidiospores: (10−) 11−13 (−14) μm long (Guzmán 1983), ventricose cheilocystidia with a cylindrical, flexuous neck, and absence of pleurocystidia. *Psilocybe moelleri* Guzmán, which is widely distributed in boreal and temperate regions (Guzmán 1978, 1983; Watling 1987; Noordeloos 1999; Cortez and Coelho 2004), has a similar basidiospore measurement but differs specifically in possessing significantly larger and taller basidiomata: 12−50 mm across in the pileus (Noordeloos 1999); 25−70 mm long in the stipe (Noordeloos 1999), typically distinct membranous annulus which is formed rather high up the stipe, subcylindrical or fusoid-ventricose cheilocystidia with a prolonged neck, an ixocutis in the pileipellis, and lacking pleurocystidia.

References 引用文献

Cortez VG, Coelho G. 2004. The Strophariooideae (Strophariaceae, Agaricales) from Santa Maria, Rio Grande do Sul, Brazil. Mycotaxon 89: 355−378.
Fries EM. 1821. Syst Mycol 1: 1−520.
Guzmán G. 1978. The species of *Psilocybe* known from Central and South America. Mycotaxon 7: 225−255.
Guzmán G. 1983. The Genus *Psilocybe*, a systematic revision of the known species including the history, distribution and chemistry of the hallucinogenic species. Beih Nova Hedwigia 74: 1−439.
Noordeloos ME. 1999. Strophariaceae. In: Bas C, Kuyper ThW, Noordeloos ME, Vellinga EC (eds). Flora Agaricina Neerlandica 4. A.A.Balkema, Rotterdam, pp 27−107.
Singer R. 1986. The Agaricales in modern taxonomy, 4th edn. Koeltz, Koenigstein.
Watling R. 1987. British Fungus Flora - Agarics and Boleti. 5. Strophariaceae and Coprinaceae p.p.: *Hypholoma, Melanotus, Psilocybe, Stropharia, Lacrymaria* and *Panaeolus*. Royal Botanic Garden, Edinburgh.

28. ハマシビレタケ（新種：高橋春樹 & 種山裕一新称）*Psilocybe definita* Har. Takah. & Taneyama, **sp. nov.**

肉眼的特徴（Figs. 219-225）：傘は径10-15（-20）mm，最初半球形のち丸山形～中高偏平，時に鈍頭の中丘を具えるかまたは中央部が浅く凹み，若いとき消失性繊維状の被膜の名残を縁部周辺に付着する；表面は粘性を欠き，吸水性，湿時半透明の条線を放射状に表し，湿時褐色，部分的に暗色を呈し，乾くと周辺部から淡色～類白色になる．肉は厚さ1 mm 以下，類白色，空気に触れても変色せず，特別な味や臭いはない．柄は10-25×1-4 mm，円柱形，時に基部に向かってやや拡大し，痩せ型，中心生，中空；表面は繊維状～やや膜質のツバより下部は綿毛状～繊維状，頂部は粉状～繊維状の筋を表し，淡褐色；根元は白色の綿毛状菌糸体に被われる．ヒダは直生～僅かに垂生，やや疎（柄に到達するヒダは17-24），1-3の小ヒダを交え，幅 3 mm 以下，幼時淡褐色，老成時暗紫褐色；縁部は長縁毛状，同色．

顕微鏡的特徴（Figs. 214-218）：担子胞子は（11.4-）13.0-14.1（-14.8）×（7.3-）8.0-8.9（-9.4）μm（n = 121，mean length = 13.52±0.57，mean width = 8.48±0.46，Q =（1.44-）1.52-1.67（-1.85），mean Q = 1.60±0.08），側面観において楕円形～長楕円形，正面観においてやや角張った六角形，平滑，水封で灰褐色，水酸化カリウム溶液において黄褐色，偽アミロイドまたは非アミロイド，頂部において切形，明瞭な発芽孔を持ち，壁は厚さ（1.1-）1.3-1.8（-2.1）．担子器は（24.1-）27.7-32.9（-37.7）×（10.3-）11.0-12.1（-12.6）μm（本体），（5.0-）5.5-6.8（-7.5）×（2.3-）2.6-3.1（-3.5）μm（ステリグマ），こん棒形，2-4胞子性．縁シスチジアは（19.5-）24.2-31.5（-34.6）×（5.4-）6.2-8.5（-11.1）μm，群生し，ヒダの縁部は不稔帯を成し，片脹れ状紡錘形，平滑，非アミロイド，水封および水酸化カリウム溶液において無色，薄壁．側シスチジアは散生，（28.1-）33.0-41.5（-46.4）×（7.9-）9.3-11.2（-13.0）μm，縁シスチジアに類似する．子実層托実質の菌糸は幅（3.8-）6.1-13.3（-17.8）μm，亜円柱形，隔壁の周囲において急激に幅が狭くなり，やや不規則に錯綜した平列型，平滑，非アミロイド，水封および水酸化カリウム溶液において淡褐色，薄壁．傘上表皮層は緩く錯綜した平行菌糸被；菌糸は幅（2.9-）3.9-6.6（-9.3）μm，亜円柱形，しばしば褐色の粒状色素が細胞壁に凝着し，水酸化カリウム溶液において無色，非アミロイド，薄壁；傘シスチジアは未分化．傘実質の菌糸細胞は（34.1-）37.3-62.9（-72.3）×（13.8-）15.6-21.8（-26.4）μm，緩く錯綜し，亜円柱形，隔壁の周囲において急激に幅が狭くなり，平滑，非アミロイド，水封および水酸化カリウム溶液において無色，薄壁．柄表皮組織は匍匐性の菌糸が緩く並列した平行菌糸被；菌糸は幅（2.4-）2.9-4.3（-5.5）μm，円柱形，しばしば水酸化カリウム溶液中で溶けやすい褐色の色素がらせん状に凝着し，水封で淡黄色，非アミロイド，薄壁；柄シスチジアはツバより上部に散在し，2形性：1）こん棒形～頂部頭状形，（14.7-）17.7-26.6（-31.4）×（7.4-）8.6-11.7（-14.3）μm，壁は厚さ 0.7-0.9（-1.1）μm；2）亜円柱形，しばしば不規則なくびれがあり，時にやや曲がりくねり，（27.1-）33.5-50.6（-61.6）×（4.4-）5.0-7.7（-9.7）μm，壁は厚さ（0.6-）0.7-1.0（-1.1）μm．柄の実質は縦に沿って配列した円柱形の菌糸からなる；菌糸細胞は（16.6-）27.6-62.3（-90.5）×（9.1-）12.3-16.8（-19.8）μm，平滑，水封で淡黄色，非アミロイド，壁は厚さ（0.6-）0.7-1.1（-1.3）μm．ツバを構成する菌糸は幅4.1-5.8（-6.8）μm，緩く錯綜し，亜円柱形，膨大せず，しばしば水酸化カリウム溶液中で溶けやすい褐色の色素がらせん状に凝着し，薄壁．全ての組織において菌糸にクランプが見られる．

生態および発生時期：孤生～散生，海岸砂浜の牛糞上に発生し，年間を通じて普通に見られる．

分布：沖縄（石垣島）．

供試標本：TNS-F-39706（正基準標本），海岸砂浜の牛糞上，沖縄県石垣市，2011年1月21日，高橋春樹採集；TNS-F-39707，同上，2011年1月24日，高橋春樹採集；TNS-F-39708，同上，2011年1月29日，高橋春樹採集；TNS-F-52280，同上，2011年1月21日，高橋春樹採集；KPM-NC0023875（複

266 —— 28. *Psilocybe definita* ハマシビレタケ

基準標本），同上，2011年1月21日，高橋春樹採集．
分子解析に用いた標本並びに GenBank 登録番号：TNS-F-39706, AB968235（ITS）．
主な特徴：子実体は小形のジンガサタケ型〜ケコガサタケ型；傘は褐色；柄は痩せ型で羊毛繊維状〜やや膜質のツバを持つ；肉は変色性を欠く；担子胞子は灰褐色，やや角張った六角形，明瞭な発芽孔を持つ；縁シスチジアおよび側シスチジアは片脹れ状紡錘形；海岸砂浜の牛糞上に発生．
コメント：変色性を欠く肉，羊毛繊維状〜やや膜質のツバを形成する痩せ型の柄，帯褐赤色で やや角張った六角形の担子胞子を持つ性質から，シビレタケ属 *Psilocybe*，カワリコシワツバタケ節 section *Merdariae* (Fr.) Singer (Singer 1986) に所属すると思われる．

　節内において温帯に広く分布するカワリコシワツバタケ *Psilocybe merdaria* (Fr.) Ricken (Fries 1821；Guzmán 1983；Watling 1987；Noordeloos 1999) は本種に近縁と考えられるが，担子胞子がやや短いこと，縁シスチジアが円柱形で曲がりくねった付属糸を持つこと，そして側シ

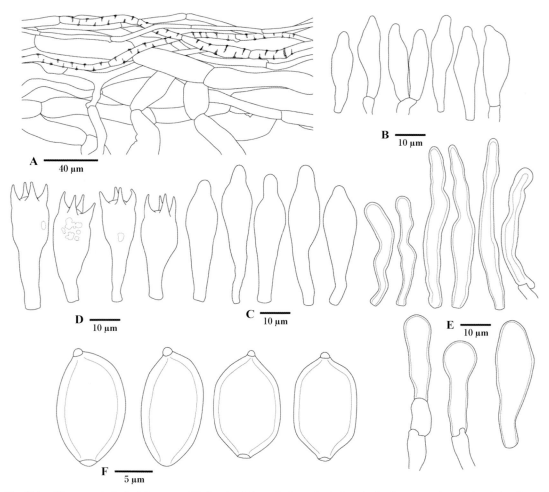

Fig. 214 – Micromorphological features of *Psilocybe definita* (holotype): **A.** Longitudinal cross section of the pileipellis and pileitrama. **B.** Cheilocystidia. **C.** Pleurocystidia. **D.** Basidia. **E.** Caulocystidia. **F.** Basidiospores. Illustrations by Taneyama, Y.
ハマシビレタケの顕微鏡図（正基準標本）：**A.** 傘表皮組織および実質の縦断面．**B.** 縁シスチジア．**C.** 側シスチジア．**D.** 担子器．**E.** 柄シスチジア．**F.** 担子胞子．図：種山裕一．

28. *Psilocybe definita* ハマシビレタケ —— 267

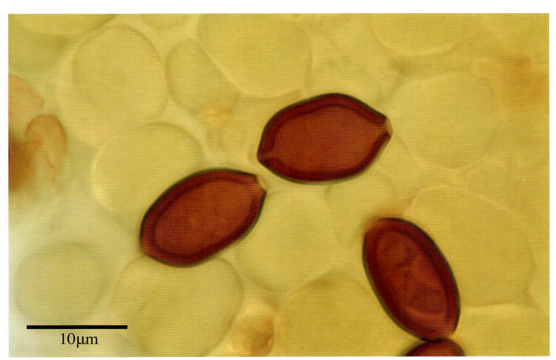

Fig. 215 – Basidiospores of *Psilocybe definita* (in Melzer's reagent, holotype). Photo by Taneyama, Y.
ハマシビレタケの担子胞子（メルツァー溶液で封入，正基準標本）．写真：種山裕一．

Fig. 216 – Cheilocystidia of *Psilocybe definita* (in 3%KOH and Congo red stain, holotype). Photo by Taneyama, Y.
ハマシビレタケの縁シスチジア（3%水酸化カリウム溶液で封入した後コンゴー赤染色，正基準標本）．写真：種山裕一．

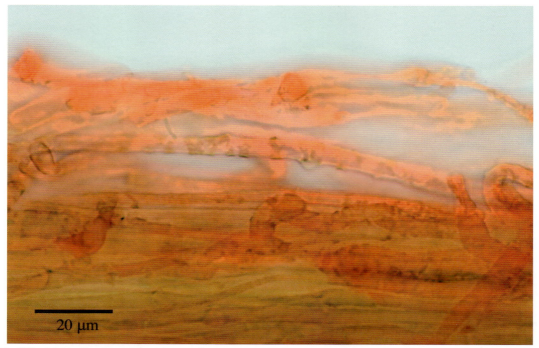

Fig. 217 – Stipitipellis of *Psilocybe definita* (in 3%KOH and Congo red stain, holotype), showing the elements spirally encrusted with brownish pigments. Photo by Taneyama, Y.
ハマシビレタケの柄表皮組織（3%水酸化カリウム溶液で封入した後コンゴー赤染色，正基準標本），褐色のらせん状凝着物を示す．写真：種山裕一．

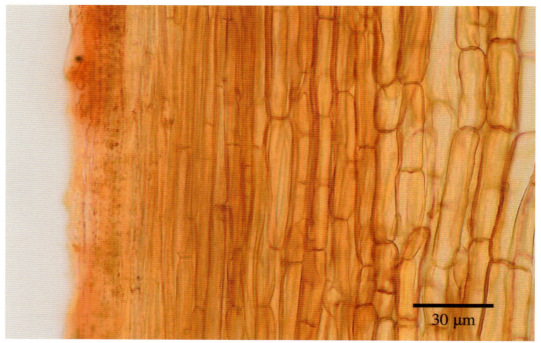

Fig. 218 – Longitudinal cross section of the stipitipellis and stipe trama in *Psilocybe definita* (in 3%KOH and Congo red stain, holotype). Photo by Taneyama, Y.
ハマシビレタケの柄表皮組織および柄実質の縦断面（3%水酸化カリウム溶液で封入した後コンゴー赤染色，正基準標本）．写真：種山裕一．

スチジアを欠くなどの相違が見られる．温帯〜寒帯に広く分布する *Psilocybe moelleri* Guzmán（Guzmán 1978, 1983；Watling 1987；Noordeloos 1999；Cortez and Coelho 2004）は担子胞子の大きさにおいて本種とほぼ一致するが，子実体がより大型：傘の径12-50 mm（Noordeloos 1999）；柄の長さ25-70 mm（Noordeloos 1999）であること，通常明瞭な膜質のツバを柄の頂部に形成すること，縁シスチジアが長い付属糸を有すること，傘表皮組織が粘性平行菌糸被であること，そして側シスチジアを欠く性質において明らかに異なる．

Fig. 219 – Mature basidioma of *Psilocybe definita* (holotype), 21 Jan. 2011, Ishigaki Island. Photo by Takahashi, H.
ハマシビレタケの成菌（正基準標本），2011年1月21日，石垣島．写：高橋春樹．

Fig. 220 – Basidioma of *Psilocybe definita* (holotype), 21 Jan. 2011, Ishigaki Island. Photo by Takahashi, H.
ハマシビレタケの子実体（正基準標本），2011年1月21日．石垣島．写真：高橋春樹．

28. *Psilocybe definita* ハマシビレタケ ── 271

Fig. 221 – Basidiomata of *Psilocybe definita* (holotype) on cow dung, 21 Jan. 2011, Ishigaki Island. Photo by Takahashi, H.
牛糞上に発生したハマシビレタケの子実体（正基準標本），2011年1月21日，石垣島．写真：高橋春樹．

Fig. 222 – Basidiomata of *Psilocybe definita* (holotype) on cow dung, 21 Jan. 2011, Ishigaki Island. Photo by Takahashi, H.
牛糞上に発生したハマシビレタケの子実体（正基準標本），2011年1月21日，石垣島．写真：高橋春樹．

28. *Psilocybe definita* ハマシビレタケ

Fig. 223 – Basidiomata of *Psilocybe definita* (TNS-F-39707) on cow dung, 24 Jan. 2011, Ishigaki Island. Photo by Takahashi, H.
牛糞上に発生したハマシビレタケの子実体（TNS-F-39707），2011年1月24日，石垣島．写真：高橋春樹．

Fig. 224 – Basidiomata of *Psilocybe definita* (TNS-F-39707) on cow dung, 24 Jan. 2011, Ishigaki Island. Photo by Takahashi, H.
牛糞上に発生したハマシビレタケの子実体（TNS-F-39707），2011年1月24日，石垣島．写真：高橋春樹．

28. *Psilocybe definita* ハマシビレタケ —— 273

Fig. 225 – Basidiomata of *Psilocybe definita* (TNS-F-39708) on cow dung, 29 Jan. 2011, Ishigaki Island. Photo by Takahashi, H.
牛糞上に発生したハマシビレタケの子実体（TNS-F-39708），2011年1月29日，石垣島．写真：高橋春樹．

29. *Pulveroboletus brunneoscabrosus* Har. Takah. ウロコキイロイグチ

Mycoscience 48 (2): 93 (2007) [MB#529990].

Etymology: The specific epithet means "*brunneo-* (brown-) + *scabrosus* (scabrous)", referring to the brownish-orange, appressed scales.

Macromorphological characteristics (Figs. 227–232): Pileus 20–50 mm in diameter, at first hemispherical, expanding to convex to broadly convex; surface wholly yellow (2A6–7 to 3A6–7) pulverulent, subviscid when wet, at first covered overall with an orange (6B7) to brownish-orange (6C7–8) veil that soon breaks up into appressed scales, ground color lemon-yellow (under scales), unchanging when bruised. Flesh up to 4 mm thick, light yellow, strongly fleshy-gelatinous, unchanging when cut; odor and taste indistinct. Stipe 60–100 × 15–25 mm, subequal or somewhat enlarged toward the base, central, terete, solid, non-reticulate; surface entirely yellow (2A6–7 to 3A6–7) pulverulent, viscid to subglutinous when wet, lemon-yellow, sheathed with orange (6B7) to brownish-orange (6C7–8), appressed scales at least over the lower part, unchanging when bruised; base covered with whitish mycelial tomentum. Partial veil yellow pulverulent, floccose-membranous to floccose-fibrillose, rarely leaving a faint fibrillose annulus. Tubes up to 6 mm deep, slightly depressed around the stipe, at first light yellow, then brown (6B7) at maturity, unchanging; pores small (2–3 per mm), subcircular, concolorous with the tubes, unchanging when bruised.

Micromorphological characteristics (Fig. 226): Basidiospores 7–10 × 4–5 μm (n = 38, Q = 1.75–2.0), inequilateral with a shallow suprahilar depression in profile, broadly ellipsoid in face view, smooth, melleous in water, thick-walled (up to 1 μm). Basidia 25–35 × 6–10 μm, clavate, four-spored. Cheilocystidia 25–40 × 5–8 μm, subfusoid to narrowly fusoid-ventricose, colorless in water, thin-walled. Pleurocystidia scattered, similar in shape to the cheilocystidia but larger (35–52 × 6–10 μm). Hyphae of hymenophoral trama 3–7 μm wide, cylindrical, parallel to each other, strongly gelatinized, smooth, pale yellow in water, thin-walled. Pileipellis consisting of repent, appressed, interwoven, gelatinized hyphal elements; constituent hyphae 2.5–6 μm wide, parallel, cylindrical, with intracellular brownish (in water) pigment, with intercellular yellow granular matter, thin-walled. Pileitrama of cylindrical, loosely interwoven hyphae 2.5–15 μm wide, strongly gelatinized, smooth, colorless in water, thin-walled. Stipitipellis composed of repent, appressed, interwoven, gelatinized hyphal elements; constituent hyphae 2.5–7 μm wide, cylindrical, with intracellular brownish (in water) pigment, with intercellular yellow (in water) granular matter, thin-walled. Stipe trama composed of longitudinally arranged, cylindrical hyphae 3–8 μm wide, occasionally branched, strongly gelatinized, smooth, colorless in water, thin-walled. Clamps absent in all tissues.

Habitat and phenology: Solitary to scattered on ground in subtropical evergreen broad-leaved forests dominated by *Quercus miyagii* Koidz. and *Castanopsis sieboldii* (Makino) Hatus. ex T. Yamaz. et Mashiba, or warm temperate evergreen broad-leaved forests dominated by *Q. gilva* Blume, May to September.

Known distribution: Okinawa (Ishigaki Island), Kagoshima (Amami-Oshima Island), Miyazaki.

Specimens examined: KPM-NC0010098 (holotype), on ground in a subtropical evergreen broad-leaved forest dominated by *Q. miyagii* and *C. sieboldii,* Banna Park, Ishigaki-shi, Okinawa Pref., 30 May 2002, coll. Takahashi, H.; KPM-NC0013148, same place, 6 Jun. 2004, coll. Takahashi, H.; KPM-NC0013147, on ground in a subtropical evergreen broad-leaved forest dominated by *Q. miyagii* and *C. sieboldii,* Omoto-dake, Ishigaki-shi, Okinawa Pref., 25 May 2003, coll. Takahashi, H.; KPM-NC0013155, on ground in an evergreen broad-leaved forest dominated by *Q. gilva,* Takaharu-cho, Miyazaki Pref., 21 Sep. 2001, coll. Kurogi, S.; KPM-NC0013156, on ground in a subtropical evergreen broad-leaved forest, Uken-son, Amami-Oshima Island, Kagoshima Pref., 22 Sep. 2002, coll. Hadano, E.

Japanese name: Uroko-kiiroiguchi.

Comments: This species is characterized by the lemon-yellow pulverulent, viscid pileus and stipe covered overall with orange to brownish-orange, appressed scales; the yellow pulverulent, floccose-membranous to floccose-fibrillose, soon collapsing annulus; the non-cyanescent (unchanging), strongly fleshy-gelatinous flesh; and the habitat in subtropical and warm temperate evergreen broad-leaved forests.

Pulveroboletus brunneoscabrosus is most similar in appearance to *Pulveroboletus ravenelii* (Berk. & M.A. Curtis) Murrill, originally described from North America (Berkeley and Curtis 1853; Singer 1947; Snell and Dick 1970; Smith and Thiers 1971; Corner 1972; Grund and Harrison 1976; Nagasawa 1989; Bessette et al. 2000), and Malaysian *Pulveroboletus frians* Corner (Corner 1972). The latter two taxa, however, differs from the present species in lacking brownish-orange, appressed scales on the pileus and stipe surfaces, and forming cyanescent flesh. *Pulveroboletus aberrans* Heinem. & Gooss.-Font. from Congo (Heinemann 1951, 1954) also bears some resemblance to *P. brunneoscabrosus*, but the former is distinct in an entirely orange-brown tomentose pileus and stipe, and has cyanescent flesh.

References 引用文献

Berkeley MJ, Curtis MA. 1853. Centuries of North American fungi. Ann Mag nat Hist, Ser. 2 12: 417–435.
Bessette AE, Roody WC, Bessette AR. 2000. North American Boletes. A color guide to the fleshy pored mushrooms. Syracuse University Press, New York.
Corner EJH. 1972. *Boletus* in Malaysia. Government Printing Office, Singapore.
Grund DW, Harrison KA. 1976. Nova Scotian boletes. Bibl Mycol 47: 1–283.
Heinemann P. 1951. Champignons récoltés au Congo Belge par Mme M. Goossens-Fontana. I. Boletineae. Bull Jard Bot Brux 21: 223–346.
Heinemann P. 1954. Boletineae. Flore Iconographique des Champignons du Congo 3: 49–80.
Nagasawa E. 1989. Boletaceae. In: Imazeki R, Hongo T (eds), Colored illustrations of mushrooms of Japan II. Hoikusha, Osaka, pp 1–44 (in Japanese).
Singer R. 1947. The Boletineae of Florida with notes on extralimital species III. Am Midl Nat 37: 1–135.
Singer R. 1986. Agaricales in modern taxonomy, 4th edn. Koeltz, Koenigstein.
Snell WH. 1936. Notes on Boletes. V. Mycologia 28: 463–475.
Snell WH, Dick EA. 1941. Notes on Boletes. VI. Mycologia 33: 23–37.
Snell WH, Dick EA. 1970. The boleti of northeastern North America. Cramer, Vaduz.
Takahashi H. 2007. Five new species of the Boletaceae from Japan. Mycoscience 48 (2): 90–99.

29. ウロコキイロイグチ *Pulveroboletus brunneoscabrosus* Har. Takah.

Mycoscience 48 (2): 93 (2007) [MB#529990].

肉眼的特徴（Figs. 227–232）：傘は径20–50 mm，最初半球形，のち饅頭形～中高偏平，表面は全体に黄色の粉質物を帯びた橙褐色～褐色の被膜に被われ，まもなく亀甲状にひび割れ，圧着した鱗片状をなし，鱗片状被膜は乾性，鱗片下部の地はレモン黄色で湿時やや粘性．肉は厚さ4 mm以下（傘中央部），淡黄色，空気に触れても変色せず，特別な味や臭いはない．柄は60–100×15–25 mm，ほぼ上下同大または下方に向かってやや太くなり，中心生，中実，網目を欠く；表面は湿時強い粘性があり，頂部はレモン黄色，下部は橙褐色の鱗片に被われる；根元は白色綿毛状菌糸体に被われる．内被膜は綿毛状膜質，レモン黄色の粉質物を帯び，まれに微かな繊維状のツバを形成する．管孔は長さ6 mm以下，柄の周囲においてやや嵌入し，最初黄色，成熟すると帯褐黄色になり，変色性を欠く；孔口は小型（2–3 per mm），管孔と同色，傷を受けても変色しない．

顕微鏡的特徴（Fig. 226）：担子胞子は7-10×4-5 μm（n = 38, Q = 1.75-2.0），広楕円形，下部側面に不明瞭でなだらかな凹みがあり（イグチ型），平坦，蜜色（水封），厚壁．担子器は25-35×6-10 μm，こん棒形，4胞子性．縁シスチジアは25-40×5-8 μm，幅の狭い紡錘形〜亜紡錘形，平滑，無色（水封），薄壁．側シスチジアは散在し，35-52×6-10 μm，形状は縁シスチジアに類似する．子実層托実質の菌糸は幅3-7 μm，並列し，円柱形，著しくゼラチン化し，平滑，淡黄色（水封），薄壁．傘表面はゼラチン化した平行菌糸被；菌糸は幅2.5-6 μm，円柱形，錯綜し，淡褐色の色素が細胞内に存在し（水封），細胞間に黄色の粒状物が散在し，薄壁．傘実質の菌糸は幅2.5-15 μm，緩く錯綜し，円柱形，著しくゼラチン化し，平滑，無色（水封），薄壁．柄表皮組織の菌糸は幅2.5-7 μm，円柱形，匍匐性，錯綜し，著しくゼラチン化し，平滑，淡褐色の色素が細胞内に存在し（水封），細胞間に黄色の粒状物が散在し，薄壁．柄実質の菌糸は幅3-8 μm，縦に沿って配列し，円柱形，時折分岐し，著しくゼラチン化し，平滑，無色（水封），薄壁．全ての組織において菌糸はクランプを欠く．

生態および発生時期：スダジイ，オキナワウラジロガシを主体とする亜熱帯性常緑広葉樹林内およびイチイガシを主体とする暖温帯性常緑広葉樹林内地上に孤生または散生，5月〜9月．

分布：沖縄（石垣島），鹿児島（奄美大島），宮崎．

供試標本：KPM-NC0010098（正基準標本），スダジイ，オキナワウラジロガシを主体とする亜熱帯性常緑広葉樹林内地上，沖縄県石垣市バンナ公園，2002年5月30日，高橋春樹採集；KPM-NC0013148，同上，2004年6月6日，高橋春樹採集；KPM-NC0013147，スダジイ，オキナワウラジロガシを主体とする亜熱帯性常緑広葉樹林内地上，沖縄県石垣市オモト岳，2003年5月25日，高橋春樹採集；KPM-NC0013155，イチイガシを主体とする暖温帯性常緑広葉樹林内地上，宮崎県高原町，2001年9月21日，黒木秀一採集；KPM-NC0013156，亜熱帯性常緑広葉樹林内地上，鹿児島奄美大島宇検村，2002年9月22日，波多野英治採集．

主な特徴：傘と柄はレモン黄色の粉質物を帯び，橙褐色〜褐色の圧着した鱗片に被われ，湿時粘性がある；ツバは消失性，綿毛状膜質，レモン黄色の粉質物を帯びる；肉は著しくゼラチン化し，変色性を欠く；亜熱帯〜温帯の常緑広葉樹林内地上に発生．

コメント：北米から新種記載されたキイロイグチ *Pulveroboletus ravenelii* (Berk. & M.A. Curtis) Murrill（Berkeley and Curtis 1853；Singer 1947；Snell and Dick 1970；Smith and Thiers 1971；Corner 1972；Grund and Harrison 1976；Nagasawa 1989；Bessette et al. 2000）並びにマレーシア産 *Pulveroboletus frians* Corner (Corner, 1972) は本種に最も類似した特徴を持つが，両種とも橙褐色の鱗片を欠き，肉は青変性を有する．本種はまたアフリカ（コンゴ）産 *Pulveroboletus aberrans* Heinem. & Gooss.-Font.（Heinemann, 1951, 1954）にもやや似るが，アフリカ産種は傘と柄の表面が全体に橙褐色を帯びた密綿毛状を呈し，肉は青変する．

29. *Pulveroboletus brunneoscabrosus* ウロコキイロイグチ —— 277

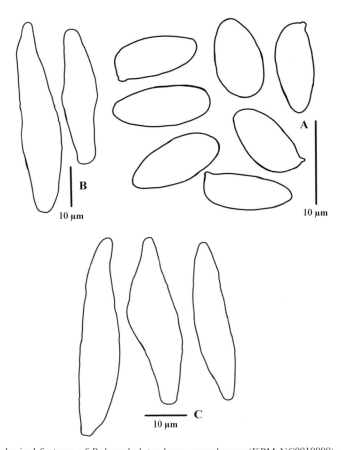

Fig. 226 – Micromorphological features of *Pulveroboletus brunneoscabrosus* (KPM-NC0010098): **A**. Basidiospores. **B**. Cheilocystidia. **C**. Pleurocystidia. Illustrations by Takahashi, H.
ウロコキイロイグチの顕鏡図（KPM-NC0010098）：**A**.担子胞子．**B**.縁シスチジア．**C**.側シスチジア．図：高橋春樹．

Fig. 227 – Immature basidioma of *Pulveroboletus brunneoscabrosus* (KPM-NC0013148), 6 Jun. 2004, Banna Park, Ishigaki Island. Photo by Takahashi, H.
ウロコキイロイグチの幼菌（KPM-NC0013148），2004年6月6日，石垣島バンナ公園．写真：高橋春樹．

29. *Pulveroboletus brunneoscabrosus* ウロコキイロイグチ —— 279

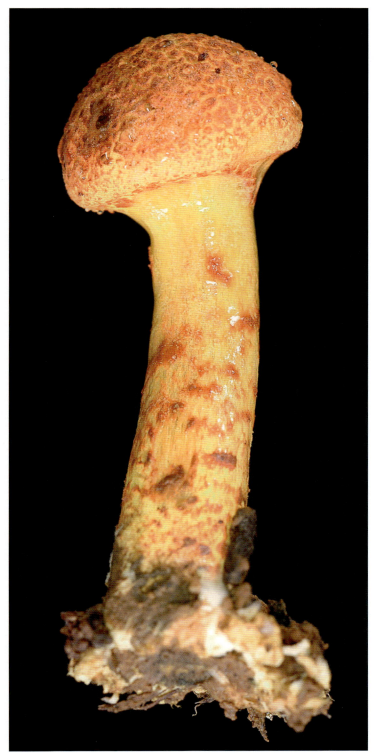

Fig. 228 – Immature basidioma of *Pulveroboletus brunneoscabrosus* (KPM-NC0013148), 6 Jun. 2004, Banna Park, Ishigaki Island. Photo by Takahashi, H.
ウロコキイロイグチの幼菌（KPM-NC0013148），2004年6月6日，石垣島バンナ公園．写真：高橋春樹．

Fig. 229 – Immature basidioma of *Pulveroboletus brunneoscabrosus* (KPM-NC0013148), 6 Jun. 2004, Banna Park, Ishigaki Island. Photo by Takahashi, H.
ウロコキイロイグチの幼菌（KPM-NC0013148），2004年6月6日，石垣島バンナ公園．写真：高橋春樹．

Fig. 230 – Mature basidioma of *Pulveroboletus brunneoscabrosus* (KPM-NC0013148), 6 Jun. 2004, Banna Park, Ishigaki Island. Photo by Takahashi, H.
ウロコキイロイグチの成熟した子実体（KPM-NC0013148），2004年6月6日，石垣島バンナ公園．写真：高橋春樹．

Fig. 231 – Mature basidioma of *Pulveroboletus brunneoscabrosus* (KPM-NC0013148), 6 Jun. 2004, Banna Park, Ishigaki Island. Photo by Takahashi, H.
ウロコキイロイグチの成熟した子実体（KPM-NC0013148），2004年6月6日，石垣島バンナ公園．写真：高橋春樹．

29. *Pulveroboletus brunneoscabrosus* ウロコキイロイグチ —— 283

Fig. 232 – Pileus surface of *Pulveroboletus brunneoscabrosus* (KPM-NC0013148), 6 Jun. 2004, Banna Park, Ishigaki Island. Photo by Takahashi, H.
ウロコキイロイグチの傘表面（KPM-NC0013148），2004年6月6日，石垣島バンナ公園．写真：高橋春樹．

30. *Resinomycena fulgens* Har. Takah., Taneyama & Oba, **sp. nov.** ギンガタケ
MycoBank no.: MB 809938.

Etymology: The specific epithet comes from the Latin word for "gleaming, shining", referring to the bioluminescent basidiomata.

Distinctive features of this species consist of small, pure white basidiomata with a pruinose pileus and stipe; inamyloid or weakly amyloid basidiospores; dimorphic cheilocystidia of slightly thick-walled oleocystidia and dendrophysis; narrowly clavate oleocystidia and dendroid elements over the pileus and stipe surfaces which are beaded with resinous drops; distinctly thick-walled, dextrinoid, short hyphal cells in the stipe trama; luminescence in the pileus, lamellae and stipe; and a lignicolous habit.

Macromorphological characteristics (Figs. 239–243): Pileus 1.5–2.5 mm in diameter, at first hemispherical, then expanding to convex to broadly convex, smooth; surface appearing pruinose under a lens, dry, not hygrophanous, pure white; margin entire, at first somewhat incurved. Flesh very thin (up to 0.1 mm), pure white, odor not distinctive, taste undetermined. Stipe 2–3 × 0.4–0.7 mm, subcylindrical but slightly thickened at the base, eccentric, terete, smooth; surface pruinose under a lens, dry, pure white; basal mycelium none. Lamellae adnate to subdecurrent, subdistant (12–15 reach the stipe), with 1–2 series of lamellulae, not intervenose, narrow (up to 0.2 mm broad), thin, pure white; edges finely fimbriate under a lens, concolorous.

Luminescence: Pileus, lamellae and stipe emit yellowish green light; mycelium luminescence unknown.

Micromorphological characteristics (Figs. 234–238): Basidiospores (5.1–) 5.8–6.8 (–7.6) × (3.0–) 3.7–4.3 (–5.0) μm (n = 102, mean length = 6.31 ± 0.52, mean width = 3.99 ± 0.32, Q = (1.39–) 1.48–1.68 (–1.90), mean Q = 1.58 ± 0.10), ellipsoid to obscurely amygdaliform, smooth, hyaline, inamyloid or weakly amyloid, colorless in KOH, thin-walled. Basidia (14.7–) 14.6–18.6 (–19.1) × (4.7–) 4.8–5.7 (–5.8) μm (n = 4, mean length = 16.60 ± 1.98, mean width = 5.26 ± 0.45) in main body, (3.1–) 3.1–3.8 (–4.2) × (1.1–) 1.1–1.4 (–1.6) μm (n = 7, mean length = 3.42 ± 0.36, mean width = 1.28 ± 0.17) in sterigmata, clavate, 4-spored, rarely 3-spored. Cheilocystidia dimorphic: 1) oleocystidia (18.1–) 22.1–30.0 (–34.8) × (5.3–) 5.7–6.8 (–7.2) μm (n = 21, mean length = 26.05 ± 3.91, mean width = 6.26 ± 0.55), abundant, narrowly clavate to subcylindrical, occasionally strangulated, covered with copious amounts of resin, smooth, hyaline, inamyloid, with walls 0.6–0.8 (–1.0) μm (n = 16, mean thickness = 0.72 ± 0.12); 2) dendroid hyphidia (dendrophysis) (16.2–) 17.4–26.1 (–26.8) × (4.3–) 4.5–5.8 (–6.0) μm (n = 5, mean length = 21.72 ± 4.34, mean width = 5.13 ± 0.64), scattered, subcylindrical, with irregularly branched, nodose, coralloid outgrowths, hyaline, inamyloid, thin-walled. Pleurocystidia absent. Hymenophoral trama subregular; element hyphae (1.8–) 1.9–2.4 (–2.6) μm wide (n = 26, mean width = 2.15 ± 0.25), cylindrical, walls thin, smooth, hyaline, dextrinoid. Pileipellis a cutis of interwoven, repent hyphae, beaded with resinous drops; constituent hyphae (1.6–) 2.0–2.6 (–2.9) μm wide (n = 23, mean width = 2.26 ± 0.30), cylindrical, at times somewhat sinuous, occasionally branched, smooth, hyaline, inamyloid, thin-walled; subpellis undifferentiated. Pilocystidia dimorphic: 1) oleocystidia (21.7–) 25.0–34.8 (–36.8) × (5.9–) 6.2–7.8 (–9.0) μm (n = 20, mean length = 29.90 ± 4.88, mean width = 6.95 ± 0.80), covered with copious amounts of resin, scattered and more common at the pileus margin, narrowly clavate, at times strangulated, smooth, hyaline, inamyloid, with walls (0.5–) 0.6–0.8 (–1.0) μm (n = 19, mean thickness = 0.72 ± 0.13); 2) dendroid elements (30.2–) 34.8–42.1 (–43.7) × (1.9–) 2.0–2.5 (–2.6) μm (n = 10, mean length = 38.46 ± 3.66, mean width = 2.24 ± 0.21), scattered, subcylindrical, with irregularly branched, nodose, coralloid outgrowths, hyaline, inamyloid, thin-walled. Elements of pileitrama (2.1–) 2.0–4.7 (–8.5) μm wide (n = 38, mean width = 3.36 ± 1.37), subparallel, cylindrical, smooth, hyaline, dextrinoid, with walls 0.5–0.7 (–0.8) μm (n = 18, mean thickness = 0.59 ± 0.09). Stipitipellis a cutis of parallel, repent hyphae (3.2–) 3.9–5.9 (–6.8) μm wide (n = 37, mean width = 4.89 ± 1.01), cylindrical, smooth, hyaline, inamyloid, with walls (0.6–) 0.7–1.1 (–1.2) μm (n = 17, mean thickness = 0.91 ± 0.19). Caulocystidia similar to the pilocystidia. Stipe trama composed of unbranched, thick-walled hyphae and occasionally branched, thin-walled hyphae; subfusiform hyphal cells (23.0–) 27.4–39.6 (–43.7) × (9.1–) 11.5–15.7 (–17.3) μm (n = 25, mean length = 33.51 ± 6.08, mean width = 13.61 ± 2.11) at the apex of stipe, cylindrical hyphae (7.2–) 8.1–11.7 (–14.7) μm wide (n = 37, mean width = 9.92 ± 1.79) at the lower portion of stipe, parallel, unbranched, smooth, dextrinoid, with pale yellow, thickened walls (1.2–) 1.6–2.4 (–3.0) μm (n = 32, mean thickness = 2.01 ± 0.39), intermixed with much narrower, occasionally branched, thin-walled hyphae (1.2–) 1.5–2.0 (–2.2) μm wide (n = 30, mean width = 1.72 ± 0.25). All tissues colorless in KOH, with clamp connections.

Known distribution: Kagoshima (Yaku Island), Tokyo (Hachijo Island), Kochi.

Habitat and phenology: Lignicolous on bark of living or dead logs of *Castanopsis sieboldii* (Makino) Hatus. ex T. Yamaz. et Mashiba, June.

Holotype: TNS-F-52273, on bark of a living log of *C. sieboldii*, Koseda, Yakushima-cho, Kagoshima Pref., 7 Jun. 2013, coll. Kurogi, S.

Extralimital specimen examined: TNS-F-52272, on bark of a living log of *C. sieboldii*, Kamogawa, Hachijo Island, Tokyo, 3 Jun. 2013, coll. Oba, Y.

GenBank accession number: AB971702.

Japanese name: Gingatake (named by Mr. Masanori Ishii).

Comments: The weakly amyloid basidiospores and the presence of dendrohyphidia and narrowly clavate oleocystidia covered with copious amounts of resin suggest that the new species may possibly be referable to the genus *Resinomycena* (Redhead and Singer 1981).

Resinomycena japonica Redhead & Nagas. from Japan (Readhead and Nagasawa 1987) closely resembles *R. fulgens* in many respects. According to the protologue, the former is described as having conchate, pleurotoid basidiomata with a minutely pubescent pileus and a poorly developed, much shorter and narrower stipe: 0.2–0.8 × 0.1–0.2 mm (Readhead and Nagasawa 1987), a small, radiating disk of silky basal mycelium, and a well differentiated subpellis made up of inflated, thick-walled cells. These characteristics show significant differences from those of *R. fulgens*. Furthermore, *R. japonica* forms a sarcodimitic stipe trama that interlaces yellowish, thick-walled, inflated sarcohyphae with much narrower, thin-walled generative hyphae, while the stipe trama in *R. fulgens* consists of mostly short, cylindrical hyphal cells that do not strictly correspond to the definition of quintessential sarcohyphae (Corner 1966; Kirk et al. 2008).

References 引用文献

Corner EJH. 1966. A monograph of cantharelloid fungi. Ann Bot Memoir 2: 1–255.
Kirk PM, Cannon PF, Winter DW, Stalpers JA. 2008. Dictionary of the fungi, 10th edn. CABI, UK.
Redhead SA, Nagasawa E. 1987. *Resinomycena japonica* and *Resupinatus merulioides*, new species of Agaricales from Japan. Can J Bot 65 (5): 972–976.
Redhead SA, Singer R. 1981. *Resinomycena* gen. nov. (Agaricales), an ally of *Hydropus, Mycena* and *Baeospora*. Mycotaxon 13 (1): 150–170.

30. ギンガタケ（新種）*Resinomycena fulgens* Har. Takah., Taneyama & Oba, **sp. nov.**

肉眼的特徴（Figs. 239-243）：傘は径1.5-2.5 mm，最初半球形のち饅頭形～中高偏平，平坦で条線または溝線を欠く；表面は粉状，非吸水性，粘性を欠き，純白色；縁部は全縁で最初やや内側に巻く．肉は厚さ0.1 mm以下，純白色，特別な匂いはなく，味は未確認．柄は2-3×0.4-0.7 mm,円柱形で根元が僅かに膨らみ，偏在生；表面は粉状，粘性を欠き，純白色，根元に発達した綿毛状菌糸体は見られない．ヒダは直生からやや垂生，やや疎（柄に到達するヒダは12-15），1-2の小ヒダを伴い，連絡脈を欠き，幅0.2 mm以下，純白色；縁部は同色，長縁毛状．

発光性：子実体全体が黄緑色に発光；菌糸体の発光性は未確認．

顕微鏡的特徴（Figs. 234-238）：担子胞子は (5.1-) 5.8-6.8 (-7.6) × (3.0-) 3.7-4.3 (-5.0) μm（n = 102, mean length = 6.31±0.52, mean width = 3.99±0.32, Q = (1.39-) 1.48-1.68 (-1.90), mean Q = 1.58±0.10），楕円形～類アーモンド形，平坦，無色，非アミロイドまたは弱アミロイド，アルカリ溶液において無色，薄壁．担子器は (14.7-) 14.6-18.6 (-19.1) × (4.7-) 4.8-5.7 (-5.8) μm（本体），(3.1-) 3.1-3.8 (-4.2) × (1.1-) 1.1-1.4 (-1.6) μm（ステリグマ），こん棒形，4胞子性，稀に3胞子性．縁シスチジアは2形性：1) オレオシスチジア (18.1-) 22.1-30.0 (-34.8) × (5.3-) 5.7-6.8 (-7.2) μm，樹脂状の油球に被われ，群生し，幅の狭いこん棒形～類円柱形，平滑，無色，非アミロイド，壁の厚さは 0.6-0.8 μm；2) 樹枝状分岐物を具えた細胞 (16.2-) 17.4-26.1 (-26.8) × (4.3-) 4.5-5.8 (-6.0) μm，散在し，類円柱形，不規則に分岐した珊瑚状～瘤状突起を具え，無色，非アミロイド，薄壁．側シスチジアはない．子実層托実質を構成する菌糸は幅 (1.8-) 1.9-2.4 (-2.6) μm，ほぼ平列し，円柱形，平滑，無色，偽アミロイド，薄壁．傘の表皮組織は樹脂状の油球が点在し，緩く配列した匍匐性の菌糸からなる：菌糸細胞は幅 (1.6-) 2.0-2.6 (-2.9) μm，円形，しばしば分岐し，平滑，無色，非アミロイド，薄壁；下表皮層は未分化．傘シスチジアは2形性：1) オレオシスチジア (21.7-) 25.0-34.8 (-36.8) × (5.9-) 6.2-7.8 (-9.0) μm，散生し，傘周縁部に分布の偏りが見られ，幅の狭いこん棒形，時に不規則にいくつかのくびれがあり，平滑，無色，非アミロイド，壁の厚さは 0.6-0.8 μm；2) 樹枝状分岐物を具えた細胞 (30.2-) 34.8-42.1 (-43.7) × (1.9-) 2.0-2.5 (-2.6) μm，散在し，類円柱形，不規則に分岐した珊瑚状～瘤状突起を具え，無色，非アミロイド，薄壁．傘実質の菌糸は径 (2.1-) 2.0-4.7 (-8.5) μm，並列し，円柱形，平滑，偽アミロイド，壁の厚さは 0.5-0.7 μm．柄表皮を構成する菌糸は平行菌糸被をなし，幅 (3.2-) 3.9-5.9 (-6.8) μm，並列し，円柱形，平滑，無色，非アミロイド，壁の厚さは 0.7-1.1 μm；柄シスチジアは傘シスチジアに類似する．柄実質は無分岐の厚壁菌糸および時折分岐する薄壁な菌糸からなる；柄の頂部（傘と柄の接続部）付近のやや膨大した類紡錘形の菌糸細胞は (23.0-) 27.4-39.6 (-43.7) × (9.1-) 11.5-15.7 (-17.3) μm，柄下半部の円柱形の菌糸は幅 (7.2-) 8.1-11.7 (-14.7) μm，平列し，無分岐，平滑，淡黄色，偽アミロイド，隔壁を有し，厚壁 (1.6-2.4 μm)；時折分岐する薄壁な菌糸は幅 (1.2-) 1.5-2.0 (-2.2) μm．全ての組織の菌糸はクランプを有し，アルカリ溶液において無色．

生態および発生時期：スダジイの生木または腐木樹皮に群生または散生，6月．

分布：日本（屋久島，八丈島，高知）．

供試標本：TNS-F-52273（正基準標本），スダジイの生木樹皮上，鹿児島県屋久島町小瀬田，2013年6月7日，黒木秀一採集．

地域外供試標本：TNS-F-52272，スダジイの生木樹皮上，東京都八丈島鴨川，2013年6月3日，大場由美子採集．

GenBank 登録番号：AB971702．

主な特徴：子実体は小型で全体に純白色を呈する；傘と柄の表面は粉状；胞子は非アミロイドまたは弱アミロイド；縁シスチジアはやや厚壁なオレオシスチジアと樹枝状分岐物を具えた細胞の

30. *Resinomycena fulgens* ギンガタケ —— 287

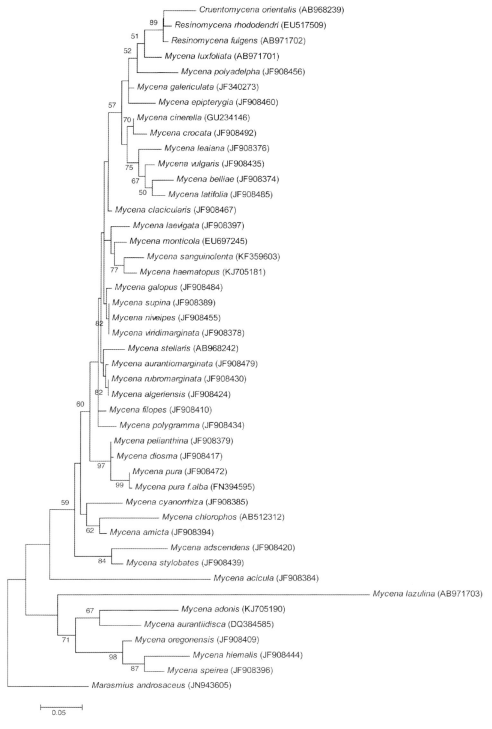

Fig. 233 – A phylogenetic tree of ITS1-5.8S-ITS2 dataset for *Mycena* species, analyzed by maximum likelihood method based on the Tamura-Nei model (Tamura and Nei 1993) in MEGA5 (Tamura et al. 2011). Numbers on the branches represent bootstrap values obtained from 2000 replications (only values greater than 50% are shown).
核リボソーム ITS1-5.8S-ITS2を用いた最尤法によるクヌギタケ属系統樹.

2形性：傘と柄の表皮組織は縁シスチジアと同様の2形性シスチジアを有し，オレオシスチジアを取り囲むように樹脂状物が点在する；柄の実質は偽アミロイドで，厚壁且つ短形の菌糸細胞からなる；子実体全体が黄緑色に発光する；スダジイの樹皮から発生．

コメント：弱アミロイドの担子胞子および傘と柄の表皮組織における樹脂状物に囲まれたオレオシスチジア並びに樹枝状細胞の存在は本種がザラメタケ属 *Resinomycena*（Redhead and Singer 1981）に近縁であることを示唆している．

　日本産ザラメタケ *Resinomycena japonica* Redhead & Nagas.（Readhead and Nagasawa 1987）は形態的に本種と共通する点が多いが，より小形の柄：0.2-0.8×0.1-0.2 mm（Readhead and Nagasawa 1987）と微毛に被われた傘からなるヒラタケ型子実体を形成し，根元を放射状に取り囲む菌糸体が存在し，そして膨大した厚壁な菌糸細胞からなる下表皮層が分化する点で明らかな有意差が認められる．またザラメタケの柄実質は非常に長く紡錘形に膨大した厚壁菌糸細胞を持つのに対し，ギンガタケのそれは一般に円柱形で短い厚壁菌糸細胞からなることも特筆される．

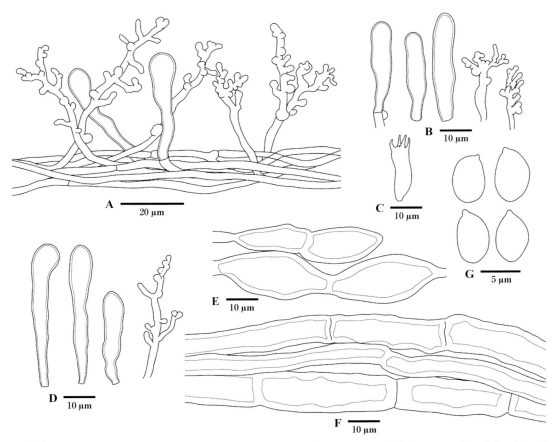

Fig. 234 – Micromorphological features of *Resinomycena fulgens* (holotype): **A**. Longitudinal cross section of the pileipellis showing the pilocystidia of two types. **B**. Cheilocystidia of two types. **C**. Basidium. **D**. Caulocystidia of two types. **E**. Subfusiform hyphal cells of the stipe trama (from the stipe apex). **F**. Cylindrical hyphal cells of the stipe trama (from the lower portion of the stipe). **G**. Basidiospores. Illustrations by Taneyama, Y.
ギンガタケの顕微鏡図（正基準標本）：**A**. 2形性の傘シスチジアを示す傘表皮組織の縦断面．**B**. 2形性の縁シスチジア．**C**. 担子器．**D**. 2形性の柄シスチジア．**E**. 柄実質の亜紡錘形の菌糸細胞（柄の頂部）．**F**. 柄実質の円柱形の菌糸細胞（柄の下半部）．**G**. 担子胞子．図：種山裕一．

30. *Resinomycena fulgens* ギンガタケ

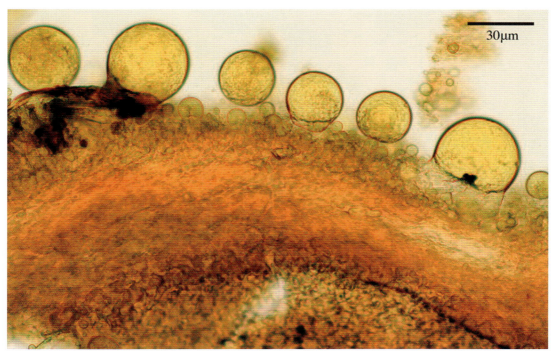

Fig. 235 – Longitudinal cross section of the pileipellis of *Resinomycena fulgens* (in Melzer's reagent, holotype), beaded with resinous drops. Photo by Taneyama, Y.
ギンガタケの傘表皮組織の縦断面（メルツァー溶液で封入，正基準標本），樹脂状物が点在する．写真：種山裕一．

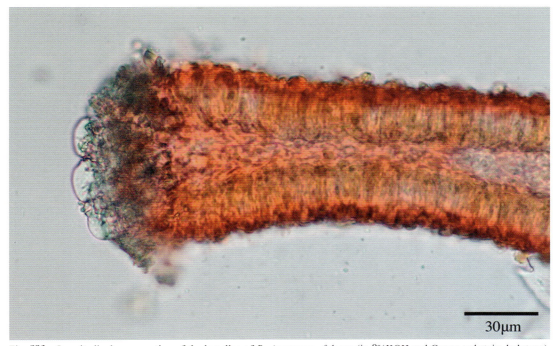

Fig. 236 – Longitudinal cross section of the lamellae of *Resinomycena fulgens* (in 3%KOH and Congo red stain, holotype), showing the lamella edge covered with copious amounts of resin. Photo by Taneyama, Y.
ギンガタケのヒダの縦断面（3%水酸化カリウム溶液で封入した後コンゴー赤染色，正基準標本），多量の樹脂状物に被われたヒダ縁部を示す．写真：種山裕一．

290 —— 30. *Resinomycena fulgens* ギンガタケ

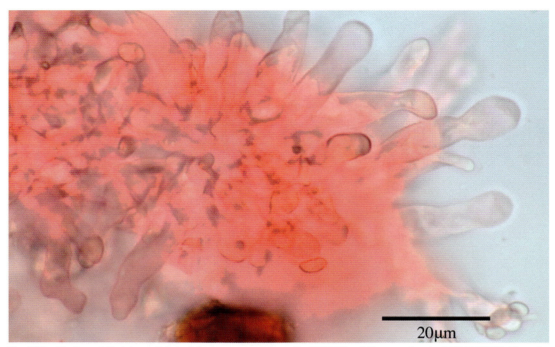

Fig. 237 – Cheilocystidia (oleocystidia) of *Resinomycena fulgens* (in 3%KOH and Congo red stain, TNS-F-52272). Photo by Taneyama, Y.
　ギンガタケのオレオシスチジア形縁シスチジア（3％水酸化カリウム溶液で封入した後コンゴー赤染色，TNS-F-52272）．写真：種山裕一．

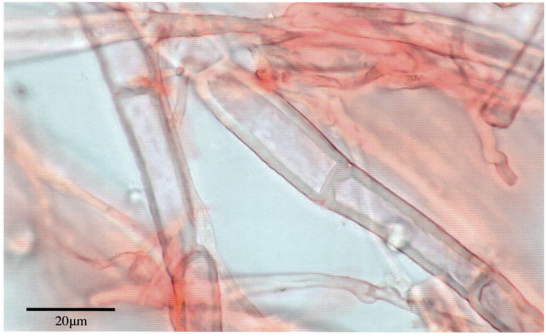

Fig. 238 – Thick-walled hyphae of the stipe trama in *Resinomycena fulgens* (in 3%KOH and Congo red stain, TNS-F-52272) from the lower portion of the stipe. Photo by Taneyama, Y.
　ギンガタケの柄の下半部実質を構成する厚壁菌糸（3％水酸化カリウム溶液で封入した後コンゴー赤染色，TNS-F-52272）．写真：種山裕一．

Fig. 240 – Pileus surface of *Resinomycena fulgens* (TNS-F-52272). 3 Jun. 2013, Hachijo Island, Tokyo. Photo by Takahashi, H.
ギンガタケの傘表面（TNS-F-52272），2013年6月3日，東京都八丈島．写真：高橋春樹．

30. *Resinomycena fulgens* ギンガタケ —— 291

Fig. 239 – Basidioma of *Resinomycena fulgens* (TNS-F-52272) on bark of a living log of *C. sieboldii*, 3 Jun. 2013, Hachijo Island, Tokyo. Photo by Takahashi, H.
　スダジイの生木樹皮上に発生したギンガタケの子実体（TNS-F-52272），2013年6月3日，東京都八丈島．写真：高橋春樹．

30. *Resinomycena fulgens* ギンガタケ —— 293

Fig. 241 – Basidiomata of *Resinomycena fulgens* (TNS-F-52272) on bark of a living log of *C. sieboldii*, 3 Jun. 2013, Hachijo Island, Tokyo. Photo by Takahashi, H.
　スダジイの生木樹皮上に発生したギンガタケの子実体（TNS-F-52272），2013年6月3日，東京都八丈島．写真：高橋春樹．

Fig. 242 – Basidiomata of *Resinomycena fulgens* (holotype) on bark of a living log of *C. sieboldii,* 7 Jun. 2013, Koseda, Yakushima-cho, Kagoshima Pref. Photo by Kurogi, S.
スダジイの生木樹皮上に発生したギンガタケの子実体（正基準標本），2013年6月7日，鹿児島県屋久島町小瀬田．写真：黒木秀一．

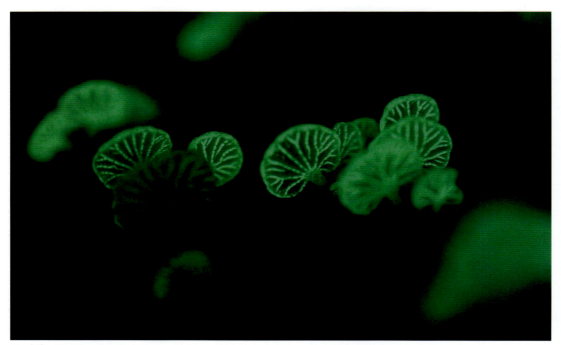

Fig. 243 – Basidiomata of *Resinomycena fulgens* (holotype) emitting yellowish green light on bark of a living log of *C. sieboldii,* 7 Jun. 2013, Koseda, Yakushima-cho, Kagoshima Pref. Photo by Kurogi, S.
スダジイの生木樹皮上において黄緑色に発光するギンガタケの子実体（正基準標本），2013年6月7日，鹿児島県屋久島町小瀬田．写真：黒木秀一．

31. *Rubinoboletus monstrosus* Har. Takah. ダルマイグチ

Mycoscience 48 (2): 95 (2007) [MB#530008].

Etymology: The specific epithet means "monstrous", referring to the peculiar habit of the basidiomata which consist of an often irregularly shaped pileus and an extremely short stipe.

Macromorphological characteristics (Figs. 245–250): Pileus 50–140 mm in diameter, at first hemispherical, expanding to broadly convex, often irregularly shaped, sometimes rimose-areolate at the center, with a slightly appendiculate margin; surface dry, subtomentose to nearly glabrous, greyish-orange (5B6) or brownish-orange (5C5–6) to yellowish-brown (5D5) overall, occasionally tinged with reddish-brown (8D7) when young and fresh. Flesh up to 12 mm thick, whitish, unchanging when exposed; odor strongly disagreeable (resembling rotten meat), taste bitter. Stipe 20–40 × 15–45 mm, very short, subequal or tapering toward the base, central, terete, hollow from the outset, non-reticulate; surface dry, almost glabrous to subtomentose, entirely yellowish-brown (5D4–5); base covered with a indistinct, whitish mycelial tomentum. Tubes up to 10 mm deep, adnate or slightly depressed around the stipe, greyish-orange (5B4–5) when young, then yellowish-brown (5D6) at maturity, unchanging when exposed; pores small (2–3 per mm), rounded to subangular, concolorous.

Micromorphological characteristics (Fig. 244): Basidiospores 5–6 × 3.5–4 μm (n = 35, Q = 1.4–1.5), ovoid-ellipsoid, smooth, pale yellow in water, walls up to 0.5 μm. Basidia 25–30 × 6–8 μm, clavate, four-spored. Basidioles clavate. Cheilocystidia gregarious, 34–52 × 8–13 μm, fusoid-ventricose, smooth, hyaline, thin-walled. Pleurocystidia scattered, 30–60 × 5–15 μm, fusoid-ventricose, smooth, with yellowish-brown to golden-yellow contents (in water), thin-walled. Elements of hymenophoral trama 6–10 μm wide, cylindrical, parallel to each other, smooth, hyaline, thin-walled. Pileipellis composed of a trichoderm formed by loosely interwoven hyphal elements, 5–8 μm wide, cylindrical, encrusted with yellowish-brown fine granules (in water), thin-walled. Pileitrama of cylindrical, loosely interwoven hyphae 4–13 μm wide, smooth, colorless, thin-walled. Stipitipellis composed of parallel, repent hyphae 2–5 μm wide, cylindrical, smooth, colorless, thin-walled; caulocystidia 12–35 × 4–10 μm, broadly clavate, smooth, hyaline, thin-walled. Stipe trama composed of longitudinally arranged, cylindrical cells 6–20 μm wide, unbranched, smooth, colorless, thin-walled. Clamps absent in all tissues.

Habitat and phenology: Solitary to scattered on ground in evergreen broad-leaved forests dominated by *Quercus miyagii* Koidz. and *Castanopsis sieboldii* (Makino) Hatus. ex T. Yamaz. et Mashiba, May.

Specimens examined: KPM-NC0013139 (holotype), on ground in an evergreen broad-leaved forest dominated by *Q. miyagii* and *C. sieboldii*, Banna Park, Ishigaki-shi, Okinawa Pref., 19 May 2004, coll. Takahashi, H; KPM-NC0013141, same location, 20 May 2003, coll. Takahashi, H.; KPM-NC0013142, same location, 25 May 2004, coll. Takahashi, H.; KPM-NC0013143, same location, 14 May 2005, coll. Takahashi, H.; KPM-NC0013140, same location, 20 May 2005, coll. Takahashi, H.

Known distribution: Okinawa (Ishigaki Island).

Japanese name: Daruma-iguchi.

Comments: Features delimiting this species include the medium to large basidiomata composed of an often irregularly shaped, brownish-orange to yellowish-brown pileus and an extremely short, non-reticulate, hollow stipe; the whitish, unchanging flesh with bitter taste and unpleasant smell; the short ellipsoid, light yellowish basidiospores; the prominent pleurocystidia containing yellowish-brown to golden-yellow contents; and the habitat of evergreen broad-leaved forests.

The unchanging, whitish flesh and the pale yellow, short ellipsoid basidiospores suggest that this species is closely related to the genera *Rubinoboletus*, *Gyroporus*, or *Tylopilus*. If greater taxonomic emphasis is placed on the non-reticulate stipe, the short ellipsoid, light yellowish basidiospores, the prominent pleurocystidia with yellowish-brown to golden-yellow contents, and the lack of clamp connections, it would be better placed in the genus *Rubinoboletus* Pilát & Dermek in the sense of Heinemman and Rammeloo (Heinemman and Rammeloo 1983; Watling and Gregory 1988; Li and Watling 1999; Watling and Li 1999).

Within the genus *Rubinoboletus*, it seems to be allied with *Rubinoboletus ballouii* (Peck) Heinemm. & Rammeloo, originally described from North America (Singer 1947; Snell and Dick 1970; Corner 1972; Heinemann and Rammeloo 1983; Bessette et al. 2000). The latter species, however, differs from *R. monstrosus* in producing a bright orange-red pileus, a much longer, solid stipe and larger pores reaching 2 mm at maturity. With respect to the taxonomy of *Tylopilus ballouii* sense Nagasawa from Japan (Nagasawa 1989), it is likely to be distinct from the North American species in that Japanese specimens have a yellowish-brown pileus.

Rubinoboletus caespitosus T.H. Li & Watling from Australia (Cleland 1924; Grgurinovic 1997; Li and Watling 1999; Watling and Li 1999) is somewhat similar in appearance, but it has a solid stipe, significantly larger basidiospores: $8–8.9 \times 5–5.5$ μm (Cleland 1924) and a caespitose habit.

References 引用文献

Bessette AE, Roody WC, Bessette AR. 2000. North American Boletes. A color guide to the fleshy pored mushrooms. Syracuse University Press, New York.
Cleland JB. 1924. Australian Fungi. Notes and descriptions 5. Trans R Soc South Australia 48: 236–252.
Corner EJH. 1972. *Boletus* in Malaysia. Government Printing Office, Singapore.
Grgurinovic CA. 1997. Larger Fungi of South Australia. The Botanic Gardens of Adelaide and State Herbarium, Adelaide.
Heinemann P, Rammeloo J. 1983. Gyrodontaceae p.p. Boletineae. Flore Illustrée des Champignons d'Afrique Central 10: 173–198.
Li T-H, Watling R. 1999. New taxa and combinations of Australian boletes. Edinb J Bot 56: 143–148.
Nagasawa E. 1989. Boletaceae. In: Imazeki R, Hongo T (eds), Colored illustrations of mushrooms of Japan II. Hoikusha, Osaka, pp 1–44 (in Japanese).
Singer R. 1947. The Boletineae of Florida with notes on extralimital species III. Am Midl Nat 37: 1–135.
Snell WH, Dick EA. 1970. The boleti of northeastern North America. Cramer, Vaduz.
Takahashi H. 2007. Five new species of the Boletaceae from Japan. Mycoscience 48 (2): 90–99.
Watling R, Gregory NM. 1988. Observations on the Boletes of the Cooloola Sandmass, Queensland and notes on their distributions in Australia. Part 2B. Smooth-spored taxa of the family Gyrodontaceae and the genus *Pulveroboletus*. Proc R Soc Queensl 99: 65–76.
Watling R, Li T-H. 1999. Australian Boletes. A preliminary survey. Royal Botanic Garden, Edinburgh.

31. ダルマイグチ *Rubinoboletus monstrosus* Har. Takah.

Mycoscience 48 (2): 95 (2007) [MB#530008].

肉眼的特徴（Figs. 245-250）：傘は径50-140 mm，最初半球形のち中高偏平になり，しばしば不規則な形状になり，時に中央部に亀甲状のひび割れを生じ，縁部は管孔側に僅かに突出する；表面は粘性を欠き，やや密綿毛状〜ほぼ平滑，全体に帯赤黄褐色〜淡黄褐色，時に若いときレンガ赤色を帯びる。肉は厚さ12 mm 以下，類白色，変色性を欠き，不快な臭気と苦みがある。柄は20-40×15-45 mm，著しく短形，ほぼ上下同大または基部に向かってやや細くなり，中心生，最初から中空，網目を欠く；表面は乾性，ほぼ平滑またはやや密綿毛状，傘より淡色〜淡黄褐色；根元は不明瞭な白色綿毛状菌糸体に被われる。管孔は長さ10 mm 以下，直性または柄の周

囲においてやや陥入し，最初灰橙色のち黄褐色，空気に触れても変色しない；孔口は小型（2-3 per mm），円形〜やや角形，管孔と同色.

顕微鏡的特徴（Fig. 244）：担子胞子は5-6×3.5-4 μm（n = 35, Q = 1.4-1.5），卵形〜短楕円形，平坦，淡黄色（水封）．担子器は25-30×6-8 μm，こん棒形，4胞子性．偽担子器はこん棒形．縁シスチジアは群生し，34-52×8-13 μm，紡錘形，平滑，無色，薄壁．側シスチジアは散在し，30-60×5-15 μm，紡錘形，黄褐色の内容物を持ち（水封），薄壁．子実層托実質の菌糸は幅6-10 μm，円柱形，並列し，平滑，無色，薄壁．傘表皮組織は緩く錯綜した菌糸からなる毛状被；菌糸は幅5-8 μm，円柱形，黄褐色の細かい粒状色素が沈着し（水封），薄壁．傘実質の菌糸は幅4-13 μm，円柱形，緩く錯綜し，平滑，無色，薄壁．柄表皮組織は平行菌糸被；菌糸は幅2-5 μm，円柱形，平滑，無色，薄壁；柄シスチジアは12-35×4-10 μm，広こん棒形，平滑，無色，薄壁．柄実質の菌糸は幅6-20 μm，縦に沿って配列し，円柱形，無分岐，平滑，無色，薄壁．全ての組織において菌糸はクランプを欠く．

生態および発生時期：スダジイ，オキナワウラジロガシを主体とする常緑広葉樹林内地上に孤生または散生，5月．

分布：沖縄（石垣島）．

供試標本：KPM-NC0013139（正基準標本），スダジイ，オキナワウラジロガシを主体とする常緑広葉樹林内地上，沖縄県石垣市バンナ公園，2004年5月19日，高橋春樹採集；KPM-NC0013141，同上，2003年5月20日，高橋春樹採集；KPM-NC0013142，同上，2004年5月25日，高橋春樹採集；KPM-NC0013143，同上，2005年5月14日，高橋春樹採集；KPM-NC0013140，同上，2005年5月20日，高橋春樹採集．

主な特徴：子実体は中〜大型で歪な形状になることが多い；傘は赤褐色〜黄褐色；柄は最初から中空で常に太短い（長さと太さがほぼ同等）；肉は類白色で変色性を欠き，不快な臭気と苦みがある；担子胞子は短楕円形，淡黄色；発達した側シスチジアが存在し，黄褐色の内容物を持つ；常緑広葉樹林内地上に発生．

コメント：類白色で変色性を欠く肉および淡黄色，淡楕円形の担子胞子はキニガイグチ属 *Rubinoboletus*，クリイロイグチ属 *Gyroporus*，またはニガイグチ属 *Tylopilus* との類縁を示唆しているが，柄に網目を欠くこと，担子胞子が短形且つ淡黄色であること，有色の内容物を持つ発達した側シスチジアを有すること，そしてクランプを欠く性質を重視すれば，Heinemann & Rammeloo（Heinemann and Rammeloo 1983；Li and Watling, 1999；Watling and Gregory, 1988；Watling and Li, 1999）の定義によるキニガイグチ属に置くのが妥当と思われる．

　属内において本種は北米産 *Rubinoboletus ballouii*（Peck）Heinemann & Rammeloo（Singer 1947；Snell and Dick 1970；Corner 1972；Heinemann and Rammeloo 1983；Bessette et al. 2000）に近い性質を示すが，北米産種は傘が明るい橙赤色を帯びること，柄はより長く，中実であること，そして孔口が成熟時大型（幅2 mmに達する）になる点で区別できる．なお日本産キニガイグチについては傘の色が黄褐色を帯びる点で北米産種とは別種の可能性が高いと思われる．

　短形で太い柄を持つ性質はオーストラリア産 *Rubinoboletus caespitosus*（Clel.）T.-H. Li & R. Watling（Cleland, 1924；Grgurinovic, 1997；Li and Watling, 1999；Watling and Li, 1999）と共通するが，オーストラリア産種は担子胞子がより大型：8-8.9×5-5.5 μm（Cleland 1924）で，子実体は一般に叢生し，柄が中実とされている．

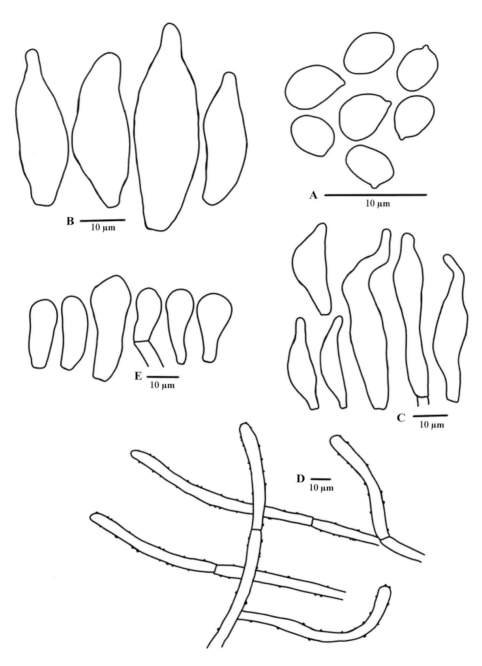

Fig. 244 – Micromorphological features of *Rubinoboletus monstrosus* (KPM-NC0013139): **A**. Basidiospores. **B**. Cheilocystidia. **C**. Pleurocystidia. **D**. Elements of the pileipellis. **E**. Caulocystidia. Illustrations by Takahashi, H.
ダルマイグチの顕鏡図（KPM-NC0013139）：**A**. 担子胞子．**B**. 縁シスチジア．**C**. 側シスチジア．**D**. 傘表皮組織．**E**. 柄シスチジア．図：高橋春樹．

Fig. 245 – Basidiomata of *Rubinoboletus monstrosus* (KPM-NC0013139) on ground in a *Castanopsis-Quercus* forest, 19 May 2004, Banna Park, Ishigaki Island. Photo by Takahashi, H.
スダジイ-オキナワウラジロガシ林内地上に発生したダルマイグチの子実体（KPM-NC0013139），2004年5月19日，石垣島バンナ公園．写真：高橋春樹．

Fig. 246 – Basidiomata of *Rubinoboletus monstrosus* (KPM-NC0013139) on ground in a *Castanopsis-Quercus* forest, 19 May 2004, Banna Park, Ishigaki Island. Photo by Takahashi, H.
スダジイ-オキナワウラジロガシ林内地上に発生したダルマイグチの子実体（KPM-NC0013139），2004年5月19日，石垣島バンナ公園．写真：高橋春樹．

31. *Rubinoboletus monstrosus* ダルマイグチ —— 301

Fig. 247 – Basidiomata of *Rubinoboletus monstrosus* (KPM-NC0013139) on ground in a *Castanopsis-Quercus* forest, 19 May 2004, Banna Park, Ishigaki Island. Photo by Takahashi, H.
スダジイ-オキナワウラジロガシ林内地上に発生したダルマイグチの子実体（KPM-NC0013139），2004年5月19日，石垣島バンナ公園．写真：高橋春樹．

Fig. 248 – Immature basidioma of *Rubinoboletus monstrosus* (KPM-NC0013139), 19 May 2004, Banna Park, Ishigaki Island. Photo by Takahashi, H.
ダルマイグチの幼菌（KPM-NC0013139），2004年5月19日，石垣島バンナ公園．写真：高橋春樹．

31. *Rubinoboletus monstrosus* ダルマイグチ

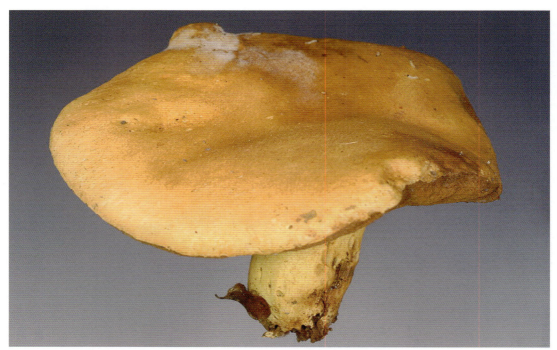

Fig. 249 – Mature basidioma of *Rubinoboletus monstrosus* (KPM-NC0013142), 25 May 2004, Banna Park, Ishigaki Island. Photo by Takahashi, H.
ダルマイグチの成熟した子実体（KPM-NC0013142），2004年5月25日，石垣島バンナ公園．写真：高橋春樹．

Fig. 250 – Underside view of the basidioma of *Rubinoboletus monstrosus* (KPM-NC0013142), 25 May 2004, Banna Park, Ishigaki Island. Photo by Takahashi, H.
ダルマイグチの成熟した子実体（KPM-NC0013142），2004年5月25日，石垣島バンナ公園．写真：高橋春樹．

32. *Strobilomyces brunneolepidotus* Har. Takah. & Taneyama, **sp. nov.** チャオニイグチ

MycoBank no.: MB 809939.

Etymology: The specific epithet comes from the Latin word, "*brunneo-* = brown, *lepidotus* = covered with small scales", referring to the reddish-brown spiny warts that cover the pileus surface.

Distinctive features of this species are found in a reddish-brown pileus covered overall with erect, soft, conical warts; a fugacious, membranous, greyish red veil remnants hanging from the margin; flesh staining reddish brown on exposure; an apically reticulated, slender stipe densely covered with reddish-brown flocci; subglobose to globose basidiospores ornamented with a constantly completed reticulum; clavate to broadly clavate or fusiform cheilocystidia; ventricose-rostrate pleurocystidia; and a habitat in a *Castanopsis-Quercus* forest.

Macromorphological characteristics (Figs. 255–258): Pileus 62–100 mm in diameter, at first hemispherical, then convex; surface dry, entirely shrouded by scattered, erect and often curved, soft, reddish brown (8D4–8E4), spiny warts 0.8–1.5 × 1.5–2.1 mm, whitish in ground color, when young with fugacious, membranous, greyish red (10D4–5) veil remnants hanging from the margin. Flesh up to 12 mm, whitish, gradually changing to reddish brown (9D4–5) where bruised; odor not distinctive, taste undetermined. Stipe ‒80 × 8–10 mm, cylindrical, often slightly thickened toward the base, slender, central, terete, solid, reticulated at the apex by a longitudinally raised, concolorous reticulum, without an annulus; surface dry, enveloped in fluffy, reddish brown (8D4–8E4) flocci; basal mycelial whitish. Tubes up to 13 mm deep, adnexed, depressed around the stipe, greyish brown (8D3–4), unchanging when bruised; pores up to 1.3 mm, angular, concolorous with the tubes.

Micromorphological characteristics (Figs. 251–254): Basidiospores (7.2–) 7.7–8.5 (–9.6) × (5.5–) 6.3–7.0 (–7.6) μm (n = 120, mean length = 8.14 ± 0.41, mean width = 6.63 ± 0.36, Q = (1.09–) 1.17–1.29 (–1.48), mean Q = 1.23 ± 0.06) including ornamentation, inequilateral with an shallow suprahilar depression in profile, subglobose to globose both in face and side view, ornamented with a constantly completed reticulum over the entire surface, brownish in water and KOH, inamyloid, thin-walled; on reticulum (1.1–) 1.2–1.6 (–1.8) × (0.4–) 0.4–0.6 (–0.7) μm (n = 27, mean length = 1.43 ± 0.19, mean width = 0.53 ± 0.08). Basidia (28.7–) 30.8–34.4 (–34.7) × (12.5–) 12.9–14.8 (–15.5) μm (n = 13, mean length = 32.60 ± 1.78, mean width = 13.83 ± 0.95) in main body, (4.6–) 5.0–5.9 (–6.0) × (2.2–) 2.5–3.1 (–3.3) μm (n = 13, mean length = 5.41 ± 0.44, mean width = 2.80 ± 0.31) in sterigmata, clavate, 4-spored. Cheilocystidia gregarious, clavate to broadly clavate or fusiform, (17.5–) 20.9–31.3 (–38.6) × (8.3–) 11.2–16.4 (–18.1) μm (n = 35, mean length = 26.09 ± 5.19, mean width = 13.78 ± 2.59) in clavate to broadly clavate type, (29.5–) 33.3–47.8 (–55.5) × (13.2–) 14.0–16.3 (–18.2) μm (n = 22, mean length = 40.53 ± 7.23, mean width = 15.11 ± 1.16) in fusiform type, smooth, thin-walled. Pleurocystidia scattered, (36.9–) 50.6–68.3 (–70.5) × (13.7–) 14.6–17.5 (–19.5) μm (n = 22, mean length = 59.44 ± 8.86, mean width = 16.08 ± 1.45), ventricose-rostrate, smooth, thin-walled. Hymenophoral trama with parallel (mature stage), cylindrical hyphae (2.9–) 3.6–5.3 (–6.7) μm wide (n = 39, mean width = 4.44 ± 0.83) in mediostratum, (7.4–) 8.9–13.0 (–16.1) μm wide (n = 37, mean width = 10.96 ± 2.03) in lateral stratum, smooth, thin-walled. Pileipellis (warts) a cutis of loosely intermingled, repent, cylindrical hyphae (13.1–) 20.0–35.9 (–45.6) × (7.8–) 9.6–14.7 (–18.8) μm (n = 60, mean length = 27.98 ± 7.95, mean width = 12.12 ± 2.56), smooth, with walls 0.8–1.3 (–1.5) μm; pilocystidia none. Pileitrama of cylindrical, loosely interwoven hyphae (6.2–) 7.6–13.3 (–19.5) μm wide (n = 85, mean width = 10.46 ± 2.84), smooth, thin-walled. Stipitipellis of a cutis with prostrate, often branched, broadly clavate to fusiform, smooth, thick-walled (1.0–1.4 μm) elements (3.5–) 4.8–7.6 (–9.9) μm wide (n = 48, mean width = 6.22 ± 1.38); edges of the reticulum mainly consisting of subcylindrical to subclavate caulocystidia (21.2–) 26.8–42.5 (–60.7) × (10.9–) 13.5–18.7 (–21.2) μm (n = 33, mean length = 34.65 ± 7.85, mean width = 16.09 ± 2.63); caulocystidia between the reticulum (19.5–) 23.4–33.6 (–43.6) × (11.9–) 13.7–17.0 (–19.2) μm (n = 28, mean length = 28.50 ± 5.13, mean width = 15.35 ± 1.61), clavate to broadly clavate; caulobasidia (31.1–) 30.3–44.5 (–53.6) × (9.9–) 10.0–13.7 (–16.6) μm (n = 10, mean length = 37.40 ± 7.10, mean width = 11.84 ± 1.89), scattered, subfusiform, 2 to 4-spored. Stipe trama composed of longitudinally running, cylindrical hyphae (6.2–) 9.0–15.6 (–20.1) μm wide (n = 42, mean width = 12.28 ± 3.28), unbranched, smooth, thin-walled. All hyphae brownish or colorless in water and KOH, inamyloid, without clamp connections.

Habitat and phenology: Solitary on the ground in *Castanopsis-Quercus* forests, June.

Known distribution: Okinawa (Ishigaki Island).

Holotype: TNS-F-48210, on the ground in an evergreen broad-leaved forest dominated by *Q. miyagii* Koidz. and *C. sieboldii* (Makino) Hatus. ex T. Yamaz. et Mashiba, Omoto-yama, Ishigaki-shi, Okinawa Pref., 9 Jun. 2012, coll. Taneyama, Y.

Japanese name: Cha-oni-iguchi (named by H. Takahashi & Y. Taneyama).

Comments: The most distinct features of the present fungus are the reddish-brown spiny warts and flocci that cover the entire surfaces of the pileus and stipe. *Strobilomyces velutinus* J.Z. Ying from south-western China (Ying and Ma 1985; Gelardi et al. 2013) has a certain similarity to the new species largely due to possessing a chestnut brown covering on the pileus. The Chinese taxon, however, forms a pileus ornamented with flat, areolate, patchy scales, unlike the erect, conical, spiny warts of *S. brunneolepidotus*. In addition, a non-reticulate, smooth stipe, black discoloration of flesh, and significantly larger, echinulate to verrucose or subcrested basidiospores: 10.1–12.6 × 8.6–10.8 μm (Ying and Ma 1985) distinguish *S. velutinus* from the *S. brunneolepidotus*.

References 引用文献

Gelardi M, Vizzini A, Ercole E, Voyron S, Wu G, Liu XZ. 2013. *Strobilomyces echinocephalus* sp. nov. (Boletales) from south-western China, and a key to the genus *Strobilomyces* worldwide. Mycol Progr 12 (3): 575–588.

Ying JZ, Ma QM. 1985. New taxa and records of the genus *Strobilomyces* in China. Acta Mycol Sin 4 (2): 95–102.

32. チャオニイグチ（新種：高橋春樹 & 種山裕一新称）*Strobilomyces brunneolepidotus* Har. Takah. & Taneyama, **sp. nov.**

肉眼的特徴（Figs. 255-258）：傘は径 62–80 (–100) mm，最初半球形のち饅頭形；表面は粘性を欠き，類白色の地に全体に赤褐色の柔らかい直立した刺状鱗片に被われる；縁部は最初膜質，赤褐色の消失性縁片膜が付着する．肉は厚さ12 mm 以下，類白色，空気に触れると次第に赤褐色に変わり，特別な匂いはなく，味は不明．柄は -80×8-10 mm，円柱形，しばしば根元に向かってやや太くなり，痩せ型で，中心生，中実，頂部に縦長の網目状隆起を表す；表面は粘性を欠き，赤褐色の綿毛に密に被われる；ツバはない；根元の菌糸体は白色．管孔は長さ13 mm 以下，上生，柄の周囲において嵌入し，灰褐色；孔口は角形，1.3 mm 以下，管孔と同色，管孔及び孔口は傷を受けても変色しない．

顕微鏡的特徴（Figs. 251-254）：担子胞子は (7.2–) 7.7-8.5 (–9.6) × (5.5–) 6.3-7.0 (–7.6) μm（表面の構造物を含む，n = 120，mean length = 8.14 ± 0.41，mean width = 6.63 ± 0.36，Q = 1.17-1.29，mean Q = 1.23 ± 0.06），下部側面に浅いなだらかな凹みがあり（イグチ型），正面観並びに側面観において類球形～球形，表面に完全な網目状隆起を表し，水封およびアルカリ溶液において帯

褐色，非アミロイド，薄壁；網目状隆起は (1.1-) 1.2-1.6 (-1.8) × (0.4-) 0.4-0.6 (-0.7) μm. 担子器は (28.7-) 30.8-34.4 (-34.7) × (12.5-) 12.9-14.8 (-15.5) μm (本体), (4.6-) 5.0-5.9 (-6.0) × (2.2-) 2.5-3.1 (-3.3) μm (ステリグマ)，こん棒形，4胞子性. 縁シスチジアは群生し，こん棒形～広こん棒形または紡錘形，(17.5-) 20.9-31.3 (-38.6) × (8.3-) 11.2-16.4 (-18.1) μm (こん棒形～広こん棒形タイプ), (29.5-) 33.3-47.8 (-55.5) × (13.2-) 14.0-16.3 (-18.2) μm (紡錘形タイプ), 平滑，薄壁. 側シスチジアは散生し，(36.9-) 50.6-68.3 (-70.5) × (13.7-) 14.6-17.5 (-19.5) μm, 頂部が嘴状に伸びた片膨れ状，平滑，薄壁. 子実層托実質は円柱形の菌糸が平列し (成熟時), 中層において幅 (2.9-) 3.6-5.3 (-6.7) μm, 側層において幅 (7.4-) 8.9-13.0 (-16.1) μm, 平滑，薄壁. 傘表皮組織（鱗片）は緩く錯綜した匍匐性の円柱形菌糸からなる平行菌糸被；菌糸は (13.1-) 20.0-35.9 (-45.6) × (7.8-) 9.6-14.7 (-18.8) μm, 平滑，壁は厚さ0.8-1.3 (-1.5) μm；傘シスチジアはない. 傘実質の菌糸は緩く錯綜し，幅 (6.2-) 7.6-13.3 (-19.5) μm, 円柱形，平滑，薄壁. 柄表皮組織は匍匐性の菌糸からなる平行菌糸被；菌糸は幅 (3.5-) 4.8-7.6 (-9.9) μm, 広こん棒形～紡錘形，しばしば分岐し，平滑，壁は厚さ1.0-1.4 μm；網目上の柄シスチジアは (21.2-) 26.8-42.5 (-60.7) × (10.9-) 13.5-18.7 (-21.2) μm, 類円柱形～類こん棒形；網目の間の柄シスチジアは (19.5-) 23.4-33.6 (-43.6) × (11.9-) 13.7-17.0 (-19.2) μm, こん棒形～広こん棒形；柄担子器は (31.1-) 30.3-44.5 (-53.6) × (9.9-) 10.0-13.7 (-16.6) μm, 散生し，こん棒形，2-4胞子性. 柄実質の菌糸は幅 (6.2-) 9.0-15.6 (-20.1) μm, 縦に沿って配列し，円柱形，無分岐，平滑，薄壁. 全ての菌糸は水封およびアルカリ溶液において無色または帯褐色で，非アミロイド，クランプを欠く.

生態および発生時期：スダジイ，オキナワウラジロガシを主体とする常緑広葉樹林内地上に孤生，6月.

分布：沖縄（石垣島）.

供試標本：TNS-F-48210（正基準標本），スダジイ，オキナワウラジロガシを主体とする常緑広葉樹林内地上，石垣市オモト山，2012年6月9日，種山裕一採集.

主な特徴：傘は赤褐色の直立した刺状鱗片に被われ，消失性の縁片膜を有する；肉は空気に触れると赤褐色に変わる；柄は痩せ型で，頂部に縦長の網目状隆起を表し，赤褐色の綿毛に密に被われる；担子胞子は球形～類球形，完全な網目状隆起を表す；縁シスチジアこん棒形～広こん棒形または紡錘形；側シスチジアは頂部が嘴状に伸びた片膨れ状；スダジイ，オキナワウラジロガシを主な構成樹種とする常緑広葉樹林内地上に発生.

コメント：最も特徴的な本種の形質は傘と柄の表面を被う赤褐色の刺状鱗片および綿毛である. 中国南西部産 *Strobilomyces velutinus* J.Z. Ying（Ying and Ma 1985）は赤褐色の鱗片に被われた傘を持つ点で本種に類似するが，傘の鱗片はチャオニイグチの刺状鱗片と異なり，平らに圧着する. また中国産種の網目状隆起を欠く平坦な柄，黒変性のある肉，そしてイボ～刺状突起に被われたより大型の担子胞子：10.1-12.6×8.6-10.8 μm（Ying and Ma 1985）はチャオニイグチには見られない特徴である.

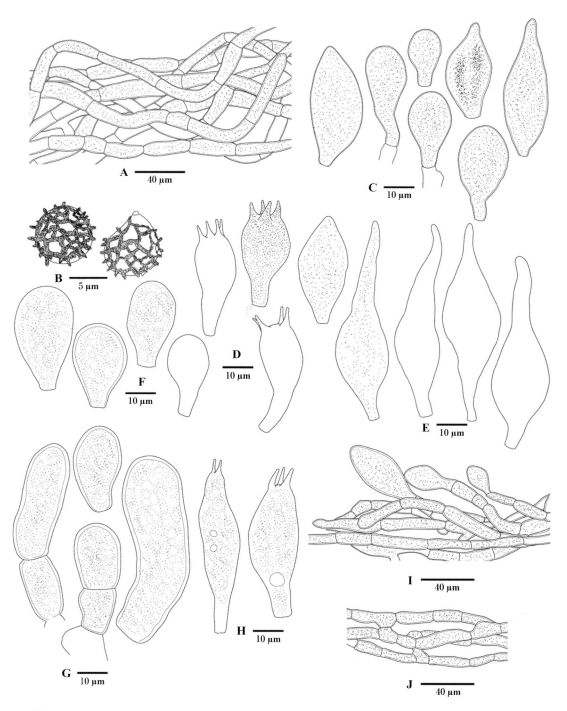

Fig. 251 – Micromorphological features of *Strobilomyces brunneolepidotus* (holotype): **A**. Longitudinal cross section of the mature pileipellis forming conical warts. **B**. Basidiospores. **C**. Cheilocystidia. **D**. Basidia. **E**. Pleurocystidia. **F**. Caulocystidia between the reticulum. **G**. Caulocystidia on the reticulum. **H**. Caulobasidia. **I**. Caulocystidia (from the middle portion of stipe). **J**. Elements of the covering of stipe. Illustrations by Taneyama, Y.

チャオニイグチの顕鏡図（正基準標本）：**A**. 刺状鱗片を形成する成熟した傘表皮組織の縦断面．**B**. 担子胞子．**C**. 縁シスチジア．**D**. 担子器．**E**. 側シスチジア．**F**. 網目の間に存在する柄シスチジア．**G**. 網目上に存在する柄シスチジア．**H**. 柄担子器．**I**. 柄中腹部の柄シスチジア．**J**. 柄表皮の綿毛鱗片を形成する菌糸．図：種山裕一．

32. *Strobilomyces brunneolepidotus* チャオニイグチ —— 307

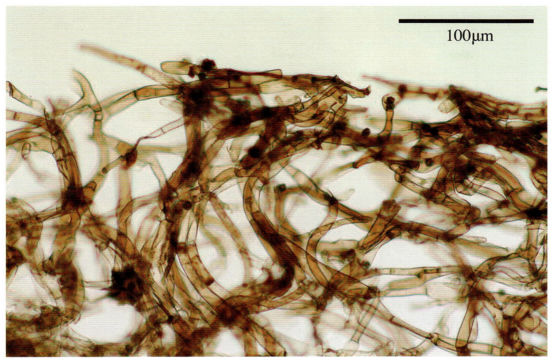

Fig. 252 – Elements of the covering layer of the pileus in *Strobilomyces brunneolepidotus* (in 3% KOH, holotype). Photo by Taneyama, Y.
チャオニイグチの傘表皮を被う菌糸（3%水酸化カリウム溶液で封入，正基準標本）．写真：種山裕一．

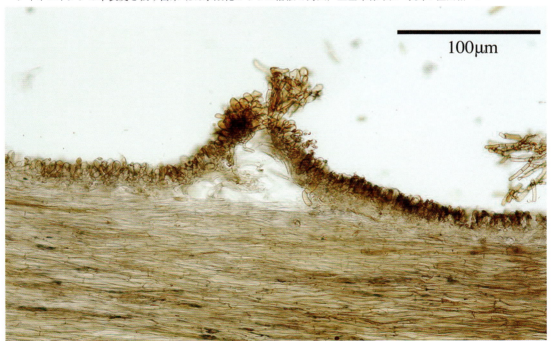

Fig. 253 – Longitudinal cross section of the stipitipellis and reticulum from upper potion of the stipe in *Strobilomyces brunneolepidotus* (in 3% KOH, holotype). Photo by Taneyama, Y.
チャオニイグチの柄頂部の表皮組織および網目の縦断面（3%水酸化カリウム溶液で封入，正基準標本）．写真：種山裕一．

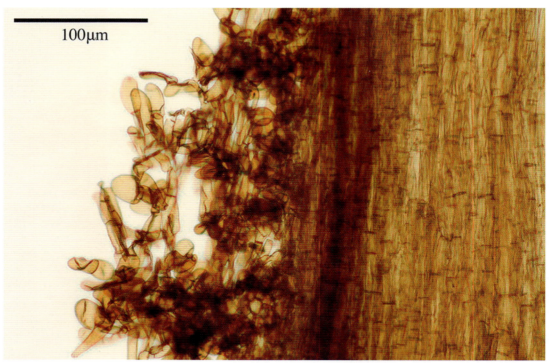

Fig. 254 – Covering layer of the stipe in *Strobilomyces brunneolepidotus* (in 3% KOH, holotype). Photo by Taneyama, Y.
チャオニイグチの柄表皮を被う綿毛状組織（3%水酸化カリウム溶液で封入，正基準標本）．写真：種山裕一．

32. *Strobilomyces brunneolepidotus* チャオニイグチ ── 309

Fig. 255 − Basidioma of *Strobilomyces brunneolepidotus* (holotype) on ground in a *Castanopsis-Quercus* forest, 9 Jun. 2012, Omoto-yama, Ishigaki Island. Photo by Taneyama, Y.
スダジイ-オキナワウラジロガシ林内地上から発生したチャオニイグチの子実体（正基準標本）．2012年6月9日，石垣島オモト山．写真：種山裕一．

Fig. 256 – Underside view of the basidioma of *Strobilomyces brunneolepidotus* (holotype), 9 Jun. 2012, Omoto-yama, Ishigaki Island. Photo by Taneyama, Y.
チャオニイグチの子実体（管孔側，正基準標本），2012年6月9日，石垣島オモト山．写真：種山裕一．

Fig. 257 – Pileus surface of *Strobilomyces brunneolepidotus* (holotype), 9 Jun. 2012, Omoto-yama, Ishigaki Island. Photo by Taneyama, Y.
チャオニイグチの傘表面（正基準標本），2012年6月9日，石垣島オモト山．写真：種山裕一．

Fig. 258 – Longitudinal cross section of the basidioma of *Strobilomyces brunneolepidotus* (holotype), 9 Jun. 2012, Omoto-yama, Ishigaki Island. Photo by Taneyama, Y.
チャオニイグチの子実体断面（正基準標本），2012年6月9日，石垣島オモト山．写真：種山裕一．

33. *Strobilurus luchuensis* Har. Takah., Taneyama & Pham, sp. nov. リュウキュウマツカサキノコ

MycoBank no.: MB 812969.

Etymology: The specific epithet refers to the specified substrate, strobilus of *Pinus luchuensis*.

This species is characterized by a pure white, finely pilose pileus; a brownish orange, white villose to strigose stipe; colorless, inamyloid basidiospores; lack of cheilocystidia; clavate to broadly clavate pleurocystidia encrusted with coarse, dextrinoid, resinous crystals in the upper portion; a hymeniform pileipellis with a few scattered, narrowly subfusiform to piliform pilocystidia; moderately thick-walled, cylindrical to subclavate or piliform caulocystidia with an occasional subcapitate apex; distinctly thick-walled elements in the stipe trama; and a habitat of a fallen or buried strobili of *P. luchuensis*.

Macromorphological characteristics (Figs. 262-266): Pileus 10-25 mm in diameter, at first convex, then becoming plano-convex to almost plane, with a deflexed to straight margin, smooth; surface subhygrophanous, without translucent striations, dry, dull, finely pilose under a lens, pure white. Flesh up to 1 mm thick, white; odor and taste indistinct. Stipe 40-80 × 1-3 mm, cylindrical, subequal but somewhat tapering toward the rooting base, central, terete, fistulose, cylindrical; surface dry, opaque, whitish at the apex, brownish orange (6C5-7) downward, villous toward the base, less so or white pruinose above; base covered with white mycelial hairs attached to the substratum. Lamellae adnexed, moderately close, L = 30-40, with 1-3 series of lamellulae, 1-3 mm broad, ventricose, white; edges even or slightly fimbriate, concolorous.

Micromorphological characteristics (Figs. 259-261): Basidiospores (3.3-) 4.0-5.1 (-6.8) × (2.1-) 2.5-3.0 (-3.6) μm (n = 144, mean length = 4.54 ± 0.56, mean width = 2.74 ± 0.27, Q = (1.39-) 1.52-1.79 (-1.95), mean Q = 1.66 ± 0.13), ellipsoid, smooth, colorless in distilled water, inamyloid, hyaline in KOH, thin-walled. Basidia (15.0-) 16.0-19.3 (-21.1) × (4.0-) 4.2-4.9 (-5.5) μm (n = 13, mean length = 17.68 ± 1.65, mean width = 4.53 ± 0.38), clavate, four-spored; sterigmata (2.6-) 3.2-4.7 (-5.5) × (1.0-) 1.0-1.3 (-1.4) μm (n = 17, mean length = 3.92 ± 0.74, mean width = 1.16 ± 0.12). Cheilocystidia not differentiated. Pleurocystidia (27.9-) 31.2-39.3 (-48.1) × (12.5-) 14.5-18.0 (-19.9) μm (n = 95, mean length = 35.28 ± 4.04, mean width = 16.22 ± 1.77), scattered, clavate to broadly clavate, usually in the upper portion encrusted with coarse, dextrinoid, resinous crystals, with thickened walls (1.2-) 1.4-1.8 (-2.0) μm (n = 21, mean thickness = 1.59 ± 0.23). Trama of lamellae composed of more or less regularly arranged, cylindrical hyphae (3.1-) 3.6-5.7 (-7.2) μm wide (n = 30, mean width = 4.66 ± 1.02), thin-walled, smooth. Surface of pileus a hymeniform layer of clavate end cells (13.0-) 16.2-22.9 (-26.1) × (5.1-) 6.3-8.7 (-10.3) μm (n = 56, mean length = 19.53 ± 3.36, mean width = 7.48 ± 1.17), thin-walled, smooth, not gelatinized; pilocystidia (32.0-) 43.8-101.1 (-176.1) × (4.4-) 6.2-8.5 (-10.9) μm (n = 71, mean length = 72.48 ± 28.63, mean width = 7.33 ± 1.12), narrowly subfusiform to cylindrical or occasionally piliform, smooth, thin-walled. Hyphae of pileus trama (1.8-) 3.2-5.6 (-6.6) μm wide (n = 74, mean width = 4.42 ± 1.17), cylindrical, loosely intricate, thin-walled, smooth. Cortical layer of stipe composed of parallel, filiform hyphae (2.4-) 3.7-5.2 (-6.0) μm wide (n = 41, mean width = 4.45 ± 0.77), thin-walled, smooth, not gelatinized; caulocystidia 4 types: 1) cylindrical caulocystidia (48.2-) 51.6-74.4 (-83.8) × (7.2-) 7.8-10.1 (-11.4) μm (n = 11, mean length = 63.02 ± 11.40, mean width = 8.99 ± 1.15) on the apex of stipe, (40.8-) 37.7-81.2 (-108.4) × (4.8-) 4.7-8.5 (-11.2) μm (n = 9, mean length = 59.44 ± 21.73, mean width = 6.58 ± 1.88) on the middle portion of stipe, occasionally with scattered, small, warty protuberances, with thickened walls (0.7-) 1.0-1.4 (-1.6) μm (n = 34, mean width = 1.23 ± 0.19); 2) subcapitate caulocystidia (25.1-) 30.7-49.6 (-69.3) × (7.6-) 9.0-12.3 (-13.9) μm (n = 24, mean length

= 40.15 ± 9.44, mean width = 10.65 ± 1.63) on the middle portion of stipe, subclavate to subcylindrical, often encrusted with coarse, resinous crystals in the upper portion, inamyloid, with thickened walls (1.1–) 1.2–1.6 (–1.8) μm (n = 22, mean width = 1.37 ± 0.19); 3) subclavate caulocystidia (30.3–) 34.0–42.8 (–50.8) × (8.7–) 10.8–14.8 (–16.7) μm (n = 19, mean length = 38.37 ± 4.41, mean width = 12.79 ± 2.00) on the apex of stipe, often encrusted with coarse, resinous crystals in the upper portion, inamyloid, with thickened walls (1.0–) 1.2–1.8 (–2.3) μm (n = 16, mean width = 1.51 ± 0.31); 4) piliform caulocystidia (127.9–) 117.2–286.1 (–359.8) × (5.5–) 6.4–10.1 (–11.7) μm (n = 14, mean length = 201.65 ± 84.50, mean width = 8.25 ± 1.84) on the apex of stipe, (242.2–) 313.8–572.6 (–708.3) × (4.6–) 6.1–9.6 (–11.4) μm (n = 26, mean length = 443.21 ± 129.37, mean width = 7.86 ± 1.71) on the middle portion of stipe, occasionally with scattered, small, warty protuberances, with thickened walls (1.1–) 1.2–1.8 (–2.3) μm (n = 33, mean width = 1.50 ± 0.27). Trama of stipe composed of longitudinally running, cylindrical hyphae (5.5–) 7.7–11.4 (–14.9) μm wide (n = 58, mean width = 9.52 ± 1.84), with thickened walls (1.2–) 1.4–2.3 (–2.3) μm (n = 7, mean width = 1.85 ± 0.43). Elements of basal mycelium up to 1000 μm long, (4.2–) 5.0–7.1 (–9.5) μm wide (n = 69, mean width = 6.02 ± 1.05), cylindrical, smooth, with walls (1.0–) 1.2–1.8 (–2.4) μm (n = 37, mean width = 1.48 ± 0.30). All tissues colorless in distilled water (except pale brownish hyphae of stipitipellis), hyaline or pale yellow in KOH, inamyloid (except dextrinoid resinous incrustations of pleurocystidia), without clamp connections.

Habitat and phenology: Solitary or scattered on fallen or buried strobili of *P. luchuensis* Mayr., November to January, common.

Known distribution: Okinawa (Ishigaki Island and Iriomote Island).

Holotype: TNS-F-52278, on fallen or buried strobili of *P. luchuensis*. Banna Park, Ishigaki-shi, Okinawa, 14 Nov. 2012, coll. Takahashi, H.

GenBank accession number: AB968234 (ITS).

Japanese name: Ryukyu-matukasakinoko (named by Y. Terashima).

Comment: The colorless, inamyloid basidiospores, the pleurocystidia encrusted with coarse, resinous crystals, the hymeniform pileipellis, and the habitat on strobili suggest the present fungus belongs to the genus *Strobilurus* Singer (Singer 1962).

Strobilurus luchuensis is macromorphologically similar to *Strobilurus stephanocystis* (Kühner & Romagn. ex Hora) Singer, which is a well-defined European species (Singer 1962; Redhead 1980; Bas et al. 1999), and *Strobilurus ohshimae* (Hongo) Hongo from Japan (Hongo 1955, 1989; Katumoto 2010). *Strobilurus stephanocystis* differs in having a usually distinctly colored pileus, a white pruinose stipe, utriform hymenial cystidia with thickened walls and lacking piliform caulocystidia with warty protuberances. *Strobilurus ohshimae* is distinct in having thick-walled cheilocystidia, capitate pilocystidia, and a habitat of dead fallen twigs of *Cryptomeria japonica* (Thunb. ex L.f.) D.Don. and *Chamaecyparis obtusa* Sieb. & Zucc.

References 引用文献

Bas C, Kuyper Th W, Noordeloos ME, Vellinga EC, Crevel R (eds). 1999. Flora Agaricina Neerlandica - Volume 4 - Critical Monographs on Families of Agarics and Boleti Occurring in the Netherlands. Monographs on the Tricholomataceae (part 3), tribus Tricholomateae and tribus Xeruleae and on the family of Strophariaceae. Rotterdam.

Hongo T. 1955. Notes on Japanese larger fungi (6). J Jpn Bot 30 (3): 73–79.

Hongo T. 1989. Tricholomataceae. In: Imazeki R, Hongo T (eds), Colored illustrations of mushrooms of Japan I. Hoikusha, Osaka, pp 56–115 (in Japanese).

Katumoto K. 2010. List of Fungi Recorded in Japan. The Kanto Branch of the Mycological Society of Japan, Kyoto.
Singer R. 1962. New genera of fungi. VIII. Persoonia 2 (3): 407-415.
Redhead SA. 1980. The Genus *Strobilurus* (Agaricales) in Canada with Notes on Extralimital Species. Can J Bot 58: 68-83.

33. リュウキュウマツカサキノコ（新種；寺嶋芳江新称）*Strobilurus luchuensis* Har. Takah., Taneyama & Pham, **sp. nov.**

肉眼的特徴（Figs. 262-266）：傘は径10-25 mm，最初丸山形，のち饅頭形，平坦；縁部は内側に巻かない；表面はやや吸水性，半透明の条線を欠き，微細な毛状，光沢を欠き，白色．肉は厚さ1 mm以下，白色，特別な味や匂いはない．柄は40-80×1-3 mm，円柱形でほぼ上下同大，時に根元は根状に長く伸びて下方に向かってやや細くなり，中心生，痩せ型，中空；表面は頂部において白色，下方に向かって褐色を帯び，乾生，光沢を欠き，頂部粉状，基部に向かって白色の長軟毛状；根元は基質につながる白色長軟毛に被われる．ヒダは上生，やや密（柄に到達するヒダは8-11），1-3の小ヒダを伴い，幅1-3 mm，丸山形，白色，縁部は同色，平坦または細鋸歯状．

顕微鏡的特徴（Figs. 259-261）：担子胞子は (3.3-) 4.0-5.1 (-6.8) × (2.1-) 2.5-3.0 (-3.6) μm (n = 144, mean length = 4.54±0.56, mean width = 2.74±0.27, Q = (1.39-) 1.52-1.79 (-1.95), mean Q = 1.66 ±0.13)，楕円形，無色，平坦，非アミロイド，アルカリ溶液において無色，薄壁．担子器は (15.0-) 16.0-19.3 (-21.1) × (4.0-) 4.2-4.9 (-5.5) μm，こん棒形，4胞子性．縁シスチジアはない．側シスチジアは (27.9-) 31.2-39.3 (-48.1) × (12.5-) 14.5-18.0 (-19.9) μm，散生し，こん棒形～広こん棒形，通常頂部において偽アミロイドに染まる粗大な樹脂状結晶に被覆され，壁は厚さ (1.2-) 1.4-1.8 (-2.0) μm．子実層托実質の菌糸は幅 (3.1-) 3.6-5.7 (-7.2) μm，ほぼ平列し，円柱形，平坦，薄壁．傘表皮組織は子実層状被；末端細胞は (13.0-) 16.2-22.9 (-26.1) × (5.1-) 6.3-8.7 (-10.3) μm，こん棒形；傘シスチジアは (32.0-) 43.8-101.1 (-176.1) × (4.4-) 6.2-8.5 (-10.9) μm (n = 71)，やや紡錘形～円柱形または毛状，平坦，薄壁．傘実質の菌糸細胞は幅 (1.8-) 3.2-5.6 (-6.6) μm，緩く錯綜し，円柱形，平坦，薄壁．柄表皮組織の菌糸は幅 (2.4-) 3.7-5.2 (-6.0) μm，並列し，糸状，平滑，薄壁；柄シスチジアは4タイプ：1）円柱形の柄シスチジアは (48.2-) 51.6-74.4 (-83.8) × (7.2-) 7.8-10.1 (-11.4) μm（柄の頂部），(40.8-) 37.7-81.2 (-108.4) × (4.8-) 4.7-8.5 (-11.2) μm（柄の中腹部），部分的に小形の疣状分岐物を具え，壁は厚さ (0.7-) 1.0-1.4 (-1.6) μm；2）頂部頭状形の柄シスチジアは (25.1-) 30.7-49.6 (-69.3) × (7.6-) 9.0-12.3 (-13.9) μm（柄の中腹部），亜こん棒形～亜円柱形，上部において粗大な樹脂状結晶物に被覆され，非アミロイド，壁は厚さ (1.1-) 1.2-1.6 (-1.8) μm；3）亜こん棒形の柄シスチジアは (30.3-) 34.0-42.8 (-50.8) × (8.7-) 10.8-14.8 (-16.7) μm（柄の頂部），上部において粗大な樹脂状結晶物に被覆され，非アミロイド，壁は厚さ (1.0-) 1.2-1.8 (-2.3) μm；4）毛状柄シスチジアは (127.9-) 117.2-286.1 (-359.8) × (5.5-) 6.4-10.1 (-11.7) μm（柄の頂部），(242.2-) 313.8-572.6 (-708.3) × (4.6-) 6.1-9.6 (-11.4) μm（柄の中腹部），部分的に小形の疣状分岐物を具え，壁は厚さ (1.1-) 1.2-1.8 (-2.3) μm．柄実質は円柱形の菌糸細胞が平列し，平坦，厚壁．根元を被う長軟毛を構成する菌糸は長さ1000 μmに達し，幅 (4.2-) 5.0-7.1 (-9.5) μm，円柱形，平滑，壁は厚さ (1.0-) 1.2-1.8 (-2.4) μm．全ての組織は水封並びに水酸化カリウム溶液において無色（淡褐色を帯びる柄表皮組織の菌糸を除く），非アミロイド（偽アミロイドに染まる側シスチジアを除く），クランプを欠く．

生態および発生時期：落下したリュウキュウマツの球果上に孤生または散生，11月～1月，普通種．
分布：沖縄（八重山諸島：石垣島，西表島）．

33. *Strobilurus luchuensis* リュウキュウマツカサキノコ

供試標本：TNS-F-52278（正基準標本），落下したリュウキュウマツの球果上，沖縄県石垣市バンナ公園，2012年11月14日，高橋春樹採集．

GenBank 登録番号：AB968234（ITS）．

主な特徴：傘表面は白色，微細な毛状；柄は帯褐色で根元に向かって白色の剛毛～長軟毛に被われる；担子胞子は無色，非アミロイド；縁シスチジアを欠く；側シスチジアはこん棒形～広こん棒形で通常頂部において偽アミロイドに染まる粗大な樹脂状結晶に被覆される；傘表皮組織は狭紡錘形～毛状傘シスチジアが散在する子実層状被を形成する；柄シスチジアはやや厚壁でしばしば頂部頭状形になり，毛状または円柱形～亜こん棒形；柄実質は厚壁の菌糸からなる；地中に埋もれたリュウキュウマツの球果から発生する．

コメント：無色で非アミロイドの担子胞子，粗大な樹脂状が付着する側シスチジア，子実層状被を成す傘表皮組織，松毬から発生する生態は本種がマツカサキノコ属 *Strobilurus*（Singer 1962）に位置することを示唆している．

　本種の子実体の外観はマツカサキノコモドキ *Strobilurus stephanocystis*（Kühner & Romagn. ex Hora）Singer（Singer 1962；Redhead 1980；Bas et al. 1999）およびスギエダタケ *Strobilurus ohshimae*（Hongo）Hongo（Hongo 1955, 1989；Katumoto 2010）にやや類似するが，マツカサキノコモドキは通常傘が有色で，柄は白色粉状を呈し，厚壁で小のう形のシスチジアが子実層に存在し，毛状の柄シスチジアを持たない．またスギエダタケは厚壁な縁シスチジアおよび頂部頭状形の傘シスチジアを有し，通常スギまたはヒノキの落枝上から発生する．

33. *Strobilurus luchuensis* リュウキュウマツカサキノコ —— 317

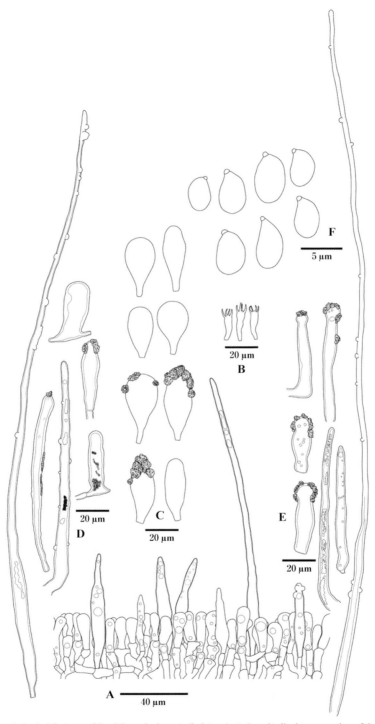

Fig. 259 – Micromorphological features of *Strobilurus luchuensis* (holotype): **A.** Longitudinal cross section of the pileipellis showing a hymeniform layer with a few scattered, narrowly subfusiform to piliform pilocystidia. **B.** Basidia. **C.** Pleurocystidia. **D**. Caulocystidia (from the apex of stipe). **E.** Caulocystidia (from the middle portion of stipe). **F**. Basidiospores. Illustrations by Taneyama, Y.
リュウキュウマツカサキノコの顕鏡図（holotype）：**A**. 傘の表皮組織の縦断面．狭紡錘形〜毛状傘シスチジアが散在する子実層状被を示す．**B**. 担子器．**C**. 側シスチジア．**D**. 柄シスチジア（柄の頂部）．**E**. 柄シスチジア（柄の中腹部）．**F**. 担子胞子．図：種山裕一．

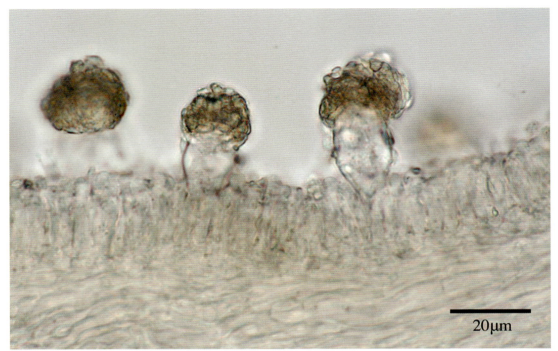

Fig. 260 – Pleurocystidia of *Strobilurus luchuensis* (in 3%KOH, holotype). Photo by Taneyama, Y.
リュウキュウマツカサキノコの側シスチジア（3%水酸化カリウム溶液で封入，holotype）．写真：種山裕一．

Fig. 261 – Longitudinal cross section of the lamellae in *Strobilurus luchuensis* (in 3%KOH, holotype), showing scattered pleurocystidia. Photo by Taneyama, Y.
リュウキュウマツカサキノコの側シスチジア（3%水酸化カリウム溶液で封入，holotype），散在した側シスチジアを示す．写真：種山裕一．

33. *Strobilurus luchuensis* リュウキュウマツカサキノコ — 319

Fig. 262 – Basidioma of *Strobilurus luchuensis* (holotype) on a fallen strobilus of *P. luchuensis*, Banna Park, Ishigaki Island, 14 Nov. 2012. Photo by Takahashi, H.
リュウキュウマツの球果から発生したリュウキュウマツカサキノコの子実体（holotype），2012年11月14日，石垣島バンナ公園．写真：高橋春樹．

Fig. 263 – Basidioma of *Strobilurus luchuensis* (holotype) on a fallen strobilus of *P. luchuensis*, 14 Nov. 2012, Banna Park, Ishigaki Island. Photo by Takahashi, H.
リュウキュウマツの球果から発生したリュウキュウマツカサキノコの子実体（holotype），2012年11月14日，石垣島バンナ公園．写真：高橋春樹．

Fig. 264 – Pileus surface of *Strobilurus luchuensis* (holotype), 14 Nov. 2012, Banna Park, Ishigaki Island. Photo by Takahashi, H.
リュウキュウマツカサキノコの傘表面（holotype），2012年11月14日，石垣島バンナ公園．写真：高橋春樹．

Fig. 265 – Underside view of the basidioma of *Strobilurus luchuensis* (holotype), 14 Nov. 2012, Banna Park, Ishigaki Island. Photo by Takahashi, H.
リュウキュウマツカサキノコの子実体（holotype），2012年11月14日，石垣島バンナ公園．写真；高橋春樹．

Fig. 266 – Stipe surface of *Strobilurus luchuensis* (holotype), 14 Nov. 2012, Banna Park, Ishigaki Island. Photo by Takahashi, H.
リュウキュウマツカサキノコの柄表面（holotype），2012年11月14日，石垣島バンナ公園．写真：高橋春樹．

34. *Termitomyces intermedius* Har. Takah. & Taneyama, sp. nov. シロアリシメジ

= ? *Termitomyces albuminosus* sensu Otani (1979).
= ? *Termitomyces eurrhizus* sensu Hongo (1987).
MycoBank no.: MB 809940.

Etymology: The specific epithet "*intermedius*" comes from the Latin word for "intermediate".

Distinctive features of this species are found in medium-sized, gymnocarpic, pluteoid basidiomata growing above fungus combs of the termite in late May to mid-June; reddish brown to greyish brown pileus with a conical perforatorium (6-12 mm high); a long pseudorrhiza (reaching to 200 mm long); ellipsoid, colorless, inamyloid basidiospores; a well differentiated pseudoparenchymatous inner layer of the pileipellis; and a dextrinoid stipe trama.

Macromorphological characteristics (Figs. 269-274): Pileus 50-90 (-120) mm in diameter, at first narrowly conical, then convex to expanding to nearly plane, with a conical to subconical perforatorium 6-12 mm high, occasionally somewhat rugulose-pitted around the perforatorium, radially striate toward the margin; margin incurved to straight, often irregularly split in places along the lines of the lamellae; surface dry, dull, radially fibrillose, greyish brown (7D3-8D3) to reddish brown (7D4-8D4), paler toward the margin, darker at the center. Veil absent. Flesh up to 5 mm thick, whitish, soft, odor and taste indistinct. Stipe 70-130 × 9-20 mm, cylindrical, somewhat thickened at the epigeal base, central, terete, solid, longitudinally striate; surface dry, subglabrous to fibrillose, often somewhat lacerate in age, whitish. Pseudorrhiza up to 200 mm long, cylindrical, tapering downward, solid. Lamellae free, subclose, 30-50 reach the stipe, with 0-1 series of lamellulae, up to 7 mm broad, at first white, then pale ochraceous or pale pinkish in age; edges even, concolorous.

Micromorphological characteristics (Figs. 267, 268): Basidiospores (5.7-) 6.6-7.7 (-8.5) × (3.3-) 4.0-4.6 (-5.0) μm (n = 112, mean length = 7.16 ± 0.56, mean width = 4.30 ± 0.33, Q = (1.39-) 1.56-1.78 (-2.00), mean Q = 1.67 ± 0.11), ellipsoid, smooth, hyaline, inamyloid, colorless in KOH, thin-walled. Basidia (18.6-) 20.7-25.1 (-26.9) × (6.9-) 7.3-8.1 (-8.6) μm (n = 17, mean length = 22.87 ± 2.19, mean width = 7.72 ± 0.42), clavate, four-spored; sterigmata (2.2-) 2.5-3.5 (-3.8) × (1.2-) 1.4-1.6 (-1.7) μm (n = 14, mean length = 3.01 ± 0.52, mean width = 1.51 ± 0.14); basidioles clavate. Cheilocystidia (22.2-) 29.4-49.2 (-59.6) × (11.2-) 14.4-22.5 (-29.4) μm (n = 48, mean length = 39.26 ± 9.90, mean width = 18.44 ± 4.06), forming a compact sterile edge, broadly clavate to pyriform, smooth, colorless, thin-walled. Pleurocystidia (33.0-) 44.6-74.1 (-97.3) × (13.2-) 17.6-29.0 (-37.3) μm, (n = 45, mean length = 59.34 ± 14.72, mean width = 23.28 ± 5.72), infrequent, broadly clavate to pyriform, smooth, colorless, inamyloid, with thin or slightly thickened walls up to 1 μm. Element of hymenophoral trama similar to those of the pileitrama. Outermost layer of pileipellis a cutis of parallel, repent hyphal cells (11.0-) 16.5-34.6 (-44.3) × (3.3-) 4.0-6.9 (-9.2) μm (n = 62, mean length = 25.56 ± 9.02, mean width = 5.44 ± 1.42), non-inflated, cylindrical, smooth, colorless or pale brownish, thin-walled. Inner layer of pileipellis (mediopellis) well differentiated from the upper stratum and subpellis, made up of voluminous, short, pseudoparenchymatous elements (13.0-) 22.1-42.7 (-59.2) × (10.6-) 13.3-22.6 (-28.6) μm (n = 40, mean length = 32.42 ± 10.30, mean width = 17.97 ± 4.63), subcylindrical to oblong, constricted at the septa, smooth, colorless or pale brownish, with walls (1.0-) 1.1-1.5 (-1.6) μm (n = 27, mean thickness = 1.28 ± 0.19). Innermost layer of pileipellis (subpellis) similar to the outermost stratum, consisting of parallel, repent hyphae (3.6-) 5.0-7.7 (-10.2) μm wide (n = 36, mean width = 6.37 ± 1.33), non-inflated, cylindrical, smooth, brownish, more deeply colored than the other strata, with walls (0.7-) 0.9-1.2 (-1.3) μm (n = 13, mean thickness = 1.05 ± 0.19). Elements of pileitrama (4.7-) 5.2-14.0 (-23.9) μm wide (n =

30, mean width = 9.60 ± 4.41), more or less parallel, loosely interwoven, cylindrical, somewhat inflated or not, smooth, colorless, with walls (0.6–) 0.7–1.0 (–1.1) μm (n = 17, mean thickness = 0.84 ± 0.16). Stipitipellis a cutis of parallel, repent hyphae (3.5–) 4.9–8.6 (–11.6) μm wide (n = 49, mean width = 6.76 ± 1.83), cylindrical, smooth, colorless, thin-walled. Stipe trama composed of longitudinally running, cylindrical hyphae (4.6–) 7.1–14.2 (–17.5) μm wide (n = 50, mean width = 10.65 ± 3.50), smooth, colorless, dextrinoid, thin-walled. All tissues hyaline in KOH, inamyloid except dextrinoid stipe trama, without clamp connections.

Habitat and phenology: Solitary to scattered, often gregarious, on fungus combs in nests of the termite *Odontotermes formosanus* Shiraki, late May to mid-June.

Known distribution: Okinawa (Ishigaki Island).

Holotype: TNS-F-48229, on a fungus comb in a termite nest, Arakawa, Ishigaki-shi, Okinawa Pref., 10 Jun. 2012, coll. Taneyama, Y.

Other specimens examined: TNS-F-48178, on a fungus comb in a termite nest, Tonoshiro, Ishigaki-shi, Okinawa Pref., 9 Jun. 2012, coll. Taneyama, Y.; TNS-F-48169, same location, 9 Jun. 2012, coll. Taneyama, Y.; KPM-NC0013154, same location, 18 Jun. 2005, coll. Takahashi, H.; TNS-F-50239 (*Termitomyces albuminosus* (Berk.) R. Heim, nom. inval.), on a fungus comb in a termite nest, Takeda, Ishigaki-shi, Okinawa Pref., 7 Jun. 1978, coll. Otani, Y.; TNS-F-50233 (*T. albuminosus*), same location, 7 Jun. 1978, coll. Otani, Y.; TNS-F-50235 (*T. albuminosus*), same location, 7 Jun. 1978, coll. Otani, Y.; TNS-F-50241 (*T. albuminosus*), same location, 7 Jun. 1978, coll. Otani, Y.; TNS-F-50237 (*T. albuminosus*), same location, 7 Jun. 1978, coll. Otani, Y.; TNS-F-50240 (*T. albuminosus*), same location, 7 Jun. 1978, coll. Otani, Y.; TNS-F-50238 (*T. albuminosus*), same location, 7 Jun. 1978, coll. Otani, Y.; TNS-F-50234 (*T. albuminosus*), same location, 7 Jun. 1978, coll. Otani, Y.; TNS-F-50236 (*T. albuminosus*), same location, 7 Jun. 1978, coll. Otani, Y.; TNS-F-50256 (*Termitomyces clypeatus* Heim), on the bank of the Shiira River, Iriomote Island, Okinawa Pref., 27 May 1981, coll. Shimizu, D., det. Otani, Y.

Extralimital specimens examined: TNS-F-50252 (*Termitomyces* sp.), Kebum Percobaan, Gekbrong, Indonesia, 28 Feb. 1979, coll. Hemmi, S., det. Otani, Y.; TNS-F-50245 (*T. albuminosus*), same location, 28 Feb. 1979, coll. Hemmi, S., det. Otani, Y.; TNS-F-50244 (*T. albuminosus*), same location, 28 Feb. 1979, coll. Hemmi, S., det. Otani, Y.; TNS-F-50243 (*T. albuminosus*), same location, 21 Feb. 1979, coll. Hemmi, S., det. Otani, Y.; TNS-F-50242 (*T. albuminosus*), same location, 15 Jan. 1979, coll. Hemmi, S., det. Otani, Y.; TNS-F-50247 (*T. albuminosus*), same location, 28 Feb. 1979, coll. Hemmi, S., det. Otani, Y.; TNS-F-50246 (*T. albuminosus*), same location, 28 Feb. 1979, coll. Hemmi, S., det. Otani, Y.; TNS-F-50248 (*T. albuminosus*), Ciburial Puncak, Bogor, Indonesia, 19 Feb. 1979, coll. Hemmi, S., det. Otani, Y.; TNS-F-50249 (*T. albuminosus*), same location, 19 Feb. 1979, coll. Hemmi, S., det. Otani, Y.

Gene sequenced specimen and GenBank accession number: TNS-F-48178, AB968241.

Japanese name: Shiroari-shimeji (named by H. Takahashi & Y. Taneyama).

Comments: Except for the dextrinoid stipe trama and the well differentiated pseudoparenchymatous inner layer of the pileipellis, the pluteoid basidiomata producing a conical perforatorium and long pseudorrhiza associated with the fungal combs of termite nests, and the ellipsoid, colorless, inamyloid basidiospores with thin walls are all characteristics of the genus *Termitomyces* (Heim 1942; Singer 1986).

Termitomyces clypeatus R. Heim, originally described from East Africa (Heim 1951, 1977; Otieno 1968; Pegler 1977; Otani and Shimizu 1981; Pegler and Vanhaecke 1994; Tibuhwa 2012), somewhat resembles the new species in appearance. African species is distinct in forming a dark brown pileus with a prominently spiniform perforatorium, and lacking a cellular layer in the pileipellis. *Termitomyces intermedius* also bears some resemblance to *Termitomyces eurrhizus* (Berk.) R. Heim from Sri Lanka (Berkeley 1847; Heim 1942; Pegler 1977, 1986; Pegler and Vanhaecke 1994; Wei et al. 2009), which differs in producing much larger basidiomata (pileus reaching more than 150 mm in diameter), an obtusely rounded perforatorium, and an evanescent membranous annulus. Taiwanese *Sinotermitomyces taiwanensis* M. Zang & C.M. Chen (Zang and Chen 1998; Wei et al. 2006, 2009), which is considered to be conspecific with *T. clypeatus* by several authors (Frøslev et al. 2003; Wei et al. 2006, 2009; Oyetayo 2012), is described as possessing a hollow stipe with a persistent annulus.

The taxonomy of Japanese *T. albuminosus* sensu Otani (Otani 1979) and *T. eurrhizus* sensu Hongo (Hongo 1987; Kinjo et al. 2005; Neda and Sato 2008) may need to be revisited.

References 引用文献

Berkeley MJ. 1847. Decades of fungi. Decade XV-XIX. Ceylon fungi. London J Bot 6: 479-514.
Frøslev TG, Aanen DK, Læssøe T, Rosendahl S. 2003. Phylogenetic relationships of *Termitomyces* and related taxa. Mycological Research 107: 1277-1286.
Heim R. 1942. Nouvelles etudes descriptives sur les agarics termitophiles d'Afrique tropicale. Arch Mus Hist Nat Paris, ser 6 18: 107-166.
Heim R. 1951. Les *Termitomyces* du Congo Belge recueillis par Medame M. Goossens-Fontana. Bull Jard bot État Brux 21: 205-222.
Heim R. 1977. Termites et Champignons. Société Nouvelle Des Éditions Boubée, France.
Hongo T. 1987. Tricholomataceae. In: Imazeki R, Hongo T (eds) Colored Illustrations of Mushrooms of Japan I (in Japanese). Hoikusha, Osaka, pp 56-115.
Kinjo K, Anucha P, Miyagi K. 2005. *Termitomyces eurrhizus* collected from the main island of Okinawa (in Japanese). Nippon Kingakukai Kaiho 46: 41-44.
Neda H, Sato H. 2008. List of agaricoid fungi reported from subtropical area of Japan (in Japanese). Nippon Kingakukai Kaiho 49: 64-90.
Otani Y. 1979. *Termitomyces albuminosus* (Berk.) Heim collected in Ishigaki Island of Ryukyu Archipelago, Japan (in Japanese). Trans mycol Soc Japan 20: 195-202.
Otani Y, Shimizu D. 1981. *Termitomyces clypeatus* Heim collected from Iriomote Island, Okinawa, Japan. Bull Natn Sci Mus Tokyo, Ser. B, 7 (4): 131-134.
Otieno NC. 1968. Further contributions to a knowledge of termite fungi in East Africa: The Genus *Termitomyces* Heim. Sydowia 22: 160-165 (published in 1969).
Oyetayo VO. 2012. Wild *Termitomyces* Species Collected from Ondo and Ekiti States AreMore Related to African Species as Revealed by ITS Region of rDNA. The ScientificWorld Journal 2012: 1-5.
Pegler DN. 1977. A preliminary agaric flora of East Africa. Kew Bull, Addit Ser 6: 1-615.
Pegler DN. 1986. Agaric flora of Sri Lanka. Kew Bull, Addit Ser 12: 1-519.
Pegler DN, Vanhaecke M. 1994. *Termitomyces* of Southeast Asia. Kew Bull 49: 717-736.
Singer R. 1986. The Agaricales in modern taxonomy, 4th edn. Koeltz, Koenigstein.
Tibuhwa DD. 2012. *Termitomyces* Species from Tanzania, Their Cultural Properties and Unequalled Basidiospores. Journal of Biology and Life Science 3 (1): 140-159.
Wei TZ, Tang BH, Yao YJ, Pegler DN. 2006. A revision of *Sinotermitomyces*, a synonym of *Termitomyces* (Agaricales). Fung Diver 21: 225-237.
Wei TZ, Tang BH, Yao YJ. 2009. Revision of *Termitomyces* in China. Mycotaxon 108: 257-285.
Zang M, Chen CM. 1998. Four new taxa of Basidiomycota from Taiwan. Fungal Science, Taipei 13 (1-2): 23-28.

34. シロアリシメジ（新種；高橋春樹 & 種山裕一新称）***Termitomyces intermedius*** Har. Takah. & Taneyama, **sp. nov.**

= ? *Termitomyces albuminosus* sensu Otani（1979）.
= ? *Termitomyces eurrhizus* sensu Hongo（1987）.

肉眼的特徴（Figs. 269-274）：傘は50-90（-120）mm，最初幅の狭い円柱形のち丸山形～ほぼ平らに開き，円錐状に突出した中丘（高さ6-12 mm）を持ち，時に中丘の周囲に小シワ状のくぼみを表し，周縁部に向かって放射状の条線を表し，しばしばヒダに沿って不規則な裂け目を生じる；縁部は最初やや内側に巻く；表面は乾性，光沢を欠き，繊維紋を放射状に表し，灰褐色～赤褐色，周縁部に向かって淡色を呈し，中央部は暗色．被膜を欠く．肉は厚さ5 mm以下，類白色，軟質，特別な味や臭いはない．柄は70-130×9-20 mm，円柱形，地上部の根元に向かってやや太くなり，中心性，中実，縦に沿って条線を表す；表面は平滑または繊維状，成熟するとしばしばささくれを生じ，類白色．偽根は長さ200 mm達し，円柱形，下方に向かって細くなり，帯黒色．ヒダは離生，やや密，柄に到達するヒダは30-50，小ヒダは0-1，幅7 mm以下，最初白色のち淡黄土色～淡紅色；縁部は全縁，同色．

顕微鏡的特徴（Figs. 267, 268）：担子胞子は (5.7-) 6.6-7.7 (-8.5) × (3.3-) 4.0-4.6 (-5.0) μm（n = 112, mean length = 7.16 ± 0.56, mean width = 4.30 ± 0.33, Q = (1.39-) 1.56-1.78 (-2.00), mean Q = 1.67 ± 0.11），楕円形，平坦，無色，非アミロイド，アルカリ溶液において無色，薄壁．担子器は (18.6-) 20.7-25.1 (-26.9) × (6.9-) 7.3-8.1 (-8.6) μm，こん棒形，4胞子性；ステリグマは (2.2-) 2.5-3.5 (-3.8) × (1.2-) 1.4-1.6 (-1.7) μm；偽担子器はこん棒形．縁シスチジアは (22.2-) 29.4-49.2 (-59.6) × (11.2-) 14.4-22.5 (-29.4) μm，不実帯をなし，広こん棒形～洋梨形，平滑，無色，非アミロイド，薄壁．側シスチジアは (33.0-) 44.6-74.1 (-97.3) × (13.2-) 17.6-29.0 (-37.3) μm，稀，広こん棒形～洋梨形，平滑，無色，非アミロイド，薄壁またはやや厚壁（1 μm以下）．子実層托実質は傘実質に類似する．傘表皮組織の最上層は平行菌糸被；菌糸細胞は (11.0-) 16.5-34.6 (-44.3) × (3.3-) 4.0-6.9 (-9.2) μm，円柱形，平滑，無色または淡褐色，薄壁．傘表皮組織の中層は偽柔組織状を成し，上層および下層から明瞭に分化する；菌糸細胞は (13.0-) 22.1-42.7 (-59.2) × (10.6-) 13.3-22.6 (-28.6) μm，著しく膨大し，楕円形～亜円柱形，隔壁において急激に幅が狭くなり，平滑，無色または淡褐色，壁は厚さ (1.0-) 1.1-1.5 (-1.6) μm．傘表皮組織の下層は上表皮層に類似し，菌糸は幅 (3.6-) 5.0-7.7 (-10.2) μm，匍匐性で並列し，円柱形，平滑，褐色を帯び，上表皮層並びに下表皮層より濃色を呈し，壁は厚さ (0.7-) 0.9-1.2 (-1.3) μm．傘実質の菌糸は幅 (4.7-) 5.2-14.0 (-23.9) μm，亜並列型，円柱形，時にやや膨大し，平滑，無色，壁は厚さ (0.6-) 0.7-1.0 (-1.1) μm．柄表皮組織は平行菌糸被；菌糸は幅 (3.5-) 4.9-8.6 (-11.6) μm，円柱形，平滑，無色，非アミロイド，薄壁．柄実質を構成する菌糸は幅 (4.6-) 7.1-14.2 (-17.5) μm，縦に沿って並列し，円柱形，平滑，無色，偽アミロイド，薄壁．全ての組織はアルカリ溶液において無色，偽アミロイドの柄実質以外は非アミロイド，クランプを欠く．

生態および発生時期：タイワンシロアリの巣の菌床から発生し，しばしば群生，5月下旬～6月中旬．

分布：沖縄（石垣島）．

供試標本：TNS-F-48229（正基準標本），タイワンシロアリの巣の菌床から発生，沖縄県石垣市荒川，2012年6月10日，種山裕一採集；TNS-F-48178，タイワンシロアリの巣の菌床から発生，沖縄県石垣市登野城，2012年6月9日，種山裕一採集；TNS-F-48169，同上，2012年6月9日，種山裕一採集；KPM-NC0013154，同上，2005年6月18日，高橋春樹採集；TNS-F-50239（*Termitomyces albuminosus* (Berk.) R. Heim, nom. inval.），シロアリの巣の菌床から発生，沖縄県石垣市嵩田，1978年6月7日，大谷吉雄採集；TNS-F-50233（*T. albuminosus*），同上，1978年6月7日，大谷吉雄採集；TNS-F-50235（*T. albuminosus*），同上，1978年6月7日，大谷吉雄採集；TNS-F-50241（*T. albuminosus*），同上，1978年6月7日，大谷吉雄採集；TNS-F-50237（*T. albuminosus*），同上，1978年6月7日，大谷吉雄採集；TNS-F-50240（*T. albuminosus*），同上，1978年6月7日，大谷吉雄採集；TNS-F-50238（*T. albuminosus*），同上，1978年6月7日，大谷吉雄採集；TNS-F-50234（*T. albuminosus*），同上，1978年6月7日，大谷吉雄採集；TNS-F-50236（*T. albuminosus*），同上，1978

年6月7日，大谷吉雄採集；TNS-F-50256（*Termitomyces clypeatus* Heim），沖縄県西表島シイラ川流域，1981年5月27日，清水大典採集，大谷吉雄同定．

地域外供試標本：TNS-F-50252（*Termitomyces* sp.），インドネシア（Gekbrong），1979年2月28日，S. Hemmi 採集，大谷吉雄同定；TNS-F-50245（*T. albuminosus*），同上，1979年2月28日，S. Hemmi 採集，大谷吉雄同定；TNS-F-50244（*T. albuminosus*），同上，1979年2月28日，S. Hemmi 採集，大谷吉雄同定；TNS-F-50243（*T. albuminosus*），同上，1979年2月21日，S. Hemmi 採集，大谷吉雄同定；TNS-F-50242（*T. albuminosus*），同上，1979年1月15日，S. Hemmi 採集，大谷吉雄同定；TNS-F-50247（*T. albuminosus*），同上，1979年2月28日，S. Hemmi 採集，大谷吉雄同定；TNS-F-50246（*T. albuminosus*），同上，1979年2月28日，S. Hemmi 採集，大谷吉雄同定；TNS-F-50248（*T. albuminosus*），インドネシア（ボゴール），1979年2月19日，S. Hemmi 採集，大谷吉雄同定；TNS-F-50249（*T. albuminosus*），1979年2月19日，S. Hemmi 採集，大谷吉雄同定．

分子解析に用いた標本並びに GenBank 登録番号：TNS-F-48178，AB968241．

主な特徴：子実体はウラベニガサ型で中型（通常傘は径12 cm 以下）で，20cm に達する長い偽根を有する；傘は灰褐色〜赤褐色，開いたとき円錐状に突出した中丘（高さ6-12 mm）を形成する；ヒダは離生し，最初白色，成熟すると淡黄褐色を帯びる；担子胞子は楕円形，無色，非アミロイド，薄壁；傘表皮組織は偽柔組織状の中層が明瞭に分化する；柄実質は偽アミロイドに染まる；5月下旬〜6月中旬にタイワンシロアリの巣の菌床から発生する．

コメント：偽アミロイドの柄実質および傘表皮組織において明瞭に分化した偽柔組織状の中層を除き，円錐状に突出した中丘とシロアリの巣に接続する長い偽根を持つウラベニガサ型の子実体，そして楕円形，無色，非アミロイド，薄壁の担子胞子は *Termitomyces* 属（Heim 1942；Singer 1986）の分類概念と一致する．

　円錐状に突出した中丘を持つ本種の外観は東アフリカから新種記載されたトガリアリヅカタケ *Termitomyces clypeatus* R. Heim（Heim 1951, 1977；Otieno 1968；Pegler 1977；Otani and Shimizu 1981；Pegler and Vanhaecke 1994；Tibuhwa 2012）に似るが，トガリアリヅカタケはより鋭く刺状に突出した中丘を形成する暗褐色の傘を持ち，傘表皮組織は細い糸状菌糸からなる．シロアリシメジはまたスリランカ産 *Termitomyces eurrhizus*（Berk.）R. Heim（Berkeley 1847；Heim 1942；Pegler 1977, 1986；Pegler and Vanhaecke 1994；Wei et al. 2009）にやや似通った性質を有する．しかしながらスリランカ産の標本は子実体がより大型（通常傘の径100-150 mm 以上）で，中丘は鈍頭になり，消失性膜質のツバを持つとされている．台湾産 *Sinotermitomyces taiwanensis* M. Zang & C.M. Chen（Zang and Chen 1998；Wei et al. 2006, 2009）は子実体の類型において *T. clypeatus* と共通する点が多く，両者を同一種と見なす見解もあるが（Frøslev et al. 2003；Wei et al. 2006, 2009；Oyetayo 2012），原記載によれば中空の柄および永存性のツバを有するとされている．

　オオシロアリタケの和名に相当する種類については，沢田兼吉（1919．台湾産菌類調査報告第1編．農事試験場特別報告 19：519-522）が台湾産の標本に対して *Collybia albuminosa*（Berk.）Petch（= *T. eurrhizus*）の学名を用いたのが最初で，その後台湾産の標本の正確な分類概念が曖昧なままオオシロアリタケの和名と *Termitomyces albuminosus* sensu Otani（Otani 1979）の学名が八重山産の標本に適用され，現在に至るまで *T. eurrhizus* sensu Hongo（Hongo 1987；Kinjo et al. 2005；Neda and Sato 2008）の学名が誤用されてきた経緯がある．少なくとも八重山産の標本（Otani 1979）はスリランカ産 *T. eurrhizus* と明らかに形態的特徴が異なり，台湾産オオシロアリタケと八重山産標本との異同が不明確な現状では，八重山産の標本に対してオオシロアリタケの和名を用いるのは不適切と考えられる．

34. *Termitomyces intermedius* シロアリシメジ —— 329

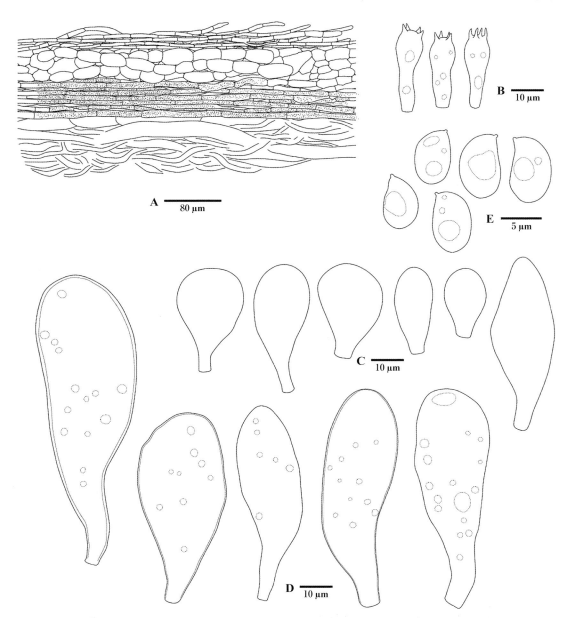

Fig. 267 – Micromorphological features of *Termitomyces intermedius* (holotype): **A**. Longitudinal cross section of the pileipellis and pileitrama. **B**. Basidia. **C**. Cheilocystidia. **D**. Pleurocystidia. **E**. Basidiospores. Illustrations by Taneyama, Y.
シロアリシメジの顕鏡図（正基準標本）：**A**. 傘表皮組織および実質の縦断面. **B**. 担子器. **C**. 縁シスチジア. **D**. 側シスチジア. **E**. 担子胞子. 図：種山裕一.

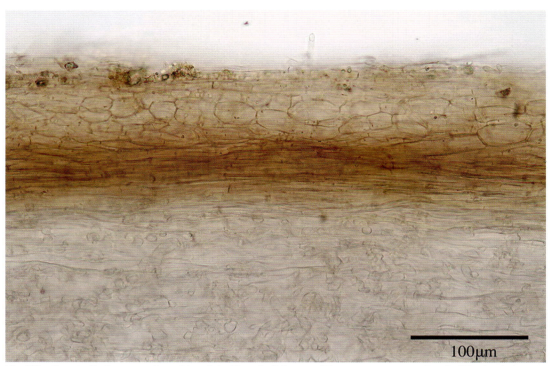

Fig. 268 – Longitudinal cross section of the pileipellis and pileitrama of *Termitomyces intermedius* (in 3%KOH, holotype), showing the deeply pigmented subpellis and the cellular mediopellis made up of highly inflated elements. Photo by Taneyama, Y. シロアリシメジの傘の表皮組織および実質の縦断面（3％水酸化カリウム溶液で封入，正基準標本）．濃色の下表皮層並びに著しく膨大した菌糸細胞からなる中表皮層を示す．写真：種山裕一．

34. *Termitomyces intermedius* シロアリシメジ ― 331

Fig. 269 – Basidiomata of *Termitomyces intermedius* (holotype), 10 Jun. 2012, Ishigaki Island. Photo by Taneyama, Y.
シロアリシメジの子実体（正基準標本），2012年6月10日，石垣島．写真：種山裕一．

Fig. 270 – Fungal comb of *Odontotermes formosanus*, 9 Jun. 2012, Ishigaki Island. Photo by Takahashi, H.
タイワンシロアリの巣の菌床，2012年6月9日，石垣島．写真：高橋春樹．

Fig. 271 – Basidiomata of *Termitomyces intermedius* (holotype), 10 Jun. 2012, Ishigaki Island. Photo by Taneyama, Y.
シロアリシメジの子実体（正基準標本），2012年6月10日，石垣島．写真：種山裕一．

Fig. 272 – Immature basidioma of *Termitomyces intermedius* (TNS-F-48178). Photo by Taneyama, Y.
シロアリシメジの未熟な子実体（TNS-F-48178）．写真：種山裕一．

Fig. 273 – Mature basidioma of *Termitomyces intermedius* (TNS-F-48169). Photo by Taneyama, Y.
シロアリシメジの成熟した子実体（TNS-F-48169）．写真：種山裕一．

Fig. 274 – Underside view of *Termitomyces intermedius* (TNS-F-48169). Photo by Taneyama, Y.
シロアリシメジのヒダ（TNS-F-48169）．写真：種山裕一．

35. *Tylopilus obscureviolaceus* Har. Takah. スミレニガイグチ

Mycoscience 45 (6): 374 (2004) [MB#370257].

Etymology: The specific epithet means "dark purple", referring to the dark purple basidiomata.

Macromorphological characteristics (Figs. 276–283): Pileus 70–100 mm in diameter, at first hemispherical, becoming convex to broadly convex, smooth, with slightly appendiculate margin; surface dry, subtomentose, persistently and evenly colored greyish Magenta (13–14D7 to 13–14E7) or deep Magenta (13–14D8 to 13–14E8). Flesh up to 10 mm thick, firm, white, unchanging when cut; odor indistinct, taste very bitter. Stipe 80–100×20–30 mm, more or less bulbous at the base, central, terete, solid, finely reticulated above or only at the extreme apex by a thin-veined, purplish reticulum; surface dry, subtomentose, sometimes longitudinally rugulose, greyish ruby (12D4–5) when fresh, greyish red (10D5 to 11D4–5) to brownish red (10D6) in age, violet brown (10E6–7) toward the base, whitish at the apex; base covered with whitish mycelial tomentum. Tubes up to 8 mm deep, adnate to slightly decurrent, white when young, dull pinkish in age, unchanging when cut; pores small (2–3 per mm), subcircular, concolorous, unchanging where handled.

Micromorphological characteristics (Fig. 275): Basidiospores 6–7.2×3.3–4 μm (n = 20 spores per two specimens, Q = 1.8), ellipsoid-subfusiform, inequilateral with a shallow suprahilar depression in profile, smooth, pinkish, thick-walled. Basidia 15–20×5–8 μm, clavate, four-spored. Basidioles clavate. Cheilocystidia gregarious, 30–40×5–10 μm, subcylindrical to subfusiform, smooth, hyaline, thin-walled. Pleurocystidia scattered, similar to the cheilocystidia. Hymenophoral trama bilateral-divergent of the *Boletus*-subtype; element hyphae 3–6 μm wide, cylindrical, smooth, colorless, thin-walled. Pileipellis a trichoderm of vertically arranged, loosely interwoven elements; terminal cells 30–60×4–7 μm, cylindrical, with reddish brown (9D7) to brownish red (10D7) intracellular pigment, thin-walled, without pilocystidia. Pileitrama consisting of loosely interwoven, cylindrical hyphae 4–11 μm wide, smooth, colorless, thin-walled. Stipitipellis hymeniform, consisting of caulocystidia; caulocystidia 13–19×5–7 μm, broadly clavate, smooth, with intracellular brownish pigment, thin-walled. Stipe trama composed of longitudinally running, cylindrical cells 4–10 μm wide, sometimes branched, smooth, colorless, with walls up to 0.5 μm. Clamps absent.

Habitat and phenology: Solitary to scattered on ground in evergreen broad-leaved forests dominated by *Quercus miyagii* Koidz. and *Castanopsis sieboldii* (Makino) Hatus. ex T. Yamaz. et Mashiba, May to June.

Known distribution: Okinawa (Iriomote Island).

Specimen examined: KPM-NC0010099 (holotype), on ground in an evergreen broad-leaved forest dominated by *Q. miyagii* and *C. sieboldii*, Otomi, Iriomote Island, Taketomi-cho, Yaeyama-gun, Okinawa Pref., 31May 2002, coll. Takahashi, H.

Japanese name: Sumire-nigaiguchi.

Comments: This species is characterized by its medium to large, boletoid basidiomata with a dark purple pileus and a brownish purple, finely reticulate, bulbous stipe; the white, bitter, unchanging flesh; the relatively small basidiospores (up to 8 μm long); the pileipellis made up of an loosely interwoven trichoderm; the hymeniform stipitipellis consisting of broadly clavate caulocystidia with intracellular brownish pigment; and the habitat of *Quercus-Castanopsis* forests.

The white, unchanging flesh and the white then pinkish pores suggest that this species belongs in the section *Tylopilus* of the genus *Tylopilus* as defined by Singer (Singer 1986). Within this section, North American *Tylopilus plumbeoviolaceus* (Snell & E. A. Dick) Singer (Bessette et al. 2000; Singer 1947; Snell 1936; Snell and Dick 1941, 1970; Wolfe 1986) and *Tylopilus neofelleus* Hongo, originally described from Japan (Hongo 1967), are similar to *T. obscureviolaceus*. The North American taxon differs in having a pileus colored more brownish or greyish when mature, much longer basidiospores: 7-11 µm (Snell and Dick 1941), and a hymeniform pileipellis composed of fusoid-ventricose to narrowly fusoid-ventricose pilocystidia (Wolfe 1986). *Tylopilus neofelleus* is distinct in forming an olive-brown to avellaneous pileus and purplish pores from the first.

References 引用文献

Bessette AE, Roody WC, Bessette AR. 2000. North American Boletes. A color guide to the fleshy pored mushrooms. Syracuse University Press, New York.
Hongo T. 1967. Notes on Japanese larger fungi (19). J Jpn Bot 42: 151-159.
Singer R. 1947. The Boletineae of Florida with notes on extralimital species III. Am Midl Nat 37: 1-135.
Singer R. 1986. Agaricales in modern taxonomy, 4th edn. Koeltz, Koenigstein.
Snell WH. 1936. Notes on Boletes. V. Mycologia 28: 463-475.
Snell WH, Dick EA. 1941. Notes on Boletes. VI. Mycologia 33: 23-37.
Snell WH, Dick EA. 1970. The boleti of northeastern North America. Cramer, Vaduz.
Takahashi H. 2004. Two new species of Agaricales from southwestern islands of Japan. Mycoscience 45 (6): 372-376.
Wolfe CB. 1986. Type studies in *Tylopilus*. III. Taxa described by Walter H. Snell, Esther A. Dick, and co-workers. Mycologia 78: 22-31.

35. スミレニガイグチ *Tylopilus obscureviolaceus* Har. Takah.

Mycoscience 45 (6): 374 (2004) [MB#370257].

肉眼的特徴（Figs. 276-283）：傘は径70-100 mm，最初半球形，のち饅頭形～ほぼ平開し，平坦，縁部は管孔側に僅かに突出する；表面は乾性，やや密綿毛状，幼菌から成菌に至るまで傘全体が一様に暗青紫色を呈する．肉は厚さ10 mm，固くしまり，白色，空気に触れても変色しない；特別な臭いはなく，強い苦みがある．柄は80-100×20-30 mm，基部が多少球根状に脹らみ，中心生，中実，上部または頂部に淡青紫色の繊細な網目模様を表す；表面は乾性，やや密綿毛状，時に縦皺を表し，傘より淡色または褐色を帯び，根元に向かって暗色になり，頂部は類白色；根本は類白色の密綿毛状菌糸体に被われる．管孔は長さ8 mm以下，直生～僅かに垂生，幼時白色，のち帯紅色，空気に触れても変色しない；孔口は小型（2-3 per mm），類円形，縁取りはなく，傷を受けても変色しない．

顕微鏡的特徴（Fig. 275）：担子胞子は6-7.2×3.3-4 µm（n = 20 spores per two specimens, Q = 1.8），楕円形～やや紡錘形，下部側面に不明瞭でなだらかな凹みがあり（イグチ型），平坦，帯紅色，厚壁．担子器は15-20×5-8 µm，こん棒形，4胞子性．偽担子器はこん棒形．縁シスチジアは群生し，30-40×5-10 µm，亜円柱形～亜紡錘形，平滑，無色，薄壁．側シスチジアは散生し，縁シスチジアに似る．子実層托実質はイグチ亜型；菌糸は幅3-6 µm，円柱形，無色，薄壁．傘表皮組織は緩く錯綜した毛状被を形成する；末端細胞は30-60×4-7 µm，円柱形，赤褐色の色素が細胞内に存在し，薄壁，傘シスチジアは分化しない．傘実質の菌糸は緩く錯綜し，幅4-11 µm，円柱形，平滑，無色，薄壁．柄表皮組織は柄シスチジアが子実層状被を成す；柄シスチジアは13-19×5-7 µm，広こん棒形，平滑，帯褐色の色素が細胞内に存在し，薄壁．柄実質の菌糸は縦に沿って配列し，幅4-10 µm，円柱形，時に分岐し，平滑，無色，壁は厚さ0.5 µm以下．全ての組織において菌糸はクランプを欠く．

生態および発生時期：スダジイ，オキナワウラジロガシを主体とする常緑広葉樹林内地上に孤生または散生，5月～6月．

分布：沖縄（西表島）．

35. *Tylopilus obscureviolaceus* スミレニガイグチ

供試標本：KPM-NC0010099（正基準標本），スダジイ，オキナワウラジロガシを主体とする照葉樹林内地上，沖縄県八重山郡竹富町西表島，2002年5月31日，高橋春樹採集．

主な特徴：子実体は中〜大型のヤマドリタケ型；傘は暗紫色；柄は基部が多少球根状に脹らみ，帯褐紫色，上部または頂部に繊細な網目模様を表す；肉は白色，強い苦みがある；担子胞子は相対的に小形（長さ8 μm以下）；傘表皮組織は緩く錯綜した毛状被からなる；細胞内に帯褐色の色素を持つ広こん棒形の柄シスチジアが柄表皮組織において子実層状被を形成する；スダジイ，オキナワウラジロガシを主体とする常緑広葉樹林内地上に発生．

コメント：白色で強い苦みがある肉そして白色のち帯紅色の管孔はSinger（Singer 1986）の分類概念によるニガイグチ属 *Tylopilus*，ニガイグチ節 section *Tylopilus* に所属することを示唆している．節内において北米産 *Tylopilus plumbeoviolaceus*（Snell & E.A.Dick）Singer（Bessette et al. 2000；Singer 1947；Snell 1936；Snell and Dick 1941, 1970；Wolfe 1986）および日本産ニガイグチモドキ *Tylopilus neofelleus* Hongo（Hongo 1967）は本種に最も近縁と思われるが，北米産種は傘の表面が成熟時褐色または灰色を帯びること，紡錘形の傘シスチジアが子実層状被を形成すること（Wolfe 1986），そして担子胞子がより長形である：7-11 μm（Snell and Dick 1941）点で有意差が認められる．ニガイグチモドキは傘の表面がオリーブ褐色〜帯紅褐色で，孔口は最初から青紫色に縁取られる．

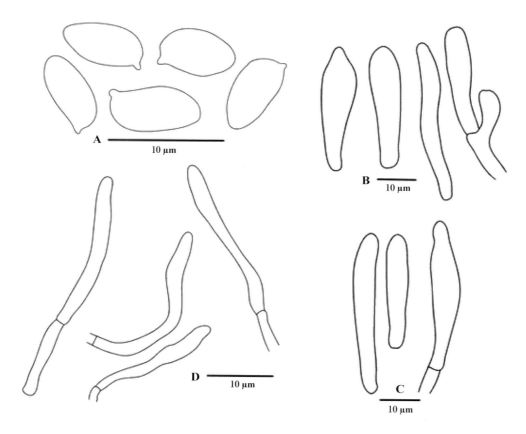

Fig. 275 – Micromorphological features of *Tylopilus obscureviolaceus* (KPM-NC0010099): **A**. Basidiospores. **B**. Cheilocystidia. **C**. Pleurocystidia. **D**. Elements of the pileipellis. Illustrations by Takahashi, H.
スミレニガイグチの顕鏡図（KPM-NC0010099）：**A**. 担子胞子．**B**. 縁シスチジア．**C**. 側シスチジア．**D**. 傘表皮組織の菌糸．図：高橋春樹．

35. *Tylopilus obscureviolaceus* スミレニガイグチ —— 339

Fig. 276 – Mature basidioma of *Tylopilus obscureviolaceus* (KPM-NC0010099), 31 May 2002, Otomi, Iriomote Island. Photo by Takahashi, H.
スミレニガイグチの成熟した子実体 (KPM-NC0010099),2002年5月31日,西表島大富. 写真:高橋春樹.

Fig. 277 – Mature basidioma of *Tylopilus obscureviolaceus* (KPM-NC0010099), 31 May 2002, Otomi, Iriomote Island. Photo by Takahashi, H.
スミレニガイグチの成熟した子実体（KPM-NC0010099），2002年5月31日，西表島大富．写真：高橋春樹．

Fig. 278 – Mature basidioma of *Tylopilus obscureviolaceus* (KPM-NC0010099), 31 May 2002, Otomi, Iriomote Island. Photo by Takahashi, H.
スミレニガイグチの成熟した子実体（KPM-NC0010099），2002年5月31日．西表島大富．写真：高橋春樹．

Fig. 279 – Underside view of the mature basidioma of *Tylopilus obscureviolaceus* (KPM-NC0010099), 31 May 2002, Otomi, Iriomote Island. Photo by Takahashi, H.
スミレニガイグチの成熟した子実体(管孔側,KPM-NC0010099),2002年5月31日,西表島大富.写真:高橋春樹.

Fig. 280 – Immature basidioma of *Tylopilus obscureviolaceus* (KPM-NC0010099), 31 May 2002, Otomi, Iriomote Island. Photo by Takahashi, H.
スミレニガイグチの幼菌（KPM-NC0010099），2002年5月31日．西表島大富．写真：高橋春樹．

Fig. 281 – Pileus surface of *Tylopilus obscureviolaceus* (KPM-NC0010099), 31 May 2002, Otomi, Iriomote Island. Photo by Takahashi, H.
　スミレニガイグチの傘表面（KPM-NC0010099），2002年5月31日，西表島大富．写真：高橋春樹．

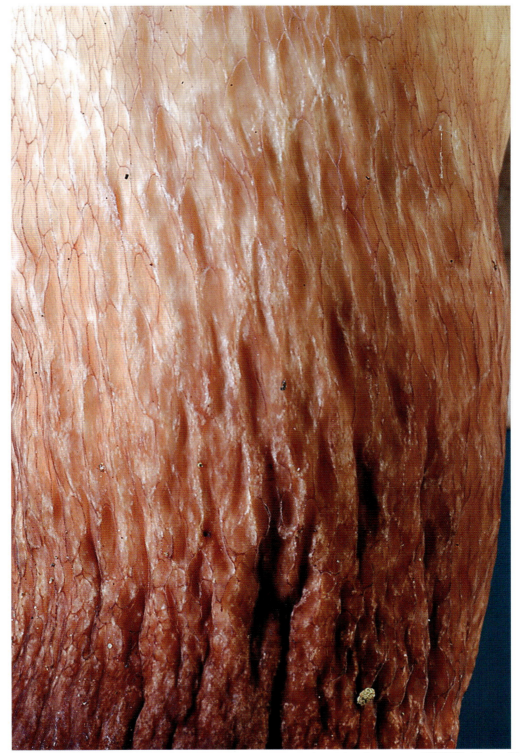

Fig. 282 – Stipe surface of *Tylopilus obscureviolaceus* (from middle portion of the stipe, KPM-NC0010099), 31 May 2002, Otomi, Iriomote Island. Photo by Takahashi, H.
スミレニガイグチの柄表面（柄中部，KPM-NC0010099），2002年5月31日，西表島大富．写真：高橋春樹．

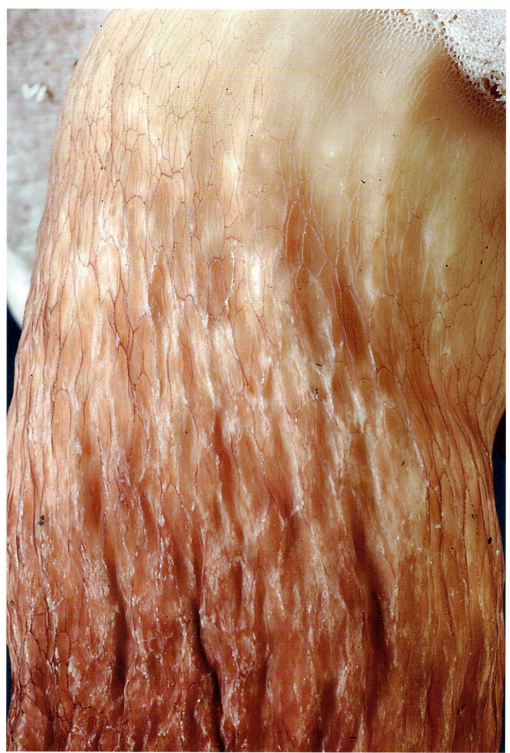

Fig. 283 – Stipe surface of *Tylopilus obscureviolaceus* (from apex of the stipe, KPM-NC0010099), 31 May 2002, Otomi, Iriomote Island. Photo by Takahashi, H.
スミレニガイグチの柄表面（柄頂部，KPM-NC0010099），2002年5月31日，西表島大富．写真：高橋春樹．

Scientific name Index 学名索引

【A】

Agaricus crocopeplus　　175-177
Agaricus trisulphuratus　　175, 177
Amanita caesarea f. *alba*　　3
Amanita caesarea var. *alba*　　3, 4
Amanita caesareoides　　7, 12
Amanita chepangiana　　3-5
Amanita hemibapha　　7, 12
Amanita hemibapha var. *ochracea*　　7, 12
Amanita javanica　　7, 12
Amanita rubromarginata　　6-16
Astrosporina humilis　　126
Aureoboletus liquidus　　17, 19- 21, 23-31
Aureoboletus longicollis　　19
Aureoboletus singeri　　19

【B】

Boletellus longicollis　　19, 20, 30
Boletellus singeri　　19, 30
Boletus bannaensis　　32-44
Boletus firmus　　33, 34
Boletus morrisii　　130, 131
Boletus quercinus　　33, 34
Boletus satanas　　32-34
Boletus virescens　　45, 46, 48-53

【C】

Chaetocalathus ehretiae　　55, 56
Chaetocalathus fragilis　　54-59
Chaetocalathus semisupinus　　55, 56
Collybia albuminosa　　328
Crinipellis actinophora　　69, 71
Crinipellis canescens　　60-66, 67
Crinipellis fragilis　　54, 55
Crinipellis hepatica　　61, 64
Crinipellis nigricaulis　　69, 71
Crinipellis omotricha　　61, 64
Crinipellis pseudostipitaria　　61, 64
Crinipellis rhizomaticola　　69, 71
Crinipellis rhizomorphica　　68-78
Crinipellis sapindacearum　　69, 71
Crinipellis septotricha　　61, 64
Crinipellis stupparia　　61; 64
Crinipellis trichialis　　69, 71
Crinipellis tucumanensis　　69, 71
Cruentomycena kedrovayae　　80, 81, 84
Cruentomycena orientalis　　79, 81, 83-89
Cruentomycena viscidocruenta　　80, 84, 85

【G】

Gymnopilus dilepis　　91, 95
Gymnopilus iriomotensis　　90-97
Gymnopilus ombrophilus　　91, 93
Gymnopilus subtropicus　　91, 93, 95
Gymnopus albipes　　98, 99, 101-104
Gymnopus oncospermatis　　105-115
Gymnopus phyllogenus　　116-118, 120-123

【I】

Inocybe fuscomarginata　　124, 125
Inocybe humilis　　126-128

【L】

Lachnella fragilis　　54, 55
Leccinellum rhodoporosum　　129, 130, 132-140
Leccinum australiense　　130
Leccinum rhodoporosum　　129, 130

【M】

Marasmiellus afer　　156, 159
Marasmiellus arenaceus　　141- 154
Marasmiellus candidus　　99, 100
Marasmiellus carneopallidus　　142, 146
Marasmiellus goossensiae　　166, 168
Marasmiellus lucidus　　155-164
Marasmiellus mesosporus　　142, 143, 146
Marasmiellus purpureoalbus　　166, 167
Marasmiellus venosus　　165, 166, 168-173
Marasmius fragilis　　54, 55
Marasmius kisangensis　　117, 119
Marasmius oncospermatis　　105, 106
Micropsalliota cornuta　　174, 175, 177-183
Moniliophthora canescens　　60, 62
Mycena arundinarialis　　210, 214
Mycena auricoma　　185, 188

Scientific name Index 学名索引

Mycena brevisetosa　　220, 223
Mycena comata　　184-196
Mycena cyanocephala　　210, 214
Mycena flammifera　　197-199, 201-208
Mycena interrupta　　210, 214
Mycena lazulina　　209-211, 213-217
Mycena luxfoliata　　219, 220, 223-226
Mycena manipularis　　198, 200
Mycena obscuritatis　　228, 231
Mycena parsimonia　　117, 119, 228, 231
mycena pocilliformis　　228
Mycena putroris　　228, 231
Mycena silvaelucens　　228, 229
Mycena stellaris　　227, 228, 231-237

【N】
Neonothopanus nambi　　239, 241

【P】
Pleurotus eugrammus　　239, 241
Pleurotus eugrammus var. *brevisporus*　　239, 241
Pleurotus eugrammus var. *radicicola*　　239, 241
Pleurotus hygrophanus　　239, 241
Pleurotus lunaillustris　　239-241
Pleurotus nitidus　　238-240, 242-249
Psilocybe capitulata　　250-252, 254-262
Psilocybe cubensis　　251, 253
Psilocybe definita　　263-273
Psilocybe magnispora　　251, 253
Psilocybe merdaria　　264, 266
Psilocybe moelleri　　264, 269
Psilocybe subaeruginascens　　251, 253
Psilocybe subannulata　　251, 253
Psilocybe subcubensis　　251, 253

Pulveroboletus aberrans　　275, 276
Pulveroboletus brunneoscabrosus　　274, 275, 277-283
Pulveroboletus frians　　275, 276
Pulveroboletus ravenelii　　275, 276

【R】
Resinomycena fulgens　　284-286, 288-294
Resinomycena japonica　　285, 288
Rubinoboletus ballouii　　296, 297
Rubinoboletus caespitosus　　296, 297
Rubinoboletus monstrosus　　295, 296, 298-302

【S】
Sinotermitomyces taiwanensis　　326, 328
Strobilomyces brunneolepidotus　　303, 304, 306-312
Strobilomyces velutinus　　304, 305
Strobilurus luchuensis　　313-315, 317-323
Strobilurus ohshimae　　314, 316
Strobilurus stephanocystis　　314, 316
Sutorius australiensis　　130, 131

【T】
Termitomyces albuminosus　　324-328
Termitomyces clypeatus　　325, 326, 328
Termitomyces eurrhizus　　324, 326, 328
Termitomyces intermedius　　324-326, 329-335
Trogia crinipelliformis　　185, 188
Tylopilus ballouii　　296
Tylopilus neofelleus　　336-338
Tylopilus obscureviolaceus　　336-346
Tylopilus plumbeoviolaceus　　336, 338

【X】
Xerocomus sulcatipes　　46, 49

Japanese name Index 和名索引

【ア】
アオアザイグチ　　45, 46, 48-53
アシグロカレハタケ　　116, 118, 120-123
アミヒカリタケ　　200

【ウ】
ウラベニヤマイグチ　　129-140
ウロコキイロイグチ　　274, 275, 277-283

【オ】
オオシロアリタケ　　328

【カ】
ガーネットオチバタケ　　79, 81, 83-89
カレハヤコウタケ　　219, 220, 223-225
カワリコシワツバタケ　　266

Japanese name Index 和名索引 ―― 349

【キ】
キイロイグチ　276
キジムナハナガサ　184, 186-196
キタマゴタケ　12
キニガイグチ　297
ギンガタケ　284, 286, 288-294

【ケ】
ケカゴタケ　56

【コ】
コガネツムタケ　93
コガネハナガサ　188
コカブラアセタケ　126, 127
コンルリキュウバンタケ　209, 211, 213-218

【サ】
ザラメタケ　288

【シ】
シビレタケモドキ（宮城仮称）　253
シラガニセホウライタケ　60, 62-67
シロアシホウライタケ　98, 99, 101-104
シロアリシメジ　324, 326, 328-335
シロスナホウライタケ　141, 143-154
シロヒカリタケ　238, 240-249

【ス】
スギエダタケ　316
スナジホウライタケ　146
スミレニガイグチ　336-346

【タ】
ダイダイツノハラタケ　174, 175, 177-183
タマゴタケ　12
ダルマイグチ　295, 296, 298-302

【チ】
チャオニイグチ　303-312

【ト】
トガリアリヅカタケ　328

【ナ】
ナガエノウラベニイグチ　34

ナンヨウウラベニイグチ　32-44
ナンヨウシビレタケ　250, 252-262
ナンヨウシロタマゴタケ　3-5

【ニ】
ニガイグチモドキ　338
ニライタケ　253

【ヌ】
ヌメリアシナガイグチ　17, 19-21, 23-31

【ハ】
ハマシビレタケ　263, 265-273

【ヒ】
ヒダフウリンタケ　54-59
ヒメヒカリタケ　165, 166, 168-173
ヒメホタルタケ　155, 157-164

【フ】
フチドリタマゴタケ　6-16
フチドリトマヤタケ　124, 125

【ホ】
ホシノヒカリタケ　227-229, 231-237

【マ】
マツカサキノコモドキ　316

【ミ】
ミドリニセホウライタケ　68, 70-78
ミナミシビレタケ　253
ミナミホタケ　90, 92-97

【モ】
モリノアヤシビ　197, 199, 201-208

【ヤ】
ヤシモリノカレバタケ　105-115

【リ】
リュウキュウマツカサキノコ　313, 315, 317-323

著　者（Authors）

波多野敦子（Atsuko Hadano）
A member of the Mycologist Circle of Japan, and Oita Mushroom Society
菌類懇話会・大分きのこ会会員

波多野英治（Eiji Hadano）
A member of the Mycological Society of Japan
日本菌学会会員

小林　孝人（Takahito Kobayashi）
Researcher of the Hokkaido University Museum
北海道大学総合博物館，研究員

黒木　秀一（Shuichi Kurogi）
Assistant Director of the Board of Education, Cultural Properties Division, Miyazaki Prefecture
宮崎県教育庁文化財課，副主幹

大場由美子（Yumiko Oba）
Chief Researcher of the Research Institute for Luminous Organisms, Specified Nonprofit Corporation Hachijojima Recreational Organization
特定非営利活動法人八丈島観光レクリエーション研究会 八丈島発光生物研究所，主任研究員

和田　匠平（Shohei Wada）
A member of the Mycological Society of Japan, Japanese Society of Mushroom and Biotechnology, Mycologist Circle of Japan, and Society of Study of Fungus, Hyogo
日本菌学会・日本きのこ学会・菌類懇話会・兵庫きのこ研究会会員

監修者（Supervisor）

寺嶋　芳江（Yoshie Terashima）
Professor of Tropical Biosphere Research Center, University of the Ryukyus, graduated from the United Graduate School of Agricultural Science, Tottori University.
鳥取大学大学院連合農学研究科 博士（農学）
琉球大学 熱帯生物圏研究センター 教授

編著者（Editors）

寺嶋　芳江（Yoshie Terashima）

高橋　春樹（Haruki Takahashi）
Mycologist
菌学者

種山　裕一（Yuichi Taneyama）
A member of the Mycological Society of Japan and Mycologist Circle of Japan,
President of the video production for commercial television
日本菌学会・菌類懇話会会員
映像制作会社　種山事務所代表

装丁：中野達彦

南西日本菌類誌　軟質高等菌類
なんせいにほんきんるいし　なんしつこうとうきんるい

2016年2月20日　第1版第1刷発行

監 修 者　寺嶋芳江
編 著 者　寺嶋芳江・高橋春樹・種山裕一
発 行 者　橋本敏明
発 行 所　東海大学出版部
　　　　　〒259-1292　神奈川県平塚市北金目4-1-1
　　　　　TEL 0463-58-7811　FAX 0463-58-7833
　　　　　URL http://www.press.tokai.ac.jp/
　　　　　振替　00100-5-46614
印 刷 所　港北出版印刷株式会社
製 本 所　誠製本株式会社

ⓒ Yoshie Terashima, 2016　　　　　　　　　　ISBN978-4-486-02085-1

Ⓡ〈日本複製権センター委託出版物〉
本書の全部または一部を無断で複写複製（コピー）することは，著作権法上の例外を除き，禁じられています。
本書から複写複製する場合は日本複製権センターへご連絡の上，許諾を得てください。
日本複製権センター（電話 03-3401-2382）